Agrometeorology

Principles and Applications of Climate Studies in Agriculture

Crop Science
Amarjit S. Basra, PhD
Senior Editor

Heterosis and Hybrid Seed Production in Agronomic Crops edited by Amarjit S. Basra

Intensive Cropping: Efficient Use of Water, Nutrients, and Tillage by S. S. Prihar, P. R. Gajri, D. K. Benbi, and V. K. Arora

Physiological Bases for Maize Improvement edited by María E. Otegui and Gustavo A. Slafer

Plant Growth Regulators in Agriculture and Horticulture: Their Role and Commercial Uses edited by Amarjit S. Basra

Crop Responses and Adaptations to Temperature Stress edited by Amarjit S. Basra

Plant Viruses As Molecular Pathogens by Jawaid A. Khan and Jeanne Dijkstra

In Vitro Plant Breeding by Acram Taji, Prakash P. Kumar, and Prakash Lakshmanan

Crop Improvement: Challenges in the Twenty-First Century edited by Manjit S. Kang

Barley Science: Recent Advances from Molecular Biology to Agronomy of Yield and Quality edited by Gustavo A. Slafer, José Luis Molina-Cano, Roxana Savin, José Luis Araus, and Ignacio Romagosa

Tillage for Sustainable Cropping by P. R. Gajri, V. K. Arora, and S. S. Prihar

Bacterial Disease Resistance in Plants: Molecular Biology and Biotechnological Applications by P. Vidhyasekaran

Handbook of Formulas and Software for Plant Geneticists and Breeders edited by Manjit S. Kang

Postharvest Oxidative Stress in Horticultural Crops edited by D. M. Hodges

Encyclopedic Dictionary of Plant Breeding and Related Subjects by Rolf H. G. Schlegel

Handbook of Processes and Modeling in the Soil-Plant System edited by D. K. Benbi and R. Nieder

The Lowland Maya Area: Three Millennia at the Human-Wildland Interface edited by A. Gómez-Pompa, M. F. Allen, S. Fedick, and J. J. Jiménez-Osornio

Biodiversity and Pest Management in Agroecosystems, Second Edition by Miguel A. Altieri and Clara I. Nicholls

Plant-Derived Antimycotics: Current Trends and Future Prospects edited by Mahendra Rai and Donatella Mares

Concise Encyclopedia of Temperate Tree Fruit edited by Tara Auxt Baugher and Suman Singha

Landscape Agroecology by Paul A. Wojkowski

Concise Encyclopedia of Plant Pathology by P. Vidhyasekaran

Testing of Genetically Modified Organisms in Foods edited by Farid E. Ahmed

Concise Encyclopedia of Bioresource Technology edited by Ashok Pandey

Agrometeorology: Principles and Applications of Climate Studies in Agriculture by Harpal S. Mavi and Graeme J. Tupper

Agrometeorology
Principles and Applications of Climate Studies in Agriculture

Harpal S. Mavi, PhD
Graeme J. Tupper, MAgSc, DipEd

CRC Press is an imprint of the
Taylor & Francis Group, an informa business

First Published by Lawrence Erlbaum Associates, Inc., Publishers
10 Industrial Avenue
Mahwah, New Jersey 07430

Reprinted 2010 by CRC Press
CRC Press
6000 Broken Sound Parkway, NW
Suite 300, Boca Raton, FL 33487

270 Madison Avenue
New York, NY 10016

2 Park Square, Milton Park
Abingdon, Oxon OX14 4RN, UK

Published by

Food Products Press®, an imprint of The Haworth Press, Inc., 10 Alice Street, Binghamton, NY 13904-1580.

Cover design by Jennifer M. Gaska.

Library of Congress Cataloging-in-Publication Data

Mavi, H. S. (Harpal Singh). 1935-
Agrometeorology : principles and applications of climate studies in agriculture / Harpal Mavi, Graeme J. Tupper.
p. cm.
Includes bibilographical references (p.) and index.
ISBN 1-56022-972-1 (hard : alk. paper)
1. Meteorology, Agricultural. I. Tupper, Graeme, J. II. Title.

S600.5.M38 2004
630'.2515—dc21

2003012333

CONTENTS

Preface ix

Acknowledgments xi

Chapter 1. Agrometeorology: Perspectives and Applications 1

Definition 1
A Holistic Science 1
Scope 2
Practical Utility 3
Chronology of Developments 5
Future Needs 8

Chapter 2. Solar Radiation and Its Role in Plant Growth 13

The Sun: The Source of Energy 13
Nature and Laws of Radiation 16
Earth's Annual Global Mean Radiative Energy Budget 18
Solar Radiation and Crop Plants 25
Solar Radiation Interception by Plants 30
Photosynthetically Active Radiation (PAR) 36
Solar Radiation Use Efficiency 38

Chapter 3. Environmental Temperature and Crop Production 43

Soil Temperature 43
Air Temperature 47
Plant Injury Due to Sudden Changes in Temperature 50
Frost: Damage and Control 55
Thermoperiodism 64
Temperature As a Measure of Plant Growth and Development 66

Chapter 4. Climatological Methods for Managing Farm Water Resources 69

Water for Crop Production 69
Making Effective Use of Rainfall 70
Evaporation and Evapotranspiration 76
Water Use and Loss in Irrigation 84

Climatological Information in Improving
Water-Use Efficiency (WUE) 85
Reducing Water Losses from Reservoirs 91

Chapter 5. Drought Monitoring and Planning for Mitigation **95**

Definition of Drought 96
Meteorological Indicators of Drought 96
Drought Monitoring in Australia 105
Drought Exceptional Circumstances 108
Overview of Drought Assessment Methods 113
Meeting the Challenge: A Drought Mitigation Plan 115
Desertification 119

Chapter 6. Climate, Crop Pests, and Parasites of Animals **123**

Role of Weather and Climate 123
Some Important Insect Pests of Crop Plants 130
Climate and Parasites of Animals 136
Helminth Parasites 136
Arthropod Parasites 138

Chapter 7. Remote-Sensing Applications in Agrometeorology **145**

Spatial Information and the Environment 145
Remote Sensing 146
Remote Sensors and Instruments 149
Image Acquisition 152
Satellite Orbits for Remote Sensing 158
Geographic Information System (GIS) 159
Global Positioning System (GPS) 160
Remote-Sensing Applications 161

Chapter 8. Role of Computer Models in Managing Agricultural Systems **179**

Modeling Biological Response to Weather Conditions 179
Models 180
Applications of Crop Models 182
Simulation Models Relevant to Australian Farming
Systems 187
Decision Support Systems (DSS) 187

Chapter 9. Agroclimatological Services **209**

Weather and Agriculture 209
Weather and Climate Forecasting 209
Tailoring Climate Information for Agriculture 211
Impacts of Weather on Specific Industries and the Role of Forecast Information 212
Agroclimatological Information Services in Australia 214
Use and Benefits of Climate Forecast Information 217
Toward Optimum Utilization of Climate Information and Forecast Products 220

Chapter 10. Using Climate Information to Improve Agricultural Systems **237**

Setting the Platform—Property Planning 238
Sustainable Production—Setting the Enterprise Mix and Production Levels 240
Making Efficient Use of Rainfall 242
Developing Resilience 252
Managing the Extremes—Droughts and Floods 254
The Decision-Making Process—Dealing with Risk and Complexity 256
Providing Climate Technology to Farmers 256
Communicating New Ideas and Practices—Creating Change Through Adult Learning 258

Chapter 11. Climate Change and Its Impact on Agriculture **263**

Climate Variability and Climate Change 263
Observed Change in Atmospheric Composition and Climate 267
Observed Impact of Climate Change 272
Future Scenarios of Climate Change 275
Impact of Climate Change on Hydrology and Water Resources 277
Impact of Climate Change on Crops 282
Impact of Climate Change on Livestock 286

References **291**

Author Index **339**

Subject Index **351**

ABOUT THE AUTHORS

Dr. Harpal S. Mavi is a climate risk management consultant in Sydney, Australia. His work involves research in climate impacts on crops, insect-pests and crop diseases, and strategies for climate-related risk management. Before going into consultancy services, he was an agroclimatologist in the New South Wales Department of Agriculture. Earlier, he was Professor of Agrometeorology at Punjab Agricultural University in India where he taught climatology and agrometeorology for over two decades. Dr. Mavi has been the research supervisor of twenty postgraduate students majoring in agrometeorology. In addition, he has been on the research advisory committee of scores of students majoring in agronomy, soil science, soil and water engineering, entomology, plant pathology, and botany. He has advised numerous individuals and organizations in climate-related risk management.

Mr. Graeme J. Tupper is Technical Specialist, Resource Information, in the NSW Department of Agriculture, Australia. His work involves research, development, and service projects in the application of spatial information technology to agriculture. Specific projects include monthly rainfall "drought" mapping, fire and flood mapping and monitoring, monitoring endemic livestock diseases, weed detection and mapping research, and water reform structural adjustment mapping. Mr. Tupper is an agricultural scientist and has worked in this capacity in research, academic, and extension institutions in Australia, Papua New Guinea, South Africa, and the United States, in environments ranging from semi-arid rangelands through temperate to tropical agriculture.

Preface

Agrometeorology is an interdisciplinary holistic science. It cuts across scientific disciplines and bridges physical and biological sciences. It has numerous applications in agricultural resource utilization and management and has progressed rapidly both in content and applications.

Currently, at the dawn of the twenty-first century, much research work has been published in scientific journals on different themes of agrometeorology, but few books are yet available in which agrometeorological principles, techniques, and applications have been presented in a systematic manner. The increasing utility of the subject, with an ample availability of literature that lies scattered in journals of various scientific disciplines, demands quality books in which the entire subject matter is dealt with in an appropriate sequence. This book is a step toward that objective—to make available a text and reference book in agrometeorology.

With numerous demonstrations that the science of meteorology has an important role to play in farm industries, agrometeorology has been declared a growth or developing field by international agencies such as the World Meteorological Organization (WMO) and the Food and Agriculture Organization (FAO). Agrometeorology has emerged as a discipline in university education. In many countries, agricultural universities and government departments of agriculture and natural resources have created separate departments or units of agrometeorology which are doing sound work in agricultural education, research, and extension. Yet very few books are available on the subject that could be assigned as texts and used as reference material. This book is intended to serve a large audience of students, teachers, researchers, and extension workers in the field of agometeorology.

Climatology has an important role to play in developing a sound understanding of the subject matter of many of the applied agricultural sciences, because weather influences their subject matter in various ways. These sciences can improve their respective techniques based on sound interpretation of meteorological knowledge. Students and teachers of the major agricultural sciences of agronomy, horticulture, soil science, animal production, entomology, and plant pathology will find this book useful in their scientific pursuits and field extension work. Students and research workers in the disciplines of geography and natural resource studies will find agrometeorological methods and techniques useful because their own subject matter overlaps with climate. It will help students of meteorology to better understand the applications of their own subject matter to various activities for agricultural development. Finally, the book will be useful to scientists and planners engaged in regional and land-use planning, soil and water conservation, risk analysis of climate hazards, harvest forecasts, and the ecological and economic implications of climate change.

Agrometeorology: Principles and Applications of Climate Studies in Agriculture is written in a simple and descriptive style. Examples of climate applications in agriculture (methods, techniques, models, and services) are mainly from Australia. Nevertheless, the majority of these applications hold true for other countries, especially those countries with climatic patterns and agricultural systems similar to those of Australia. The book is relevant to global agriculture and is documented with the latest literature from international research journals. A range of topics has been covered that could generate the interest of a large cross section of people. Care has been taken that the material covered is a blend of different views of faculties of physical and biological sciences. It covers material that is taught in several disciplines of scientific education. In addition to use as a text in the discipline of agrometeorology, this book is applicable to several courses taught across other disciplines at the college and university level.

Acknowledgments

The authors are grateful to Darren Bayley and David Brouwer, Education Officers, NSW Agriculture, C.B. Alexander Agricultural College ("Tocal"), Australia, for contributing Chapter 10 to the book.

For their contributions and suggestions in the contents of the book we are indebted to

Andrew Kennedy, Technical Specialist, Natural Resources, NSW Agriculture;
Bernie Dominiak, Coordinator, Queensland Fruit Fly, NSW Agriculture;
Damien O'Sullivan, Senior Climate and Pasture Officer, Queensland Department of Primary Industries;
David George, Senior Education Officer, Queensland Department of Primary Industries;
Darren Bayley, Education Officer, NSW Agriculture, Tocal;
Ian McGowen, Senior Research Officer, NSW Agriculture;
Jason Crean, Technical Specialist, Economic Research, NSW Agriculture;
John Crichton, Research Officer, NSW Agriculture;
Lee Cook, Veterinary Officer, NSW Agriculture;
Paul Carberry, Climate Education Officer, NSW Agriculture;
Penny Marr, Editor, NSW Agriculture;
Peter Hayman, Coordinator, Agroclimatology, NSW Agriculture;
Rendle Hannah, Agricultural Coordinator, Water Reform Implementation, NSW Agriculture;
Samsul Huda, Senior Lecturer, Agroclimatology, University of Western Sydney; and
Tarjinder Mavi, Veterinarian, Technical Services, Australia

Figures 2.8 to 2.11, 3.1, 4.2, and 4.3 are adapted from articles published in various issues of *Agricultural and Forest Meteorology*. Permission given by Elsevier Science to use these figures is acknowledged with thanks. Permission granted by MicroImages, Inc., for adapting figures in Chapter 6 is also acknowledged with thanks.

Chapter 1

Agrometeorology: Perspectives and Applications

DEFINITION

Agrometeorology, abbreviated from agricultural meteorology, puts the science of meteorology to the service of agriculture, in its various forms and facets, to help with the sensible use of land, to accelerate the production of food, and to avoid the irreversible abuse of land resources (Smith, 1970). Agrometeorology is also defined as the science investigating the meteorological, climatological, and hydrological conditions that are significant to agriculture owing to their interaction with the objects and processes of agriculture production (Molga, 1962).

The definition of biometeorology adopted by the International Society of Biometeorology (ISB) states, "Biometeorology is an interdisciplinary science dealing with the application of fields of meteorology and climatology to biological systems" (Hoppe, 2000, p. 383). The general scope includes all kinds of interactions between atmospheric processes and living organisms—plants, animals, and humans. By this definition, it becomes evident that there are roughly three subbranches of biometeorology: plant, animal, and human biometeorology (Hoppe, 2000). The domain of agrometeorology is the plant and animal subbranches. The third subbranch, human biometeorology, is outside the scope of agrometeorology.

A HOLISTIC SCIENCE

Agrometeorology is an interdisciplinary science in which the main scientific disciplines involved are atmospheric sciences and soil sciences, which are concerned with the physical environment, and plant sciences and animal sciences (including their pathology, entomology, and parasitology, etc.), which deal with the contents of the biosphere.

The interdisciplinary nature of agrometeorology is both its greatest strength and its greatest weakness (Hollinger, 1994). The strength is obtained from an agricultural meteorologist's understanding of the interactions of physical and biological worlds. The weakness is due to the political reality that agricultural meteorology is not fully appreciated by the more traditional practitioners of the physical and biological sciences. Current academic structures do not foster interactions between biological and physical scientists. As a result, neither group fully understands the other, leading to the mistrust of each other's scientific methods. The perspective of agrometeorology is more holistic than that of the climatology or biology disciplines (Hatfield, 1994). An agrometeorologist is fluent in both the biological and physical sciences and looks at the world from a different and wider perspective than the physical or biological scientist does.

Though interdisciplinary in nature, agrometeorology is a well-defined science. It has a set approach in theory and methodology. Its subject matter links together the physical environment and biological responses under natural conditions. An agrometeorologist applies every relevant meteorological skill to help farmers make the most efficient use of their physical environment in order to improve agricultural production both in quality and quantity and to maintain the sustainability of their land and resources (Bourke, 1968). Using a four-stage approach, an agrometeorologist first formulates an accurate description of the physical environment and biological responses. At the second stage, he or she interprets biological responses in terms of the physical environment. Third, he or she makes agrometeorological forecasts. The final goal is to develop agrometeorological services, strategies, and support systems for on-farm strategic and tactical decisions and to implement them in collaboration with specialists in agriculture, livestock, and forestry.

SCOPE

For optimum crop growth, specific climatic conditions are required. Agrometeorology thus becomes relevant to crop production because it is concerned with the interactions between meteorological and hydrological factors on the one hand and agriculture, in the widest sense including horticulture, animal husbandry, and forestry, on the other (Figure 1.1). Its objective is to discover and define such effects and to apply knowledge of the atmosphere to practical agricultural use. The field of interest of an agrometeorologist extends from the soil surface layer to the depth down to which tree roots penetrate. In the atmosphere he or she is interested in the air layer near the ground in which crops and higher organisms grow and animals live, to the highest levels in the atmosphere through which the transport of seeds, spores, pollen, and insects may take place. As new research uncovers the se-

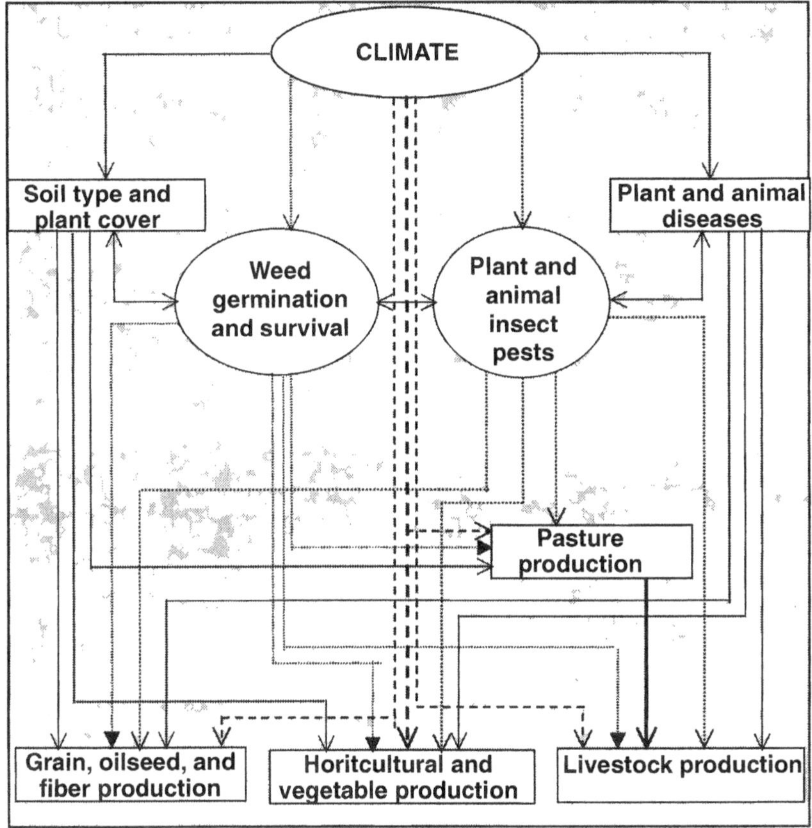

FIGURE 1.1. Climate and agricultural production

crets of meteorological phenomena, there is increasing interest in remote sensing and interactions between oceans and the atmosphere in shaping seasonal conditions.

PRACTICAL UTILITY

The dangers to the natural resource base, crops, and livestock that have a meteorological component include pollution of soil and air; soil erosion from wind or water; the incidence and effects of drought; crop growth; animal production; the incidence and extent of pests and diseases; the incidence, frequency, and extent of frost; the dangers of forest or bush fires; losses during storage and transport; and all farm operations. Agrometeorology offers practical solutions for harnessing climate potential and for protection against or avoidance of climate-related risks.

The role of agrometeorology is both strategic and tactical. The strategic role is involved in the assessment of long-term utilization of natural resources in the development of crop diversity. The tactical role is more concerned with the short-term and field-scale decisions that directly influence crop growth and development. If communicated to the right client and applied, agrometeorological information can help farmers practice sustainable, high-quality, more profitable agriculture, with fewer risks, lower costs, and less environmental pollution and damage (Rijks and Baradas, 2000).

Of the total annual crop losses in world agriculture, a large percentage is due to direct weather effects such as drought, flash floods, untimely rains, frost, hail, and storms. Losses in harvest and storage, as well as those due to parasites, insects, and plant diseases, are highly influenced by the weather (Mavi, 1994). When specifically tailored weather information is readily available to the needs of agriculture, it greatly contributes toward making short-term adjustments in daily agricultural operations, which minimize losses resulting from adverse weather conditions and improve the yield and quality of agricultural products. Tailored weather information also provides guidelines for long-range or seasonal planning and the selection of crops most suited to anticipated climatic conditions (Newman, 1974; Ogallo, Boulahya, and Keane, 2000). Most decisions in livestock enterprises involve a considerable lag between decisions and their effects. Some decisions affect the product three to four years in the future (Plant, 2000). A long-range forecast is a very good climate risk-management tool which helps increase livestock production (Anonymous, 2000). Seasonal climate forecasts can play an important role in shaping the economic polices of governments. For example, with a forecast of a major drought, economic growth would be less than expected. By taking serious note of the forecast, monetary policy could be relaxed to maintain growth targets (White, 2000).

Other applications of agrometeorology are through improvement in techniques based on sound interpretation of meteorological knowledge. These include irrigation and water allocation strategies; shelter from wind or cold; shade from excessive heat; antifrost measures, including the choice of site; antierosion measures; soil cover and mulching; plant cover using glass or plastic materials; artificial climates of growth chambers or heated structures; animal housing and management; climate control in storage and transport; and efficient use of herbicides, insecticides, and fertilizers. Agrometeorological models can be used in efficient land-use planning; determining suitable crops for a region; risk analysis of climatic hazards and profit calculations in farming; production or harvest forecasts; and the adoption of farming methods and the choice of effective farm machinery.

CHRONOLOGY OF DEVELOPMENTS

Attempts to relate agricultural production to weather go back at least 2,000 years and are still evolving. The twentieth century can be termed a very progressive and fertile one in respect to meteorology and agrometeorology (Fleming, 1996). Qualitative studies in the nineteenth century were followed by statistical analyses, then by microclimatic measurements, and most recently by modeling (Decker, 1994; Monteith, 2000).

Quantification of Crop-Weather Relationships

Visual observations of the microclimate and its impact on crop plants have been going on for several centuries. However, measurement of the characteristics of the microclimate in laboratory and experimental fields was strengthened during the first half of the twentieth century. The role of water in soil climate was recognized, and the link between the physical properties of soil, heat exchange, and water movement was investigated.

It is well recognized that year-to-year variations in yields and regional commodity production are associated with variations in climate. Efforts were made to describe this relationship through statistical analysis of the correlation of yields with monthly rainfall. These analyses were a first attempt to use statistics to describe the nature of the relationship between variable yields, production, and climate. Later, the relationship between yields and rainfall was studied using multiple correlation methods. Since the 1920s (with some refinements in techniques), correlation and regression analyses have become the favorite tools for describing yield-weather relations.

Energy Balance Quantification

The first half of the twentieth century saw great contributions toward the quantification of water loss and use by plants. Research studies on the measurement and analysis of energy fluxes above crops and on crop evapotranspiration were stimulated.

It was in this period that Bowen proposed a method (Bowen ratio) of partitioning the energy used in evaporation and heating the air. Penman published a rational method for using meteorological observations to estimate evaporation from a free water surface and vaporization from a plant canopy.

The Penman method for estimating evapotranspiration has become a standard tool for estimating the need for irrigation water by agricultural engineers, agronomists, and meteorologists throughout the world. At the same time, the Thornthwaite method for estimating potential evapotranspiration was published. Further research into the energy balance resulted in the development of the eddy correlation method for estimating latent and sensible heat transfer. Later, Blaney and Criddle developed the consumptive use principle for irrigation scheduling. This technique has been widely used by agriculturists in the semiarid regions of the world.

In more recent years, advanced and reliable instrumentation has made possible continuous measurements of biometeorological exchange processes, such as measurements of mass and energy exchange to assess the plant community's response to atmospheric variables.

Biological Studies in Controlled Climates

By the middle of the twentieth century, technology was available to build facilities in which biological responses to environmental conditions could be measured quantitatively. These facilities provided a way to study the responses of plants and animals to diurnal variations in weather conditions. One such facility was at the California Institute of Technology. At about the same time, a large animal facility at the University of Missouri, called the Missouri Climatic Laboratory, was established for studies dealing with the physiological response and production of dairy cattle to variations in temperature, humidity, wind, and radiation loads. As a result of these facilities, several excellent studies have contributed to a better understanding of climate and weather effects on plants and animals. These two laboratories served as precursors to the development of the growth chambers used today in nearly every part of the world.

Modeling Biological Response to Climate

By the late 1960s and early 1970s, an extensive literature was available that documented the response of plant growth and development to environmental conditions. This information paved the way for work on mathematical models of plant response and yields to varying environmental conditions. The comprehensive development and use of plant and animal dynamic simulation models started with the availability of computers in the early 1970s. By the close of the twentieth century, several thousand computer-based plant and animal dynamic simulation models had been developed to expand scientific insights into complex biological and environmental systems, and their use has resulted in huge economic benefits.

The application of crop simulation models and simulation-based decision support systems became more acceptable to the agricultural community during the final decade of the twentieth century (Hoogenboom, 2000). Increases in the sophistication of computers and decreased costs are further fueling rapid advances in modeling. Interest has arisen in the topic of scaling from the leaf level to the global scale, to examine the global nature of climate change and its impact (Paw U, 2000).

Remote Sensing of the Environment and Vegetation

Remote sensing detects and measures the characteristics of a target without being in physical touch with it. Information about the object is derived through electromagnetic energy. Aircraft and satellites are the main platforms for remote-sensing observations. Aerial photographs are the original form of remote sensing and remain the most widely used method. Infrared thermometry provides a way to determine the surface temperatures of plants and animals. Precise handheld infrared thermometers are commercially available to provide these measurements. The technology allows the measurement of the surface temperature with a resolution of a few square centimeters.

The development and deployment of earth satellites in the 1970s brought a revolution in remote sensing. Remote sensing now provides a sequence of reliable and irreplaceable information for agriculture planning and management (Maracchi, Pérarnaud, and Kleschenko, 2000).

Weather and Climate Information for Agriculture

The agriculture industry is the most sensitive to variability in weather and climate. Throughout the world, efforts have been made to provide agriculture with a specifically focused weather service. Most countries of the world have developed programs to provide agroclimatological services.

Unfortunately, in many countries there is a lack of coordination and cooperation to link agencies representing agriculture and meteorology in their efforts to advise farmers of weather-related risk management. This lack of cooperation has adversely affected improvements and further development in agro-advisory services. Furthermore, due to a lack of financial support, the network of meteorological stations does not adequately cover various agrometeorological zones to meet potential needs. Conflicts within and between countries often halt the collection and exchange of weather data. This has a detrimental impact on projects in which analysis of weather and climate data is attempted.

Crop-Related Climate Data Archives

Early studies dealing with relationships between yields and climate were accomplished using limited climatic data and primitive computational procedures. The advent of the computer era saw the development of new methods for storing historical data. Computer programs are now available that can electronically archive huge amounts of climatic data. These archives have further enhanced the evaluations of weather and climate risk for agriculture. The World Meteorological Organization (WMO) supports the sharing of computerized climate data in all countries of the world (Decker, 1994).

Climate Change and Impact

Over the past 100 years, human activities have significantly altered the earth's atmosphere. Increases in the concentrations of greenhouse gases have led to warming of the earth's surface. An accumulating body of evidence suggested that by the last decade of the twentieth century global warming had already made significant negative impacts in a large number of regions. The menace of global climate change became a central issue of investigation in the 1990s and beyond. The investigations considered the effects of global warming on individual plants, plant stands, and entire vegetation units from regional to global scales (Overdieck, 1997). The investigations were not confined only to plants; the impact on hydro-resources, livestock, insect pests, and diseases has also been investigated.

FUTURE NEEDS

Agroclimatological Database

The availability of a proper meteorological and agrometeorological database is a major prerequisite for studying and managing the processes of agricultural and forest production. Historical data and observations during the current growing season will play a critical role in increased applications of crop models and model-generated output by farmers, consultants, and other policy- and decision makers. A major and inevitable priority is to build a database of meteorological, phenological, soil, and agronomic information. The acquisition of pertinent climate and agrometeorological data, their processing, quality control, and archiving, and timely access and database management are important components that will make the information valuable to agrometeorological research and operational programs. The major concerns for the availability of climate and agrometeorological data will continue to be in the areas of data collection and database management (Sivakumar, Stigter, and Rijks, 2000; Stigter, Sivakumar, and Rijks, 2000).

Research

The most important development for science in general and for agrometeorology in particular is the rapid advances in electronics and their impact on computer, communication, and measurement technologies. The potential to handle data by computers and exchange them globally via the Internet is growing daily. Computers have opened the gates to the ability to store huge amounts of data and to process them through more computationally intensive statistical techniques (Serafin, Macdonald, and Gall, 2002). In agrometeorology, in which a vast amount of atmospheric data must be linked with complex sets of biological data, the availability of data in a uniform file format and the vanishing of data processing limitations result in a strong momentum for research.

Agrometeorological Models

Agrometeorological models have many potential uses for answering questions in research, crop management, and policy. As society becomes more computerized and technology oriented, there will be a greater possibility for the application of crop simulation models and decision support systems to help provide guidance in solving real-world problems related to agricultural sustainability, food security, the use of natural resources, and protection of the environment.

Environmental Management

A major area for future research is the response of environmentally sensitive agricultural practices to weather events (De Pauw, Göbel, and Adam, 2000). As the public becomes more concerned about the environment, greater pressure will be put on the agricultural community to document and prove that chemical applications are not harming the environment. This will require a better understanding of the role weather plays in the fate of agricultural chemicals during application, their persistence and movement after application, and their effect on natural organisms. To gain this understanding, research will require a more extensive interdisciplinary approach than is employed today. Adaptive research is required under on-farm or at least close to on-farm conditions, ideally with farmers participating (Olufayo, Stigter, and Baldy, 1998).

Climate Change Impact

One of the most prominent current problems of humankind is global warming and its impact on the environment, water resources, agriculture, and human health. Agrometeorology has to play a leading role in the assessment of climate change, its impact on the biosphere, and adaptation strategies to increasing climate variability and climate change.

Investigation of the effects of global warming on animals will be another challenge. Future animal agrometeorologists have to search deeper into animal responses to specifically defined factors of the environment. These findings will permit the development of more adaptive, more tolerant, and more productive animals in stressful environments (Salinger, Stigter, and Das, 2000).

Pest and Disease Management

Increasing environmental, population, and economic pressures are creating difficulties in solving agricultural pest and disease management problems. Future climate change and increased variability will further complicate pest and disease management problems. This will require improved analyses of the weather to develop new pest management techniques and strategies. Agrometeorologists trained in weather-pest and weather-disease relationships and in the basics of pest management disciplines need to play a key role in developing pest and disease management strategies (Strand, 2000).

Education and Training

Neither education nor training is a one-time effort. The acquisition of knowledge and skills should be viewed as a continuous process throughout one's career (Lomas, Milford, and Mukhala, 2000). The need for continued training in agrometeorology was demonstrated by a survey on education and training requirements by the WMO (Olufayo, Stigter, and Baldy, 1998). The study revealed that the national meteorological and hydrological services in many countries do not have adequately trained personnel. Furthermore, in many countries there are neither facilities nor sufficient national resources available to train personnel in the home country or abroad.

In the academic setting, there is a need for creative educational programs in schools. These programs will educate younger generations on the importance of agriculture and how weather affects the food supply (Blad, 1994). Development of such programs requires professionals with an understanding of the interactions of weather and climate with agriculture. Unfortunately, fewer professionals are being trained in this discipline because of limited independent programs or departments in universities. Future training of agricultural meteorologists at the university level will require cooperative efforts between the WMO and member countries.

A second area of education that has been lacking is that of the public and agricultural producers. Perry (1994) points out that agricultural producers need to learn how to better use weather-driven models in their daily decision making. More important, they need to be taught how weather affects the various decisions they make and how their productivity or profit can be improved by using this information. The perception is that both long- and short-range forecasts are not sufficiently reliable to use in decision making (Jagtap and Chan, 2000). Research programs are needed to improve and quantify the reliability of forecasts and show how these forecasts can be used to improve decision making. Subsequently, extension programs are needed to transfer these findings to agricultural producers.

Services

Information has value when it is disseminated in such a way that end users receive the maximum benefit from applying it (Weiss, Van Crowder, and Bernardi, 2000). Areas of agricultural expertise that have prospered throughout the years are those with a product that is appreciated and used by farmers. Plant breeding, soil science, entomology, and plant pathology are areas that have been particularly successful. Each has some specific products that attract agricultural producers. The opportunity for agrometeorological services will grow dramatically if the importance and economic benefits of agrometeorological services are demonstrated. A major challenge to agricultural meteorologists is to educate agricultural producers to use weather data in various management decisions (Seeley, 1994). Demonstration of successful uses of the climate and weather through case studies is a useful example to begin discussion and to transfer potential applications to adopters of new technology.

Agrometeorology has a broad number of perspectives and applications. Computer usage has brought rapid advances in this science and has opened new doors in perspectives and applications that were not available before. The twenty-first century offers a challenge for the development of applications, risk analysis, crop and forest models, and assessments of production under global warming.

Chapter 2

Solar Radiation and Its Role in Plant Growth

THE SUN: THE SOURCE OF ENERGY

The sun is the nearest star to the earth, and its radiant energy is practically the only energy source to the earth. Very small and insignificant quantities of energy are available from other sources such as the interior of the earth, the moon, and other stars. The mean sun-earth distance, also known as one astronomical unit (1 AU), is 1.496×10^8 km or, more accurately, 149, 597, 890 ± 500 km. The earth revolves round the sun in an elliptical orbit. The minimum sun-earth distance is about 0.983 AU and the maximum approximately 1.017 AU. The earth is at its closest point to the sun (perihelion) on approximately January 3 and at its farthest point (aphelion) on approximately July 4. The visible disk or photosphere has a radius of 6.599×10^5 km, and the solar mass is 1.989×10^{30} kg (Goody and Yung, 1989; Iqbal, 1983).

The sun is a completely gaseous body. The chemical composition of the outer layers is (by mass) 71 percent hydrogen, 26.5 percent helium, and 2.5 percent heavier metals. Its physical structure is complex, although several regions, including the core, photosphere, reversing layer, chromosphere, and corona, are well recognized.

The innermost region, the *core,* is the densest and hottest part of the sun. It is composed of highly compressed gases at a density of 100 to 150 $g{\cdot}cm^{-3}$. The core temperature is in the range of 15×10^6 to 40×10^6°C. Outside the core is the *interior* which contains practically all of the sun's mass. The core and interior are thought to be a huge nuclear reactor in which fusion reactions take place. These reactions supply the energy radiated by the sun. The most important reaction is the process by which hydrogen is transformed to helium. The energy is first transferred to the surface of the sun and then radiated into space. The radiation from the core and interior of the sun is thought to be in the form of X rays and gamma rays.

The surface of the sun, called the *photosphere,* is the source of most of the visible radiation arriving at the earth's surface. The photosphere is the crust that is visible to the naked eye when looking at the sun through a blue glass. It is composed of very low density gases. The temperature in this region is 4,000 to 6,000°C. In spite of the fact that it has low density (10^{-4} that of air at sea level), the photosphere is opaque because it is composed of strongly ionized gases. The photosphere is the source of radiation flux to space because it has the capability to emit and absorb a continuous spectrum of radiation.

Outside the photosphere is the *solar atmosphere,* which is several hundred kilometers deep and almost transparent. This solar atmosphere is referred to as the *reversing layer.* This layer contains vapors of almost all of the known elements found on the earth. Outside the reversing layer is the *chromosphere,* which is about 25,000 km deep. It is seen from the earth only during a total eclipse when it appears as a rosy color layer. It is in this zone that the short-lived, brilliant solar flares occur in the clouds of hydrogen and helium. These flares are a source of intense bursts of ultraviolet (UV) and radio wave radiation. The solar flares also eject streams of electrically charged particles called *corpuscles,* which, on reaching the earth's surface, disturb its magnetic field. The temperature in the chromosphere is several times higher than that of the photosphere.

The outermost portion of the sun is the *corona,* which is composed of extremely rarefied gases known as the *solar winds.* These winds are believed to consist of very sparse ions and electrons moving at very high speeds and are thought to extend into the solar system. The corona can be seen during a total eclipse. It has a temperature on the order of 1,800,000°K. There is no sharp boundary to this outermost region.

These zones suggest that the sun does not act as a perfect black body radiator at a fixed temperature. The radiation flux is the composite result of its several layers. For general purposes, however, the sun can be referred to as a black body at a temperature of 5,762°K. The sun rotates at a rate that is variable in depth and latitude. As measured by the motion of sunspots, the synodic period (as seen from the earth) is $26.90 + 5.2 \sin^2$ (latitude) days.

The sun is a variable star. It is estimated to be about 5×10^9 years old. Theories of climatic changes on geological time scales indicate definite changes that must have taken place during the lifetime of the sun. According to widely accepted theories, when the sun was formed it was 6 percent smaller and 300°K cooler, and its irradiance was 40 percent lower than present-day values (Goody and Yung, 1989).

Some of the variations occurring in the sun are monitored on a regular basis. These variations are associated with magnetic activity resulting from interactions between convective motions, the solar rotation, and the general magnetic field of the sun. Magnetic fields and electric currents penetrate the chromosphere and corona, where magnetic variations have far greater influence because of the low densities.

The most striking visual disturbances are on the photosphere, and these are known as *sunspots.* These are patches varying in diameter from a few thousand to 100,000 kilometers, with an emission temperature in the center about 1,500°K lower than that of the undisturbed photosphere. The fraction of the photosphere covered by spots is never more than 0.2 percent, and their average persistence is about a week. For most of the period for which the observations are available, a *sunspot cycle* averages 11.04 years. The number of spots is only one characteristic feature of the sun that changes in this rhythmic manner. Just after the minimum, spots first appear near 27° latitude in both hemispheres. As the cycle proceeds, they drift equatorward and disappear close to 8° latitude. They are rarely observed at latitudes higher than 30° or lower than 5°.

When a sunspot is near to the extremity it can be seen to be surrounded by a network of enhanced photospheric emission, patches which are called *faculae.* These photospheric emissions have longer lifetimes than the associated sunspot group, appearing before and disappearing after the spots themselves.

Flocculi or *plages* are other disturbances that are typical features in hydrogen light (H-alpha). Flocculi are the most prominent features, and they occur at high latitudes, where spots do not. Occasionally, a hydrogen flocculus near a spot will brighten up. In extreme cases, the brightening is visible to the eye. These brightenings are known as *solar flares,* and they are associated with great increases of Lyman alpha and other ultraviolet radiations that influence the upper atmosphere.

Prominences are photospheric eruptions extending into the chromosphere. Many different forms occur, but a typical prominence might be 30,000 km high and 200,000 km long, with a temperature of 5,000°K.

Large changes in the corona are well established. Coronal ultraviolet emission is the heat source for levels in the upper atmosphere where the density is very low. The thermosphere, above 150 km, is greatly influenced by variable conditions on the sun. Coronal disturbances are closely related to the sunspot cycle. In visible light the corona appears more jagged at the sunspot maximum than at the minimum. Solar radio emission from the corona shows a marked variation with the sunspot cycle and is also correlated with shorter period changes in sunspot number.

Solar Constant

The sun is the source of more than 99 percent of the thermal energy required for the physical processes taking place in the earth-atmosphere system. The solar constant is the flux of solar radiation at the outer boundary of the earth's atmosphere, received on a surface held perpendicular to the sun's direction at the mean distance between the sun and the earth. The value of the solar constant is 1,370 W m^{-2} (about 2 cal·cm^{-2}·min^{-1}), giving an average flux of solar energy per unit area of the earth's surface equal to 350 W m^{-2}. The solar constant is only approximately constant. Depending on the distance of the earth from the sun, its value ranges from approximately 1,360 to 1,380 W m^{-2}.

Of this energy, approximately 31 percent is scattered back to space, 43 percent is absorbed by the earth's surface, and the atmosphere absorbs 26 percent. The ratio of outward to inward flux of solar radiation from the entire earth's surface (termed *albedo*) is about 0.31, leaving an average around 225 W m^{-2} (range 220 to 235 W m^{-2}) that is available for heating, directly and indirectly, the earth-atmosphere system (Goody and Yung, 1989; Kiehl and Trenberth, 1997; Roberto et al., 1999). The irradiation amount at the earth's surface is not uniform, and the annual value at the equator is 2.4 times that near the poles. The solar energy incident upon a surface depends on the geographic location, orientation of the surface, time of the day, time of the year, and atmospheric conditions (Boes, 1981).

NATURE AND LAWS OF RADIATION

The behavior of electromagnetic radiation may be summed up in the following simplified statements:

- Every item of matter with a temperature above absolute zero emits radiation.
- Substances that emit the maximum amount of radiation in all wavelengths are known as black bodies. Such bodies will absorb all radiation incident upon them. A black body is thus a perfect radiator and absorber.
- Substances absorb radiation of wavelengths, which they can emit.
- The wavelengths at which energy is emitted by substances depend on their temperature—the higher the temperature, the shorter the wavelength.
- Gases emit and absorb radiation only in certain wavelengths.
- The amount of radiation absorbed by a gas is proportional to the number of molecules of the gas and the intensity of radiation of that wavelength.

Wavelength

The wavelength of electromagnetic radiation is given by the equation

$$\lambda = c/v \tag{2.1}$$

where λ is the wavelength, the shortest distance between consecutive crests in the wave trans; c is the constant equal to the velocity of light, 3×10^{10} $cm \cdot sec^{-1}$; and v is the frequency, the number of vibrations or cycles per second.

Planck's Law

Electromagnetic radiation consists of the flow of quanta or particles, and the energy content (E) of each quantum is proportional to the frequency given by the equation

$$E = hv \tag{2.2}$$

where h is Planck's constant (having a value of 6.625×10^{-27} $erg \cdot sec^{-1}$) and v is the frequency. The equation indicates that the greater the frequency, the greater is the energy of the quantum.

Kirchoff's Law

Any gray object (other than a perfect black body) that receives radiation disposes of a part of it in reflection and transmission. The absorptivity, reflectivity, and transmissivity are each less than or equal to unity.

This law states that the absorptivity a of an object for radiation of a specific wavelength is equal to its emissivity e for the same wavelength. The equation of the law is

$$a\,(\lambda) = e\,(\lambda). \tag{2.3}$$

Stefan-Boltzman Law

This law states that the intensity of radiation emitted by a radiating body is proportional to the fourth power of the absolute temperature of that body:

$$\text{Flux} = \sigma T_a^4 \tag{2.4}$$

where σ is the Stefan-Boltzman constant (5.67×10^{-5} erg·cm^{-2}·sec^{-1}·K^{-4}) and T_a is the absolute temperature of the body.

Wein's Law

The wavelength of maximum intensity of emission from a black body is inversely proportional to the absolute temperature (T) of the body. Thus,

$$\text{Wavelength } (\lambda) \text{ of maximum intensity } (\mu\text{m}) = 2897\ T^{-1}. \qquad (2.5)$$

For the sun the wavelength of the maximum emission is near 0.5 μm and is in the visible portion of the electromagnetic spectrum.

Lambert's Law

This law states the permeability of the atmosphere to solar radiation. The intensity of solar radiation on a vertical irradiation at the earth's surface is given by the equation

$$I_m = I_o qm \qquad (2.6)$$

where I_o represents the solar constant, q is the transmission factor for the layer thickness 1 (solar angle 90°), and m represents distance of the air transversed. When the transmission factor q is replaced by the extinction coefficient a $(a = In \cdot q)$, the equation takes the form

$$I_m = I_o e^{-a \cdot m}. \qquad (2.7)$$

About 95 percent of the sun's radiation is contained between 0.3 and 2.4 μm, 1.2 percent in wavelengths < 0.3 μm, and 3.6 percent in wavelengths > 2.4 μm (Iqbal, 1983). A systematic division of solar radiation according to frequency and wavelength is given in Tables 2.1 to 2.3. An approximation of energy content in various segments of shortwave radiation is given in Table 2.2. A more detailed picture of the energy content and nature of the solar radiation spectrum is given in Table 2.3.

EARTH'S ANNUAL GLOBAL MEAN RADIATIVE ENERGY BUDGET

The global annual mean energy budget is determined by the net radiation flow of energy through the top of the atmosphere and at the earth's surface.

TABLE 2.1. Broad bands of the solar spectrum

Color	λ (μm)	Irradiance W m^{-2}	% of solar constant
Ultraviolet	< 0.4	109.81	8.03
Visible	0.390-0.770	634.40	46.41
Infrared	> 0.77	634.40	46.40

Source: Adapted from Iqbal, 1983.

TABLE 2.2. Electromagnetic spectrum energy content in various color bands

Color	λ (μm)	Irradiance W m^{-2}	% of solar constant
Violet	0.390-0.455	108.85	7.96
Blue	0.455-0.492	73.63	5.39
Green	0.492-0.577	160.00	11.70
Yellow	0.577-0.597	35.97	2.63
Orange	0.597-0.622	43.14	3.16
Red	0.622-0.770	212.82	15.57

TABLE 2.3. Partition of solar irradiation, 0.2 to 5.0 μm wavelength

Wavelength (λ μm)	% Irradiance	Wavelength ((λ μm)	% Irradiance
0.20	0.003	0.55	6.675
0.22	0.024	0.60	6.300
0.24	0.102	0.65	5.585
0.26	0.477	0.70	4.972
0.28	0.817	0.80	3.882
0.30	1.873	0.90	3.031
0.32	2.520	1.0	2.418
0.34	3.031	1.10	1.975
0.36	3.542	1.20	1.635
0.38	4.019	1.4	1.090
0.39	4.257	1.6	0.715
0.40	4.904	1.8	0.511
0.42	6.198	2.0	0.368
0.44	6.879	2.5	0.167
0.46	7.356	3.0	0.89
0.48	7.390	4.0	0.031
0.50	7.152	5.0	0.014

Source: Adapted from Goody and Yung, 1989.

At the top of the atmosphere, the net energy output is determined by the incident shortwave radiation from the sun minus the reflected shortwave radiation. This difference determines the net shortwave radiation flux at the top of the atmosphere. To balance this inflow of shortwave energy, the earth-atmosphere system emits longwave radiation to space.

Satellite observations of the top of the atmosphere have made fairly accurate estimates of the global mean energy budget. According to these estimates, the global mean annual outgoing longwave radiation is 235 W m^{-2} and the annual mean absorbed shortwave flux is 238 W m^{-2}. Hence, the measured top-of-atmosphere budget balances to within 3 W m^{-2}. A part of this imbalance could be associated with the buildup of greenhouse gases and a part is probably associated with El Niño events (Kiehl and Trenberth, 1997).

Incoming Shortwave Radiation

Solar radiation that encounters matter, whether solid, liquid, or gas, is called incident radiation. Interactions with matter can change the following properties of incident radiation: intensity, direction, wavelength, polarization, and phase. Radiation intercepted by the earth is absorbed and used in energy-driven processes or is returned to space by scattering and reflection (Figure 2.1). In mathematical terms, this disposal of solar radiation is given by the equation

$$Qs = Cr + Ar + Ca + Aa + (Q + q)(1 - a) + (Q + q)\,a \qquad (2.8)$$

where Qs is the incident solar radiation at the top of the atmosphere; Cr is reflection and scattering back to space by clouds; Ar is reflection and scattering back by air, dust, and water vapors; Ca is absorption by clouds; Aa is absorption by air, dust, and water vapors; $(Q + q)\,a$ is reflection by the earth; $(Q + q)(1 - a)$ is absorption by the earth's surface, where Q and q are, respectively, direct beam and diffused solar radiation incident on the earth and a is albedo. The global disposal of shortwave radiation (W m^{-2} per year) is given in Table 2.4.

About a quarter of the solar radiation is reflected back to space by clouds. On average, the reflection is greatest in middle and high latitudes and least in the subtropics. A small portion of the incident radiation is scattered back to space by the constituents of the atmosphere, mainly air molecules, dust particles, and water vapors. About 30 percent of the radiation is scattered downward. Atmospheric scattering results from multiple interactions between light rays and the gases and particles of the atmosphere. The two

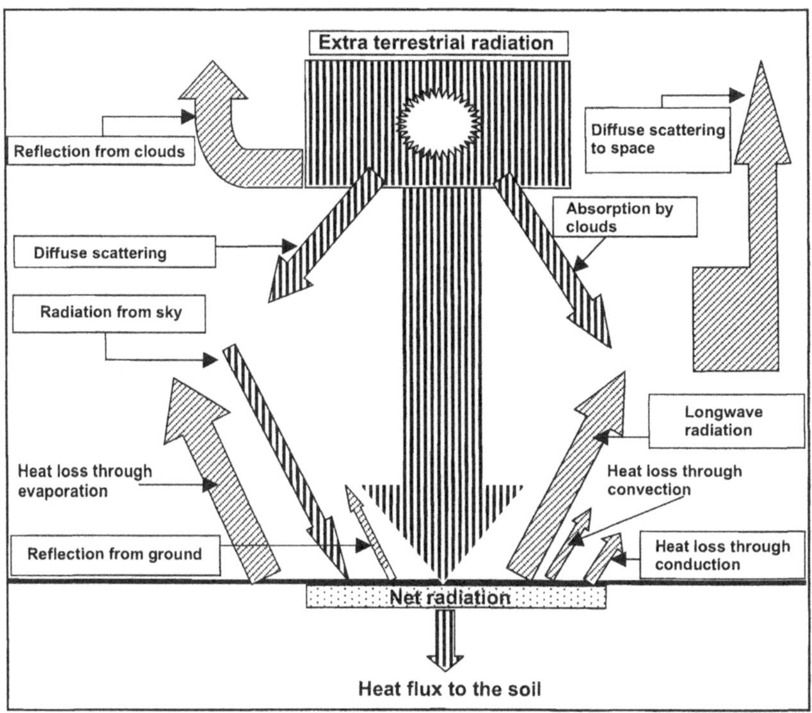

FIGURE 2.1. Daytime radiation balance over the earth's surface

TABLE 2.4. Disposal of solar radiation

Solar energy	W m^{-2}
Incident on the top of the atmosphere	342
Reflected by clouds, aerosols, and atmosphere	77
Reflected from the earth	30
Total reflected	107
Absorbed by the atmosphere	67
Absorbed by the earth	168
Total absorbed by earth-atmosphere system	235

Source: Adapted from Kiehl and Trenberth, 1997.

major processes, selective scattering and nonselective scattering, are related to the size of the particles in the atmosphere. In selective scattering, the shorter wavelength of UV energy and blue light are scattered more severely than that in longer wavelengths (red) and infrared (IR) energy. Selective scattering is caused by fumes and by gases such as nitrogen, oxygen, and carbon dioxide. This is known as Rayleigh scattering and is the primary cause of the blue color of the sky. For larger sizes of particles, scattering is independent of the wavelength, i.e., white light is scattered. The phenomenon is known as Mei scattering. As the path length increases, the percentage of solar energy in the visible part decreases. Within the visible part itself, the ratio of the blue to the red part decreases with increased path length. This is because the part of the spectrum with higher frequency is scattered to a greater extent than the part with lower frequency. The red color of the sky at sunrise and sunset is because of increased path length in the atmosphere which scatters blue and green wavelengths so only red light reaches the viewer (Sabins, 1997).

The atmosphere absorbs about 20 percent of the solar radiation. The constituents of the atmosphere that absorb the solar radiation significantly are oxygen, ozone, carbon dioxide, and water vapors. This absorption is of great importance to life on the earth's surface, because only a very small amount of this radiation can be tolerated by living organisms.

Oxygen and ozone: Solar radiation in the wavelengths <0.3 μm is not observed on the ground. It is absorbed in the upper atmosphere. Energy of <0.1 μm is highly absorbed by the atomic and molecular oxygen and also by nitrogen in the ionosphere. Energy of 0.1 to 0.3 μm is absorbed efficiently by ozone in the ozonosphere. Further but less complete ozone absorption occurs in the 0.32 to 0.36 μm region and at minor levels around 0.6 μm (visible part) and 4.75 μm, 9.6 μm, and 14.1 μm (infrared part).

Carbon dioxide: This gas is of chief significance in the lower part of the atmosphere. Carbon dioxide has a weak absorption band at about 4 μm and 10 μm and a very strong absorption band around 15 μm.

Water vapor: Among the atmospheric gases, water vapors absorb the largest amount of solar radiation. Several weak absorption bands occur below 0.7 μm, while important broad bands of varying intensity exist between 0.7 and 0.8 μm. The strongest water absorption is around 6 μm, where almost 100 percent of longwave radiation may be absorbed if the atmosphere is sufficiently moist (Barrett, 1992).

Thus, after reflection, scattering, and absorption in the atmosphere, about half of the solar radiation reaches the earth's surface. Out of this, about 6 percent is reflected back to outer space. This is known as albedo. The albedo is defined as the fraction of incoming shortwave radiation that is reflected by the earth's surface. The albedo varies with the color and composition of the earth's surface, the season, and the angle of the sun's rays. The values are higher in winter as well as at sunrise and sunset. The albedo also varies with the wavelength of the incident radiation (Roberto et al., 1999). Very small values have been recorded in the ultraviolet part of the spectrum and higher values in the visible part. The albedo values of some selected surfaces are given in Table 2.5.

Outgoing Longwave Radiation

The surface of the earth after being heated by the absorption of solar radiation becomes a source of radiation itself (Figure 2.2). Because the average temperature of the earth's surface is about 285°K, 99 percent of the radiation is emitted in the infrared range from 4 to 120 μm, with a peak near 10 μm, as indicated by Wein's displacement law. This is longwave radiation and is also known as *terrestrial radiation.* The average annual global disposal of infrared radiation is represented by equations 2.9, 2.10, and 2.11.

$$I(e) = Ia + Is \tag{2.9}$$

$$I(a) = I\downarrow + I(a)s \tag{2.10}$$

$$I = I(e) - I\downarrow \tag{2.11}$$

where $I(e)$ is infrared radiation emitted by the earth's surface; Ia is infrared radiation from the earth's surface absorbed by the atmosphere; Is is infrared radiation from the earth lost to space; $I(a)$ is infrared radiation from the atmosphere; $I\downarrow$ is counter radiation; $I(a)s$ is infrared radiation from the atmosphere lost in space; and I is the effective outgoing radiation from the earth. The quantitative disposal of longwave radiation (W m^{-2} per year) from the earth-atmosphere system is summarized in Table 2.6.

The earth's atmosphere absorbs about 90 percent of the outgoing radiation from the earth's surface. Water vapors absorb in wavelengths of 5.3 to 7.7 μm and beyond 20 μm; ozone in wavelengths of 9.4 to 9.8 μm; carbon dioxide in wavelengths of 13.1 to 16.9 μm; and clouds in all wavelengths. Longwave radiation escapes to space between 8.5 and 11.0 μm, known as the *atmospheric window.* A large part of the radiation absorbed by the atmosphere is sent back to the earth's surface as counter radiation. This counter radiation prevents the earth's surface from excessive cooling at night.

TABLE 2.5. Albedo of shortwave radiation

Surface	Albedo (%)	Surface	Albedo (%)
Fine sandy soil	37	Alfalfa	2-5
Dark black soil	14	Cotton	20-22
Moist black soil	8	Grass (dry)	31-33
Deciduous forest	17	Grass (green)	26
Pine forest	14	Lettuce	22
Prairie	12-13	Lucerne	23-32
Desert scrubland	20-29	Maize	16-23
Ice sheet with water	26	Rice	11-21
Sea ice	36	Sugar beet	18
Dense clean dry snow	86-95	Rye	11-21
Water surface at 30° latitude	6-9	Wheat	16-23

Source: Adapted from Barrett, 1992; Iqbal, 1983.

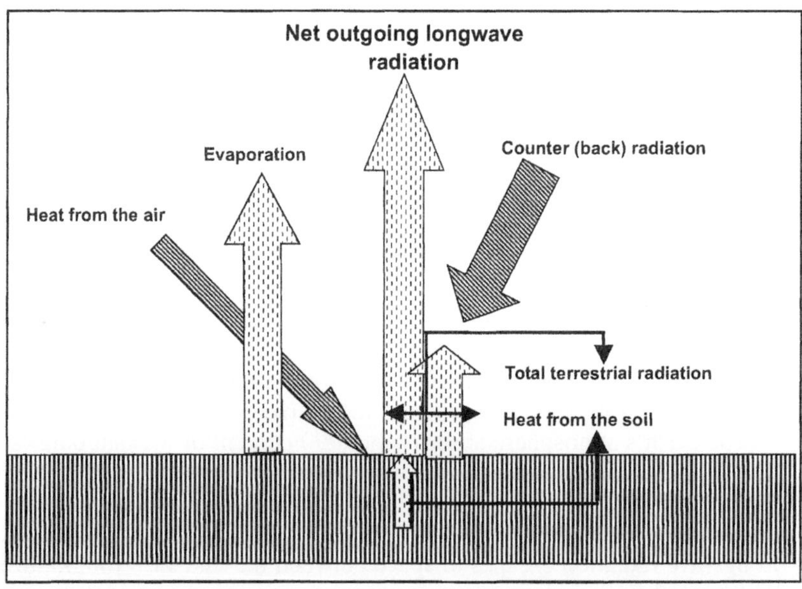

FIGURE 2.2. Outgoing longwave radiation balance at night

TABLE 2.6. Disposal of longwave radiation

Longwave radiation	**W m^{-2}**
Emitted by the earth's surface	390
Lost to space	40
Absorbed by the atmosphere	350
Emitted by the atmosphere and clouds	519
Lost to space from atmosphere	195
Back radiation from atmosphere absorbed by earth	324
Total outgoing longwave radiation	235

Source: Adapted from Kiehl and Trenberth, 1997.

Radiation Balance in the Earth-Atmosphere System

When averaged over the globe, the earth's surface absorbs about 168 W m^{-2} of solar radiation every year and effectively radiates 66 W m^{-2} of longwave energy to the atmosphere. The difference, +102 W m^{-2}, is the net radiation gain of the earth's surface. Likewise, the net radiation balance of the atmosphere comes to –102 W m^{-2} per year. Thus, the atmosphere loses as much radiative energy in a year as the earth's surface gains. To keep the thermal balance in equilibrium, energy is transferred from the earth's surface to the atmosphere. This vertical heat exchange occurs mainly through the evaporation of water from the surface of the earth (heat loss), through condensation in the atmosphere (heat gain), and by the conduction of sensible heat from the surface and transfer to the atmosphere through convection.

SOLAR RADIATION AND CROP PLANTS

Solar radiation is the energy source that sustains organic life on earth. Crop production is in fact an exploitation of solar radiation. The three broad spectra of solar energy described in this section are significant to plant life.

The shorter-than-visible wavelength radiation segment in the solar spectrum is chemically very active. When plants are exposed to excessive amounts of this radiation, the effects are detrimental. However, the atmosphere acts as a regulator in this type of solar radiation, and none of the cosmic, gamma, and X rays reach the earth (Evans, 1973). The ultraviolet radiation of this segment reaching the earth's surface is very low and is normally tolerated by plants.

Solar radiation in the higher-than-visible wavelength segment, referred to as infrared radiation, has thermal effects on plants. In the presence of water vapors, this radiation does not harm plants; rather, it supplies the necessary thermal energy to the plant environment.

The third spectrum, lying between the ultraviolet and infrared, is the visible part of solar radiation and is referred as light. This segment of solar radiation plays an important part in plant growth and development through the processes of chlorophyll synthesis and photosynthesis and through photosensitive regulatory mechanisms such as phototropism and photoperiodic activity. Light of the correct intensity, quality, and duration is essential for normal plant development. Poor light availability is frequently responsible for plant abnormalities and disorders. Virtually all plant parts are directly or indirectly influenced by this part of the spectrum. It affects the production of tillers; the stability, strength, and length of the culms; the yield and total weight of plant structures; and the size of leaves and root development (Rodriguez et al., 1999). The length of day or the duration of the light period determines flowering and has a profound effect on the content of soluble carbohydrates present. The majority of plants flower only when exposed to certain specific photoperiods. It is on the basis of this response that the plants have been classified as short-day plants, long-day plants, and day-neutral plants. When other environmental factors are not limiting it, photosynthesis increases with longer duration of the light period (Salisbury, 1981).

Reflection, Transmission, and Absorption

Reflection and transmission from the leaves have similar spectral distributions as shown in Figures 2.3 and 2.4. The maxima for both are in the green light as well as in the infrared region. The impression of the green color of the plants depends on the high reflectivity, the relatively high intensity of solar radiation, and the greater sensitivity of the human eye for green light. The strong infrared reflection from plants is an important natural device for protection of plant life against damage due to overheating. On average, the plant canopy absorbs about 75 percent of the incident radiation, with about 15 percent reflected and 10 percent transmitted.

Due to their chemical components or physical structures, plants absorb selectively in discrete wavelengths (Figure 2.5). The transparent epidermis allows the incident sunlight to penetrate into the mesophyll, which consists of two layers: (1) the palisade parenchyma of closely spaced cylindrical cells and (2) the spongy parenchyma of irregular cells with abundant interstices filled with air. Both types of mesophyll cells contain chlorophyll,

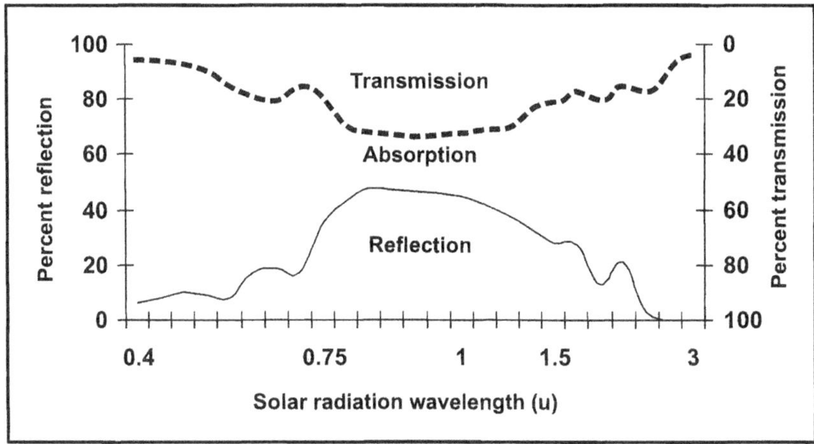

FIGURE 2.3. A generalized pattern of reflection, absorption, and transmission of solar radiation through a green leaf

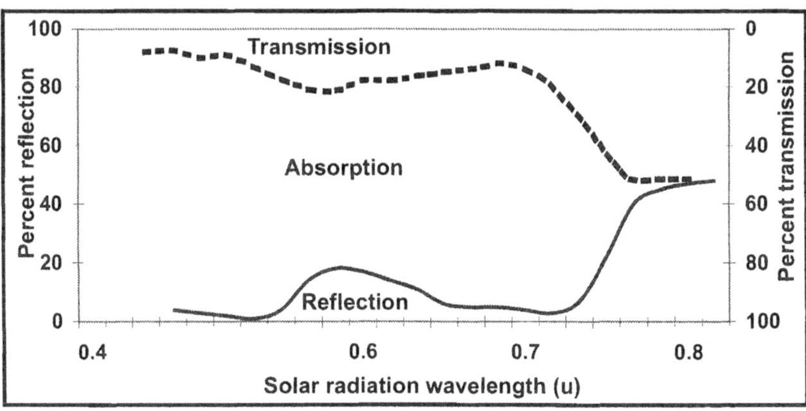

FIGURE 2.4. A generalized pattern of reflection, absorption, and transmission of light through a green leaf

which reflects part of the incident green wavelengths and absorbs all the blue and red energy for photosynthesis (Sabins, 1997). Chlorophyll absorption is maximum in the blue (0.45 μm) and in the red (0.65 μm) regions (Table 2.7). The longer wavelengths of photographic IR energy penetrate into the spongy parenchyma, where the energy is strongly scattered and reflected by the boundaries between the cell walls and air spaces. The high

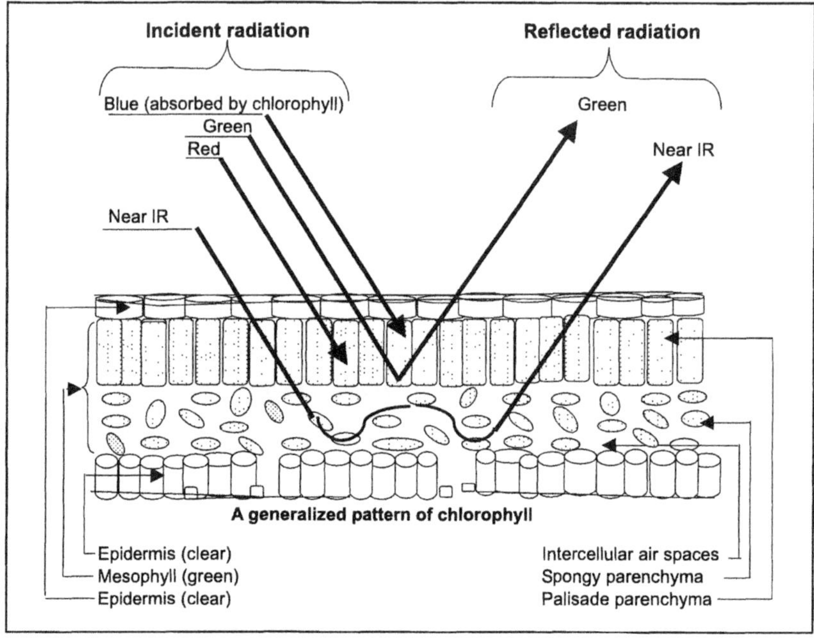

FIGURE 2.5. Interactions of incident solar radiation in a leaf cross section

TABLE 2.7. Green leaf response to spectral radiation components

Wavelength (μm)	Reflection (%)	Transmission (%)	Absorption (%)
0.34	9	0	91
0.44	11	2	87
0.51	14	10	76
0.58	14	10	76
0.64	13	9	78
1.0	45	50	5
2.4	7	28	65

Source: Adapted from Baumgartner, 1973.

near-infrared (near-IR) reflectance of leaves is caused not by chlorophyll but by the internal cell structure. Near the border of visible light, absorption by the plant decreases but then increases again in the infrared. Infrared radiation greater than 3 μm is completely absorbed by the plants.

It can be summed up that the plant leaf strongly absorbs blue and red wavelengths, less strongly absorbs the green, very weakly absorbs the near infrared, and strongly absorbs in the far-infrared wavelengths. Because the absorption of the near-infrared wavelengths (which contain the bulk of energy) by the leaf is limited, by discarding this energy it prevents the internal temperature from becoming lethal. At the infrared wavelengths, the plant leaf is an efficient absorber, but in these wavelengths the energy at the surface is small, with the result that the plant is a good absorber in the far-infrared. It is an equally a good radiator at these wavelengths.

The quality of radiation affects flowering, germination, and elongation. Red light with a wavelength of 0.66 μm is by far the most effective inhibitor of flowering in the case of long-day plants. Red light helps mature apples to turn red. Germination of seeds is inhibited when they are exposed to green, blue, and other short wavelength colors. However, germination is induced when seeds are exposed to the red portion of the spectrum. The red and infrared parts of the spectrum have reversible effects on seed germination. Stem elongation is promoted by exposure to far-red wavelengths, whereas the red part of the spectrum suppresses the elongation (Butler and Roberts, 1966; Takaichi et al., 2000).

The visible part of the spectrum also influences the orientation of shoots, phenomenon known as *phototropism* (Stowe-Evans, Luesse, and Liscum, 2001; Koller, Ritter, and Heller, 2001; Jin, Zhu, and Zeiger, 2001). When shoots turn toward the light, the phenomenon is known as *positive phototropism*. With increasing intensity of light, positive phototropism turns into *negative phototropism*. The strongest influence on phototropism is by the blue part of the spectrum (0.5 μm) and the weakest influence is by red rays. The phototropism action of the visible spectrum increases from the red to the blue part; subsequently, it declines again in the ultraviolet part. However, Ruppel, Hangarter, and Kiss (2001) have demonstrated that, in addition to the previously described blue-light-dependent negative phototropic response in roots, roots of wild-type and mutant (ACG 21) *Arabidopsis thaliana* display a previously unknown red-light-dependent positive phototropic response.

The ultraviolet and gamma part of the spectrum has only a slight effect on the plant. This may be partly because very little of this part of the spectrum reaches the earth's surface. However, it is well known that these rays have biological effects (Skorska, 2000; Predieri and Gatti, 2000). These rays may kill microorganisms, disinfect the soil, and eradicate diseases (Sharp and Polavarapu, 1999). Ultraviolet rays also influence the germination and quality of seeds. These rays lead to many irregularities in the growth and development of plants (Caldwell, 1981). Ultraviolet radiation leads to a strong inhibition of photosynthesis and metabolism (Karsten et al., 1999; Correia et al., 2000).

The solar spectrum can be divided into the following eight broad bands on the basis of the physiological response of plants:

1. Wavelength greater than 1.000 μm: Most of this radiation absorbed by plants is transformed into heat without interfering with the biochemical processes.
2. Wavelength 1.000 to 0.700 μm: Elongation effects on plants.
3. Wavelength 0.700 to 0.610 μm: Very strong absorption by chlorophyll, the strongest photosynthetic activity, and in many cases strong photoperiodic activity.
4. Wavelength 0.610 to 0.510 μm: Low photosynthetic effectiveness in the green segment and weak formative activity.
5. Wavelength 0.510 to 0.400 μm: Strong chlorophyll absorption, strong photosynthetic activity, and strong formative effects.
6. Wavelength 0.400 to 0.315 μm: Produces fluorescence in plants and a strong response by photographic emulsions.
7. Wavelength 0.315 to 0.280 μm: Significant germicidal action. Practically no solar radiation of wavelengths shorter than 0.29 μm reaches the earth's surface.
8. Wavelength shorter than 0.280 μm: Very strong germicidal action. It is injurious to eyesight and when below 0.26 μm can kill some plants. No such radiation reaches the earth's surface.

SOLAR RADIATION INTERCEPTION BY PLANTS

Three aspects of solar radiation are biologically significant. The first is the intensity of radiation, the amount of radiant energy falling on a unit of surface area in a unit of time. The second is the spectral distribution of radiation that governs the photochemical process of photosynthesis. The third aspect is the radiation distribution in time, which is important for photoperiodic phenomenon.

Quantification of intensity and spectral distribution of radiation within crop canopies is important because of its control of the photosynthetic process and the microclimate of the plant community. The rate of photosynthesis is dependent on the availability of photosynthetically active radiation intercepted by the leaves. The rate of transpiration taking place from the plant canopy is also controlled to a great extent by the radiation energy. Thus, knowledge of radiation transmission through the elements of a plant community is necessary to know the quality and quantity of incident radiation used by the plants.

The capture of radiation and its use in dry matter production depends on the fraction of the incident photosynthetically active radiation (PAR) that is intercepted and the efficiency with which it is used for dry matter production. Intercepted radiation (*Si*) is often estimated as the difference between the quantity of incident radiation (*S*) and that transmitted through the canopy to the soil (*St*). However, this approach has inherent technical and theoretical difficulties since is does not account for the reflection of incident radiation from the canopy surface (typically 5 to 20 percent depending on surface characteristics and moisture content), or for radiation intercepted by nonphotosynthetic canopy elements. As a result, interception by photosynthetically competent tissues may be greatly overestimated, particularly for canopies which are senescing or which contain numerous woody structural elements.

The quantity of radiation intercepted depends on the amount received by the canopy, canopy size, duration, and fractional interception (*f*). The seasonal time course of *f,* defined as *Si/St,* varies greatly depending on canopy architecture and the phenology of the vegetation involved. As such, *f* increases more rapidly in cereals such as sorghum than in legumes such as groundnut. Furthermore, mean *f* values calculated over the duration of the crop are generally lower in short-duration cereals and legumes than in perennial species, largely because of the differing duration of ground cover (Squire, 1990).

Factors Affecting the Distribution of Solar Radiation Within the Plant Community

The distribution of radiation in a plant canopy is determined by several factors, such as the transmissibility of the leaf, leaf arrangement and inclination, plant density, plant height, and the angle of the sun (Vorasoot, Tienroj, and Apinakapong, 1996; Cohen et al., 1999; Courbaud, Coligny, and Cordonnier, 2003). Leaves of deciduous trees, herbs, and grasses (including cereals) have transmissibility ranging from 5 to 10 percent. The broad leaves of evergreen plants have a value of 2 to 8 percent. Transmissibility varies slightly with the age of the leaf. The transmissibility of a young leaf is relatively high. With the maturing of the leaf, it declines but then rises again as the leaf turns yellow.

The transmissibility of a leaf is directly related to its chlorophyll content. The logarithm of transmissibility decreases linearly with an increase in the chlorophyll content. If the leaves that transmit 10 percent of the radiation were horizontally displayed in continuous layers, only 1 percent of light, mostly in the green region, could penetrate the second layer. However, leaves are rarely displayed horizontally. The relative light interception of horizontal and erect foliage is calculated in the ratio 1 to 0.44. Therefore, the actual light gradient within the canopy is not as steep as the transmissibility will suggest. On average, when the total leaf area equals the area at the ground, the mean transmissibility is around 75 percent for the more upright leaves and 50 percent for the more horizontal leaves. In weak light, any departure of the leaves from the horizontal position reduces the net photosynthesis. In full sunlight, the optimum leaf inclination for efficient light use is 81° (Figure 2.6). At full sunlight, a leaf placed at the optimum inclination is 4.5 times as efficient in using light as a horizontal leaf (Figure 2.7). For more efficient use of light, the upper leaves in a plant canopy should have a near-vertical orientation, whereas the lower foliage should be almost horizontal. An ideal arrangement of the plant canopy is for the lower 13 percent of the leaves to be oriented at an angle of 0 to 30°, the middle 37 percent of the leaves should be at 30 to 60°, and the upper 50 percent leaves should be at 60 to 90° with the horizontal (Chang, 1968).

In the case of young plants, the percentage of light interception is not only small but also variable with the time of day. It is at a minimum at noon and at maximum during the morning and evening hours. When the plant height increases, the interception of light by the canopy also increases, with only a small variation at different times of the day.

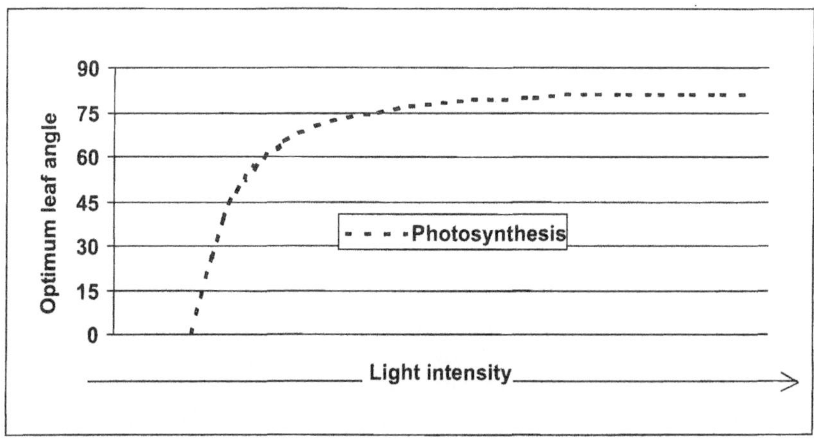

FIGURE 2.6. Light intensity and leaf angle for optimum photosynthesis (*Source:* Mavi, 1994.)

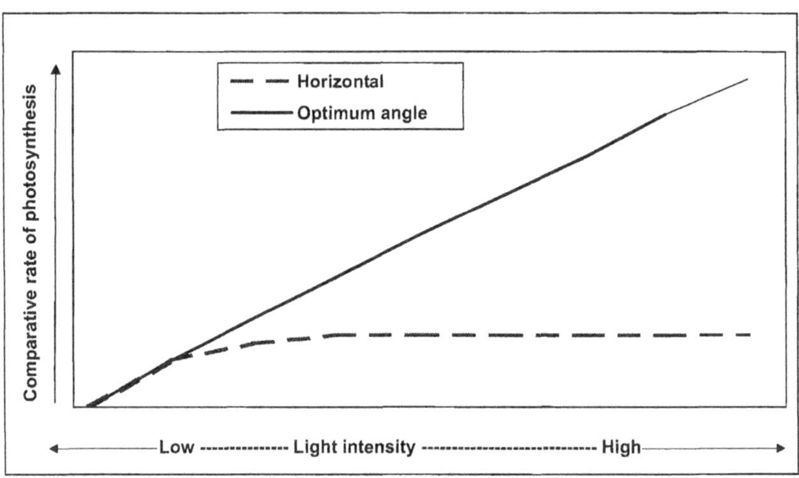

FIGURE 2.7. Rate of photosynthesis in a leaf placed at two different angles (*Source:* Mavi, 1994.)

Numerous investigators have studied the radiation distribution in a plant canopy and put forward equations for determining light at a particular height in a canopy (Monteith and Elston, 1983; Kull and Kruijt, 1998; Mariscal, Orgaz, and Villalobos, 2000; Marques, Filho, and Dallarosa, 2000). So far, the equation for Beer's law is thought to be the most appropriate. The equation of the law is written as

$$I = Ia\ e^{-kf} \tag{2.12}$$

where I is the intensity of light at a particular height within the canopy, Ia is the intensity at the top, k is the extinction coefficient of the leaf, f is the leaf-area index (LAI), and e is the base of natural log. The extinction coefficient can be defined as the ratio between the light loss through the leaf to the light at the top of the leaf. The extinction coefficient varies with the orientation of the leaf. Its value is low in stands with upright leaves and high in stands with more or less horizontal leaves.

Roujean (1999) made actual measurements of solar radiation profiles in black spruce canopies on typical summer days and compared those with Beer's law values (Figure 2.8). He observed certain deviations from the Beer's law extinction and assigned those to seasonal effects, such as the angle of the sun's rays.

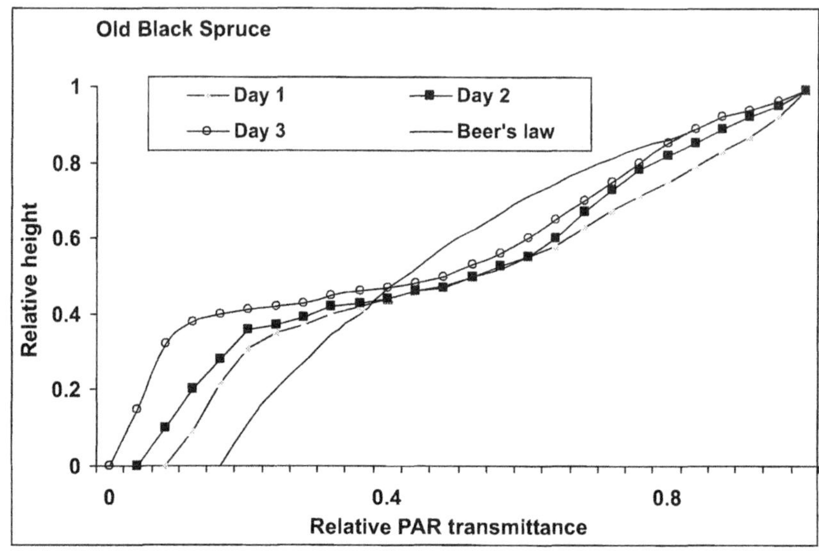

FIGURE 2.8. Solar radiation profile in a spruce forest (*Source:* Reprinted from *Agricultural and Forest Meteorology,* 93, J. L. Roujean, Measurement of PAR transmittance within boreal forest stands during BOREAS, pp. 1-6, 1999, with permission from Elsevier Science.)

Spectral Composition of Radiation in a Plant Canopy

As solar radiation penetrates the canopy, its quality undergoes transformation in different layers (Baumgartner, 1973). After every reflection and transmission, red and infrared radiation increases relative to the other wavelengths. In the interior of the canopy there is a relatively greater decrease of light in the chlorophyll absorption bands at 0.45 μm and 0.65 μm, and a relatively small decrease in green at 0.55 μm and infrared at 0.80 μm. In less tall crops such as alfalfa, about 30 percent of the total radiation and 20 percent of light reaches the ground. For a tall maize crop, the transmission of infrared radiation to the ground is 30 to 40 percent. In the visible part of the spectrum, the transmission is only 5 to 10 percent.

Flint and Caldwell (1998) measured global (total) and diffuse solar radiation in canopy gaps of a semideciduous tropical forest in Panama. Compared to unobstructed measurements taken outside the forest, the sunlit portions of gaps were depleted in the proportion of UV-B relative to PAR, especially at midday. Shaded areas, in contrast, were always richer in UV-B relative to PAR, but the magnitude of the change varied greatly. It was suggested that this variation was due to the differences in the directional nature of diffuse solar UV-B radiation as compared to diffuse PAR. Measurements in the gaps showed substantial reductions in the proportion of radiation in the diffuse components of both the UV-B and PAR wavebands. However, because of the greater proportion of UV-B that is diffuse, it tended to predominate in shaded areas. Similar patterns were seen in measurements taken at temperate latitudes.

The composition of solar radiation changes with the angle of the sun. The maximum visible spectrum penetration is at noon. Penetration of infrared radiation is comparatively high soon after sunrise and just before sunset. The early morning and evening values are higher because of the greater amount of diffused light. Anisimov and Fukshansky (1997) measured the spectral composition of incident solar and diffuse sky PAR as well as the spectral scattering coefficient of PAR for a green leaf. The results are shown in Figures 2.9 and 2.10.

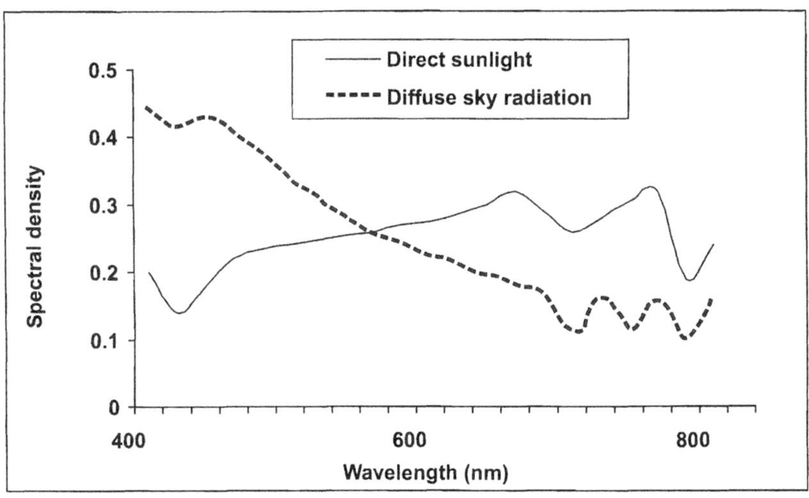

FIGURE 2.9. Daylight spectra of direct sunlight and diffuse sky radiation (*Source:* Reprinted from *Agricultural and Forest Meteorology,* 85, O. Anisimov and L. Fukshansky, Optics of vegetation: Implications for the radiation balance and photosynthetic performance, pp. 33-49, 1997, with permission from Elsevier Science.)

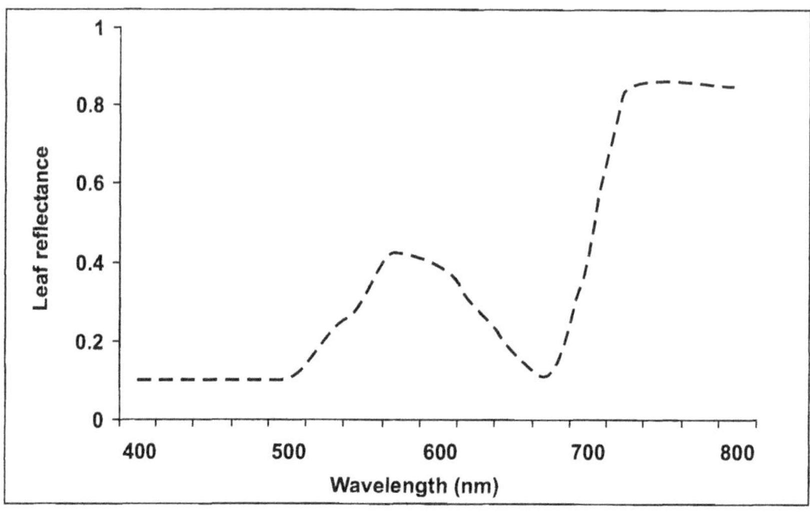

FIGURE 2.10. Spectral scattering coefficient of photoelements (*Source:* Reprinted from *Agricultural and Forest Meteorology,* 85, O. Anisimov and L. Fukhansky, Optics of vegetation: Implications for the radiation balance and photosynthetic performance, pp. 22-49, 1997, with permission from Elsevier Science.)

PHOTOSYNTHETICALLY ACTIVE RADIATION (PAR)

The visible region (approximately 0.385 to 0.695 μm) of the solar spectrum is generally referred to as photosynthetically active radiation. Although the global radiation is expressed in terms of W m^{-2}, the unit of PAR measurement is μE m^{-2} s^{-1} . Photosynthetic photon flux density (PPFD) is the number of photons in the photosynthetically active band of solar radiation. It is usually defined in moles of photons per unit surface and per unit time (mol m^{-2} s^{-1}). 1 μmol photons m^{-2} s^{-1} = 6.022×10^{17} photons m^{-2} s^{-1} = 1 μE m^{-2} s^{-1}. For conversion sake, 2.02 μmol photons m^{-2} s^{-1} of PAR is treated as equivalent to 1 W m^{-2} of global radiation (Berbigier and Hassika, 1998; Alados et al., 2002).

PAR is often calculated as a constant ratio of the broadband solar irradiance. Many reports are available in the literature to estimate PAR from the more routinely measured parameters of solar radiation, light intensity, and cloud amount. Several of these reports indicate the desirability of local calibration for the relationship between PAR and solar irradiance to account for local climatic and geographic differences such as cloudiness, day length, and diurnal pattern of solar radiation. A wide range of values has been quoted for the ratio (f_e) of PAR (W m^{-2}) to global solar radiation (W m^{-2}). Several researchers suggest that this variation can be ascribed to differences in the waveband limits chosen to define PAR and in part to the different methods used to measure or calculate f_e (Olesen, 2000). On the other hand, many people argue that different lower and upper waveband limits have no significant influence on the ratio received at the earth's surface.

At higher and middle latitudes, the daily average value of f_e is little affected by atmospheric and sky conditions. Systematic differences from day to day are largely a function of cloudiness. Even in the tropics, f_e should be a conservative quantity on clear days. For a clear day, $f_e = 0.51$, and for very cloudy skies, $f_e = 0.63$ have been measured in tropical countries.

Udo and Aro (1999) made measurements of global solar radiation (R_s) and global photosynthetically active radiation for a period of 12 months at Ilorin, Nigeria, to find the relationship between them. The results of the measurements showed that the average ratio of PAR to R_s for the year was 2.08 E MJ^{-1}, with the dry and rainy season values of 2.02 and 2.12 E MJ^{-1}, respectively. The minimum monthly mean daily ratio of 1.92 E MJ^{-1} was in January, representing a typical dry season month, while the maximum was 2.15 E MJ^{-1} in May, representing a rainy season month. The ratio values in the rainy season months and even dry season months remain constant at about 2.1 E MJ^{-1}. On a daily basis, the maximum and minimum ratios were 1.86 and 2.31 E MJ^{-1}, respectively. Hourly values of the ratio increased as the sky conditions changed from "clear" to "cloudy."

Hassika and Berbigier (1998) made continuous measurements of global and diffuse PAR throughout the year, within and above a forest. On clear sky days, roughly 65 percent of the incident PAR was absorbed by the needles, stems, and branches, 20 percent was reflected, and the understory absorbed the remaining 15 percent (Figure 2.11).

PAR interception in actively growing wheat crops was studied by Prasad and Sastry (1994). Two wheat varieties were grown with irrigation during the 1985-1986 winter season and assessed for total solar radiation interception, PAR, net radiation, and albedo. Maximum solar radiation and PAR interception was at 100 days after sowing (milk ripeness stage). For high values of crop net photosynthesis, the number of rows is more important at high light than at low light, whereas crop height is more important at low light than at high light (Thornley, Hand, and Wilson, 1992). The distribution of leaf angles (more vertical than horizontal angles) is important for maximizing whole-plant photosynthesis (Herbert, 1991).

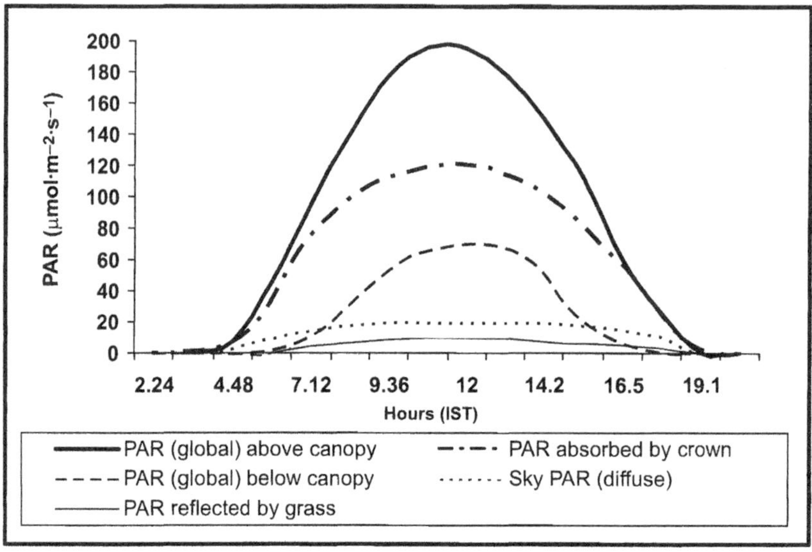

FIGURE 2.11. Cycle of PAR above and within a pine forest (*Source:* Reprinted from *Agricultural and Forest Meteorology,* 90, P. Hassika and P. Berbigier, Annual cycle of photosynthetically active radiation in maritime pine forest, pp. 157-171, 1998, with permission from Elsevier Science.)

SOLAR RADIATION USE EFFICIENCY

The conversion coefficient, defined as the quantity of biomass produced per unit of intercepted radiation (g MJ^{-1}), provides a measure of the efficiency *e* with which the captured radiation is used to produce new plant material. The alternative term, radiation use efficiency (RUE), is also commonly used (Black and Ong, 2000). Corlett and colleagues (1992) measured the *e* values for a millet crop under varying agronomic practices (Table 2.8).

As the values in Table 2.8 indicate, solar radiation use efficiency under the current crop production systems is very low. It is much below the theoretically estimated (8 to 10 percent) upper limit (Mavi, 1994). The efficiency of the conversion of photosynthetically active radiation by C3 plants falls off with increasing intensity. This decrease is caused by finite resistance to diffusion of CO_2 through the leaf to the chloroplast. However, an increase in the productivity of direct solar energy can be achieved if, by redistribution, it is intercepted at more uniform and lower intensity by a greater portion of the leaf area of a crop. Aikman (1989) developed a model which

TABLE 2.8. Intercepted solar radiation, aboveground biomass production, and biomass production per unit of intercepted radiation (*e*)

Season/crop	Intercepted radiation (MJ m^{-2})	Biomass (t ha^{-1})	*e*(g MJ^{-1})
Rainy season (July-August 1986)			
Sole millet	581	4.7	0.81
Alley millet	300	3.1	1.03
Sole *L. leucocephala*	520	4.0	0.77
Alley *L. leucocephala*	510	4.0	0.77
Total alley system	810	7.1	0.81
Rainy season (July-August 1987)			
Sole millet	504	5.0	0.98
Alley millet	180	0.9	0.050
Sole *L. leucocephala*	861	7.1	0.82
Alley *L. leucocephala*	748	6.4	0.86
Total alley system	928	7.3	0.79

Source: Adapted from Corlett et al., 1992.

predicts that redistributing direct solar radiation over twice the leaf area at half the intensity would give an increase of 22 percent of annual productivity. The model gives reasonable values for the reduction in productivity reported for shade regimes. The results of this study suggest that in protected cultivation, screens of partially reflective material could be used to redistribute solar radiation from leaves exposed to high intensities onto shaded leaves and so raise the photosynthetic efficiency. Assuming absorption of direct light by the screens of 0.1, the increase in productivity is estimated to be 17 percent. Li, Kurata, and Takakura (1998) also demonstrated that solar radiation enhancement through reflected radiation on the cultivated area could be achieved to raise the photosynthetic productivity throughout the winter.

When water or nutrient supplies do not limit growth, the quantity of biomass produced by monocrops is limited primarily by the quantity of radiation captured, and seasonal biomass accumulation for a given species may be expressed as the time integral of the product (Monteith, 1990, 1994). Numerous studies of annual crops, and some with perennial species, have demonstrated the existence of close correlations between dry matter production and cumulative intercepted radiation. For example, Stirling and colleagues (1990) examined the impact of artificial shade imposed on groundnut between the onset of peg initiation and pod filling, and final harvest using bamboo screens. A close linear correlation between aboveground biomass and cumulative intercepted radiation was found in all treatments, although the quantity of biomass produced per unit of intercepted radiation was substantially greater when shading was imposed from peg initiation onward. In the absence of stress, *e* is often conservative, typically ranging between 1.0 and 1.5 g MJ^{-1} for C3 species in temperate environments, 1.5 to 1.7 g MJ^{-1} for tropical C3 species, and up to 2.5 g $MJ^{-1}$for tropical C4 cereals under favorable conditions (Squire, 1990). However, the work of Stirling and colleagues (1990) showed that *e* may vary substantially within a single season between 0.98 g MJ^{-1} in the unshaded control and 2.36 g MJ^{-1} in crops shaded from peg initiation onward. Thus, plants in the latter treatment intercepted approximately one-quarter of the radiation received by the unshaded control but converted this to dry matter 2.4 times more efficiently (Monteith and Elston, 1983; Russell, Jarvis, and Monteith, 1988). Choudhury (2000) also observed a strong linear relationship between RUE and diffuse fraction of the incident solar radiation.

The observed variability in experimentally determined *e* values contrasts with earlier views that *e* is highly conservative except during severe water stress but complies with more recent suggestions that the assumption of a constant value within species or cultivars may be incorrect (Demetriades Shah et al., 1994; Sumit and Kler, 2000; Bonhomme, 2000).

This leads to criticism of the concept that biomass accumulation may be linked directly with cumulative intercepted radiation, and those meaningful *e* values may be derived from such correlations. It is argued that the concept of radiation use efficiency is oversimplistic, cannot improve our understanding of crop growth, and is of limited value in predicting yield. This argument concludes little evidence exists that incident radiation is a critical limiting factor determining crop growth under normal field conditions. Demetriades Shah and colleagues (1992) advocated that analysis of crop growth in terms of cumulative intercepted radiation and the conversion efficiency of solar energy during dry matter production should be approached with caution. A major plank in this argument was that photosynthesis, and hence crop growth rate, depends on numerous soil, atmospheric, and biological factors, of which radiation is only one component. They suggested that good correlations would always be found between radiation interception and any growing object, even when radiation is not the limiting variable. So a close correlation between crop growth and radiation interception may be expected even when light is not a major limiting factor. Therefore, although solar energy may be the most fundamental natural resource for crop growth from a physical viewpoint, from a biological viewpoint it is no more important than water, nutrients, CO_2, or any other essential commodity. As such, analysis of crop growth in terms of its radiation conversion coefficient may be inappropriate when variables other than radiation are the primary limiting factor. Further experimental support for this view was provided by Vijaya Kumar and colleagues (1996), when they showed that the conversion coefficient for rainfed castor beans *(Ricinus communis)* was less stable than previously suggested. The values obtained varied from year to year and were influenced by sowing date, decreasing with lateness of planting within the range 0.79 to 1.10 g MJ^{-1}. Values recorded prior to flowering were more stable than those obtained after flowering began. Campbell and colleagues (2001) also demonstrated that RUE steadily declined during growth of the rice crop and suggested that when RUE is used as a model parameter, it must be changed for differing LAI and for pre- and postanthesis periods.

Monteith (1994), however, defends the validity, generality, and robustness of correlations between intercepted radiation and growth and the conservativeness of *e*. Monteith concludes that few of the arguments advanced against conversion coefficient *e* are not convincing, and errors involved in measuring intercepted radiation can be minimized. In contrast to the view of Demetriades Shah and colleagues (1992), he saw no reason to abandon the concept, but instead highlighted the need to test and improve methodology as new information becomes available.

Monteith's arguments are supported by Kiniry (1994) and Arkebauer and colleagues (1994), who suggested that Demetriades Shah and colleagues (1992) had overlooked the fact that many environmental stresses that limit growth act through physiological pathways directly involving the photosynthetic process and its products. Arkebauer and colleagues (1994) argued that *e* cannot be expected to be constant, even within a single species or genotype, in the face of changes in other environmental variables. They argued that the definition of *e* involves three separate factors. First, the type and energy content of the carbon involved, i.e., net CO_2 uptake by the canopy, total aboveground dry matter production, or total plant dry matter including roots and storage organs. Second, the way in which radiation is characterized, i.e., total incident solar radiation, intercepted shortwave radiation, intercepted PAR, or absorbed PAR. Third, the time scale over which *e* is calculated is extremely important and may range from instantaneous to hourly, daily, weekly, or seasonal estimates. Because widely differing definitions of *e* have been adopted, the values obtained may be expected to show substantial variation.

Weighing arguments for and against the concept of solar radiation use efficiency, it can be concluded that RUE is likely to remain as a tool in understanding and predicting crop growth and yield.

Chapter 3

Environmental Temperature and Crop Production

Solar radiation is the main source of heat energy to the biosphere. Temperature is the intensity aspect of heat energy, and it is of paramount importance for organic life. Temperature governs the physical and chemical processes that in turn control biological reactions within plants. It controls the diffusion rate of gases and liquids within plants, and solubility of plant nutrient substances is dependent on temperature. As such, environmental temperature has a primary role in plant growth and its geographical distribution over the earth.

SOIL TEMPERATURE

Soil temperature is an important environmental factor in plant growth and distribution. In comparison to air temperature, the amplitude of variation in soil surface temperature is much more pronounced because of the varying characteristics and composition of soil.

Factors Affecting Soil Temperature

- *Aspect and slope:* These factors are of great importance in determining soil temperature outside the tropics. In the Northern Hemisphere, a south-facing slope is always warmer than a north-facing slope or a level plain. The reverse is the case in the Southern Hemisphere. The difference in soil surface temperature exceeds the difference in air temperature.
- *Tillage:* By loosening topsoil and creating mulch, tillage reduces the heat flow between the surface and subsoil. Because a mulched surface has a greater exposed area and the capillary connection with moist layers below is broken, cultivated soil has greater temperature amplitude than uncultivated soil. At noon, the air temperature 2.5 cm above the soil surface can be 5 to 10°C higher in cultivated soil as compared to uncultivated soil.

- *Soil texture:* Because of lower heat capacity, sandy soils warm up and cool down more rapidly than clay soils; hence, they are at a higher temperature during the day and a lower temperature at night.
- *Organic matter:* Organic matter reduces the heat capacity and thermal conductivity of soil, increases its water-holding capacity, and has a dark color which increases its solar radiation absorptivity. In humid climates, because of a large water content, peat and marsh are much cooler than mineral soils in spring and warmer in winter. However, when organic soils are dry, they become warmer than mineral soils in summer and cooler in winter.

Soil Temperature and Crop Germination

Soil temperatures influence the germination of seeds, the functional activity of the root system, the incidence of plant diseases, and the rate of plant growth (Singh, Singh, and Rao, 1998). Living tissues of many temperate plants are killed when they are exposed to a surface temperature of about 50°C (Chaurasia, Mahi, and Mavi, 1985). Excessively high soil temperatures are also harmful to roots and cause lesions on the stem. Extremely low temperatures are equally detrimental. Low temperatures impede the intake of nutrients. Soil moisture intake by plants stops when they are at a temperature of 1°C. Root growth is generally more sensitive to temperature than that of aboveground plant parts, meaning that the range between maximum and minimum temperature for roots is less than for shoots and leaves.

In numerous cases soil temperature is more important than air temperature to plant growth. In Canada, sowing of agronomically important crops takes place during the early months of spring when temperatures are well below the optimum. This often results in reducing the rate and success of germination, slow, asynchronous seedling emergence, and poor stand establishment (Nykiforuk and Flanagan, 1998).

Several rice varieties do not emerge as long as the soil temperature is below 11°C (Kwon, Kim, and Park, 1996). Germination of warm-season grasses is very poor during the winter season. Slower germination rates during cooler seasons require long periods of soil water availability at the surface to enable germination (Roundy and Biedenbender, 1996). Figure 3.1 shows that cassava plants of variety MAus 10 did not emerge below 14.8°C or above 36.6°C, whereas those of variety MAus 7 did not emerge below 12.5°C or above 39.8°C (Yin et al., 1995). Germination of sunflower, maize, and soybean is very poor when day/night soil temperature is above 21/12°C and soil water content is too low during the first week after sowing (Helms, Deckard, and Gregoire, 1997; Hernandez and Paoloni, 1998).

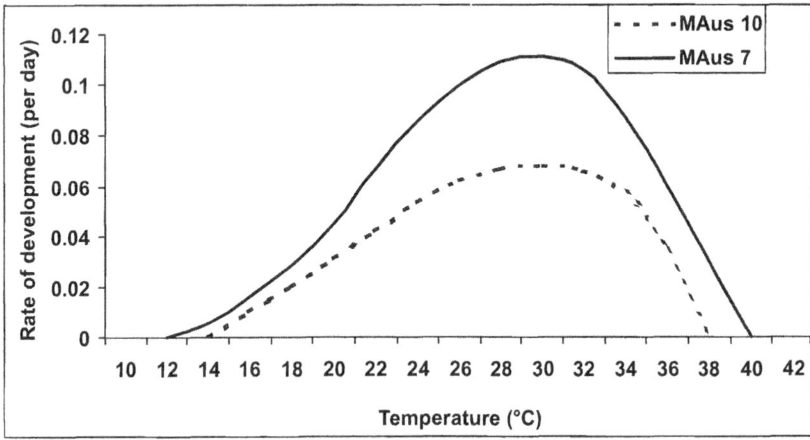

FIGURE 3.1. Soil temperature and rate of development from sowing to emergence in two cassava cultivars (*Source:* Reprinted from *Agricultural and Forest Meteorology*, 77, X. Yin et al., A nonlinear model for crop development as a function of temperature, pp. 1-16, 1995, with permission from Elsevier Science.)

In the tropics, high soil temperature causes degeneration of the tuber in potato. Optimum soil temperature for this crop is 17°C. Tuber formation is practically absent above a soil temperature of 29°C. Preconditioning of potato seed under specific temperatures has an important impact on germination. Seed stored at 27°C showed the best germination, while that stored at 45°C failed to germinate even after eight days of lowering the temperature in the germination environment to 17°C (Pallais, 1995).

Impact of Soil Temperature on Plant Growth

After germination, soil temperature is important for the vegetative growth of crops. For each species, a favorable soil temperature is needed for ion and water uptake. The daytime soil temperature is more important than the nighttime temperature, because it is necessary to maintain a favorable internal crop water status to match the high evaporation rate.

Maize yield is closely related to soil temperature at planting. Some cultivars sown at soil temperatures above 30°C show reduced final seedling emergence (Arachchi, Naylor, and Bingham, 1999). Soil temperature controls the rate of maize development while the meristem is underground. Increased soil temperature accelerates the rates of leaf tip appearance and full leaf expansion, enabling the crop to more rapidly attain maximum green leaf-area index. This enables a better synchrony between the time of peak radiation interception and peak radiation incidence. The extent to which soil temperature affects yield will therefore vary with sowing time and the latitude of the crop's location (Stone, Sorensen, and Jamieson, 1999).

Tomato seed germination, plant growth, and fruit yields are governed by the prevailing soil temperature conditions. Germination is completely inhibited at low temperatures (up to 5°C) as well as high temperatures (40°C). Germination is highest at 25 to 30°C. At 10°C, plant growth is slow, almost no fruit formation occurs, and plants start to die off prematurely. At 18°C, the highest growth rates and earliest fruit formation are recorded (Sakthivel and Thamburaj, 1998; Nieuwhof, Keizer, and Van Oeveren, 1997).

Studies on the effect of temperature on root yield and quality of sugar beet show that soil temperature correlated positively with root yield and negatively with sugar content (Hayasaka and Imura, 1996).

The root zone temperature significantly affects the quality and yield of sweet pepper. Growth is more inhibited by low temperature than high temperature. Sugar content is influenced by root zone temperature. Phosphate content is lower at 13 and 33°C root zone temperatures than at other temperatures. Higher numbers of fruits are obtained at 18 to 28°C, and higher yields are obtained at 23 to 28°C than at other root zone temperatures. A 23°C root zone temperature is considered optimal for economic production of sweet pepper (Kim et al., 2001).

Optimal soil temperature for growth of wheat plant roots during the vegetative stage is below 20°C and is lower than that for the shoots. Temperatures higher than 35°C have been shown to reduce terminal root growth and accelerate its senescence. Root growth may cease altogether if soil temperatures drop below 2°C. Studies have shown (Porter and Gawith, 2000) that an air temperature of –20°C is lethal for root survival, although this must be translated into a soil surface temperature, which would, in most cases, be higher.

Cardinal Temperatures

Three temperatures of vital plant activity have been recognized, which are often termed *cardinal points.*

1. A minimum temperature below which no growth occurs: For typical cool-season crops, it ranges between 0 and 5°C, and for hot-season crops between 15 and 18°C.
2. An optimum temperature at which maximum plant growth occurs: For cool-season crops, it ranges between 25 and 31°C, and for hot-season crops between 31 and 37°C.

3. A maximum temperature above which the plant growth stops: For cool-season crops, it ranges between 31 and 37°C, and for hot-season crops between 44 and 50°C.

The cardinal temperatures for germination of some plants are given in Table 3.1. The cardinal points can be measured only approximately because their position is related to external conditions, the duration of exposure, the age of the plant, and its previous treatment.

AIR TEMPERATURE

Air temperature is the most important climatic variable that affects plant life. The growth of higher plants is restricted to a temperature between 0 and 60°C, and crop plants are further restricted to a narrower range of 10 to 40°C. However, each species and variety of plants and each age group of plants has its own upper and lower temperature limits. Beyond these limits, a plant becomes considerably damaged and may even be killed. It is therefore the amplitude of variations in temperature, rather than its mean value, that is more important to plant growth.

The midday high temperature increases the saturation deficit of plants. It accelerates photosynthesis and ripening of fruits. The maximum production

TABLE 3.1. Cardinal temperatures for the germination of some important crops

	Cardinal temperature (°C)		
Plant	**Minimum**	**Optimum**	**Maximum**
Wheat	3-4.5	25	30-32
Barley	3-4.5	20	38-40
Maize	8-10	32-35	40-44
Rice	10-12	30-32	36-38
Tobacco	13-14	28	35
Sugar beet	4-5	25	28-30
Peas	1-2	30	35
Oats	3-4	25	30
Sorghum	8-10	32-35	40
Lentils	4-5	30	36
Carrot	4-5	8	25
Pumpkin	12	32-34	40

Source: Adapted from Bierhuizen, 1973

of dry matter occurs when the temperature ranges between 20 and 30°C, provided moisture is not a limiting factor. High temperature can devernalize cryophytes, especially the buds of sun-exposed deciduous trees. When high temperature occurs in combination with high humidity, it favors the development of many plant diseases. High temperature also affects plant metabolism.

High night temperature increases respiration. It favors the growth of the shoot and leaves at the cost of roots, stolons, cambium, and fruits. It governs the distribution of photosynthates among the different organs of the plants, favoring those which are generally not useful for human consumption. High night temperature also affects plant metabolism. It accelerates the development of noncryophytes.

Most crop plants are injured and many are killed when the night temperature is very low. Tender leaves and flowers are very sensitive to low temperature and frost. Plants that are rapidly growing and flowering are easily killed. Low temperature interferes with the respiration of plants. If low temperature coincides with wet soil, it results in the accumulation of harmful products in the plant cells. Frost also interferes with plant metabolism.

Spring wheat grain yields generally decline as temperature increases. Temperature stress intensity is severe under late sowing, causing a reduction in the duration of later growth phases. Grain test weight, spikelets/spike, and grains/spike under hot (normal sowing) environments and spike length and spikes/m^2 under very hot (late sowing) environments are adversely affected. Other factors that significantly contribute to yield under high temperatures are tiller numbers and reduced height (Hanchinal et al., 1994; Frank and Bauer, 1996; Chowdhury, Kulshrestha, and Deshumukh, 1996).

Grain yield of rice is highly correlated with minimum temperature. A prediction model in the Philippines (Pamplona et al., 1995) showed that the high yield observed especially during the dry season is due to lower minimum temperature. Higher grain yield corresponds with a seasonal minimum temperature of 22.5°C, compared to an average seasonal minimum temperature of 24.2°C.

Reasons for low and variable cotton yields are associated with extremes of temperatures (Oosterhuis, 1997). Yield and fiber characteristics respond to variations of daily mean and amplitude of temperature (Liakatas, Roussopoulos, and Whittington, 1998). Mean temperature reduction improves yield components, but high temperatures, particularly high day temperatures, increase fiber length, uniformity, and strength. Large daily temperature amplitude produces an intermediate number of flowers and the lowest retention percentage. Fruiting and yield are increased by a reduction in temperature down to the threshold mean temperature of 22°C. An adverse effect of low minimum temperature on lint and fiber properties was also observed.

Sowing date, reflecting temperature conditions, significantly affected phenology (time to emergence, flowering, and maturity) and pod yield of groundnut. The observed responses appear to have been due to the effect of temperature differences on partitioning during the pod-filling phase (Ntare, Williams, and Ndunguru, 1998).

Temperature and Photosynthesis

The rate of photosynthesis and respiration increases with an increase in temperature, until a maximum value of photosynthesis is reached. This value is maintained over a broad range of temperatures (Figures 3.2 and 3.3). Then, at considerably higher temperatures, when the enzyme becomes inactivated and various reactions are disturbed, photosynthesis decreases and ultimately stops.

The range of temperature in which photosynthesis is more than 90 percent of the maximum obtainable can be regarded as optimum. This range is narrower for net photosynthesis than for gross photosynthesis, because while gross photosynthesis is still operating at top speed in the optimum

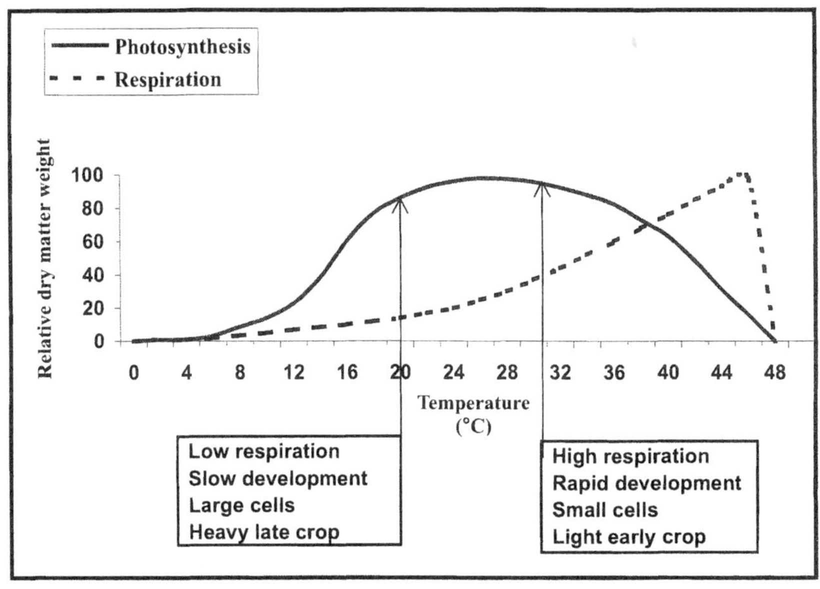

FIGURE 3.2. Effect of temperature on photosynthesis and respiration of potato (*Source:* Mavi, 1994.)

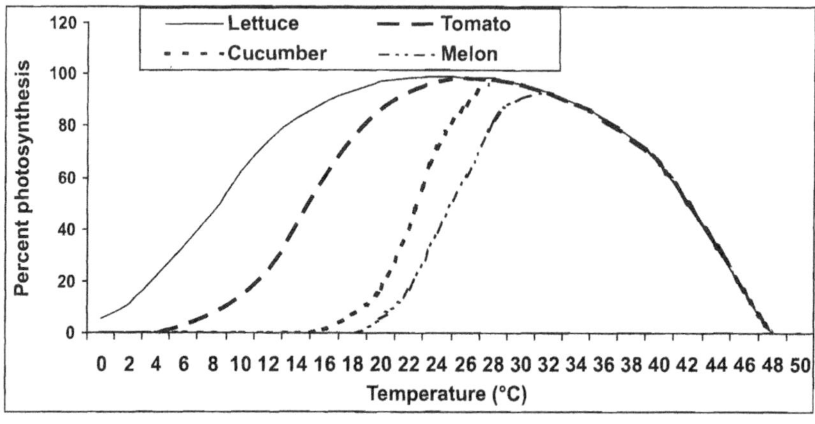

FIGURE 3.3. Temperature limits for photosynthesis (*Source:* Mavi, 1994.)

range of temperatures, the rate of respiration increases, diminishing the net photosynthetic yield. The temperature limits for net photosynthesis for some plant groups are given in Table 3.2.

PLANT INJURY DUE TO SUDDEN CHANGES IN TEMPERATURE

Living organisms receive and transfer thermal energy through radiation, conduction, and convection. Transpiring water to the surrounding atmosphere also transfers thermal energy from growing plants. Through these processes, they remain in equilibrium with the surrounding environment and maintain normal growth and development. However, with the passage of weather systems, changes in atmospheric temperature are often very sudden, and plants cannot adjust to these severe variations and are damaged beyond recovery.

Leaf Temperature versus Air Temperature

Under normal conditions, leaf temperature remains around the ambient temperature but differs under certain situations. At a temperature of about 33°C, there is a tendency for equality between air and leaf temperature. Below this temperature, leaves tend to be warmer than the air and vice versa.

TABLE 3.2. Temperature dependence of net photosynthesis during the growing season under conditions of natural carbon dioxide availability and light saturation

	Temperature limits (°C) for carbon dioxide intake		
Plant group	**Lower limit**	**Optimum**	**Upper limit**
C4 plants of hot habitats	–5 to 7	35 to 45	50 to 60
C3 crop plants	–2 to over 0	20 to 30	40 to 50
Alpine plants	–7 to –2	10 to 20	30 to 40
Evergreen tropical trees	0 to 5	25 to 30	45 to 50
Deciduous trees of temperate zone	–3 to –1	15 to 25	40 to 45
Evergreen conifers	–5 to –3	10 to 25	35 to 42
Shrubs of tundra	–8	15 to 25	40 to 45
Lichens of cold regions	–25 to –10	5 to 15	20 to 30

Source: Adapted from Larcher, 1980.

Where the temperature exceeds 33°C, leaves appear to suffer from water stress.

- *Leaves exposed to sun:* Thick leaves that are not transpiring actively in still air are several degrees warmer than the air. Under intense radiation and high humidity, some leaves may be at a temperature 15°C higher than the air. Likewise, under very hot and low humidity conditions, leaf temperature can be as much as 10°C higher than the air. Where plants do not suffer for want of moisture, the difference between leaf and air temperature is very small.
- *Leaves under shade:* Leaves shaded from direct sunlight are usually somewhat warmer than the surrounding air.
- *Leaves exposed to a clear night:* At night when the sky is clear, leaf temperature is usually lower than the air temperature. During a cold and clear night, a leaf may be around 2°C cooler than the surrounding air.
- *Leaves exposed to a cloudy night:* With cloud cover, the difference in air and leaf temperature is small. In certain cases the leaf temperature may be slightly higher than air temperature.

High-Temperature Injury to Plants

Thermal death point of active cells ranges from 50 to 60°C for most plant species, but it varies with the species, the age of tissue, and the length of time of exposure to high temperature. It has been reported (Chang, 1968) that most plant cells are killed at a temperature of 45 to 55°C, but some tissues withstand a temperature of up to 105°C.

For aquatic and shade plants the lethal limit is 40°C, and for most xerophytes it is 50°C, when the plants are exposed to a saturated atmosphere for about half an hour. High temperature results in the desiccation of the plant and disturbs the balance between photosynthesis and respiration. Once the temperature exceeds the maximum up to which growth takes place, plants enter a state of quiescence. When the temperature becomes extremely high, a lethal level is reached. At temperatures higher than the optimum cardinal, the physiological activity declines as a consequence of inactivation of enzymes and other proteins. Leaf functions are disturbed at about 42°C, and lethal effects on active shoot tissues generally occur in the range of 50 to 60°C.

Many rice varieties subjected to high temperature just before and just after flowering result in more than 20 percent sterility. High temperature just before or during flowering decreases pollen size, causes a shortage of starch in pollen, and increases the proportion of anthers that did not dehisce. High temperature during ripening decreases grain weight. In wheat crops, a major effect of high temperature appears to be the acceleration of senescence, including cessation of vegetative and reproductive growth, deterioration of photosynthetic activities, and degradation of proteinaceous constituents (Xu et al., 1995).

Serious damage to fruit and vegetable crops resulting from excessively high temperature has also been recorded (Muthuvel et al., 1999; Atta-Aly and Brecht, 1995; Chen, Lin, and Chang, 1994; Oda et al., 1994; Inaba and Crandal, 1988). Apart from desiccation and disturbed photosynthesis and respiration balance, plants are injured in several ways, such as excessive respiration from seeds, sun scald, and stem girdle.

The higher the temperature, the greater is the rate of respiration, which results in the rapid exhaustion of food reserves of seeds. Temperatures on the sunny side of the bark on stems during hot afternoon and late night undergo great fluctuations. The injury inflicted because of this short period fluctuation in temperature is known as *sun scald.* Stem girdle is another injury associated with high temperature. Exceptionally high temperature at the soil surface and the adjoining laminar sublayer of the air frequently scorches the short stems. The scorching of the stem is known as *stem girdle.* This type of injury is most common in young seedlings of cotton in sandy soils where the temperature of the soil surface during summer afternoons may be as high as 60 to 65°C (Chaurasia, Mahi, and Mavi, 1985). Stem girdle injury is first noted through a discolored band a few millimeters wide. This is followed by shrinkage of the discolored tissues. It appears that stem girdle causes the death of plants by destroying the conductive tissues or by an injury that helps the establishment of pathogens.

Low-Temperature Injury to Plants

Exposure to extremely low temperatures and heavy snowfall damages the plant in several ways including suffocation, desiccation, heaving, chilling, and freezing.

Suffocation

Small plants may suffer from deficient oxygen when covered with densely packed snow. When suffocated, certain toxic substances accumulate in contact with roots and crowns and tend to inhibit the diffusion of carbon dioxide.

Physiological Drought and Desiccation

Spring drought sometimes occurs in coniferous trees in cool temperate climates. This results from excessive transpiration and a time lag in absorption of moisture from the soil, caused by a warm period when the soil is still frozen. The result is an internal moisture deficit sufficient to cause death of the twigs. The decreased water absorption by plants at low temperatures is the combined effect of the decreased permeability of the root membrane and increased viscosity of water. This results in increased resistance to water movement across the living cells of the roots.

Heaving

Injury to a plant is caused by the soil layer lifting upward from the normal position and causing the root to stretch or break at a time when the plant is growing. Sometimes the roots are pushed completely above the soil surface. After thawing, it is difficult for the roots to become firmly established, and the plants may die because of this mechanical damage and desiccation.

Chilling

Plants of tropical origin are damaged by exposure to mild chilling for two to three days. Plants of temperate origin withstand chilling for long periods without suffering any injury. Rice, cotton, and cowpea are killed when exposed to temperatures near 0°C for about two to three days. Sudan grass and peanuts are injured by short exposure to chilling temperatures but recover if favorable temperature conditions return shortly afterward. Short duration mild chilling does not seriously injure corn, sorghum, and pumpkin plants. Plants of cool climate origin such as wheat and soybean are injured when exposed to prolonged chilling but recover with the return of favorable conditions.

In temperate climates, two types of injuries occur because of low temperature. These are delayed growth and sterility. For example, rice yield decreases due to insufficient grain maturation caused by low temperatures during the ripening period. When flowering is delayed by low temperatures at a certain stage before heading, insufficient time is available to the grains to ripen fully before frost occurs in autumn.

In the sterility type of injury, rice yields decrease due to sterile spikelets caused by low temperatures at the booting stage or at anthesis. The observed injury in developmental order is a stoppage of anther development; pollen unripeness; partial or no dehiscence; pollen grains remaining in anther loculi; little or no shedding of pollen grains on the stigma; and failure of germination on the stigma.

Chilling injury of fruits is of particular interest because they are often stored and shipped at low temperatures. Symptoms of chilling injury to fruits include surface pitting, lesions, discoloration, susceptibility to decay organisms, and shortening of storage life. Fruits subjected to chilling injury do not ripen normally. The critical temperature at which chilling injury occurs is 8 to 12°C for tropical fruits such as banana, avocado, and mango, and 0 to 4°C for temperate zone fruits such as apple (Kozlowski, 1983).

Freezing Injury

Plant parts or an entire plant may be killed or damaged beyond repair as a result of actual freezing of tissues. Freezing damage is caused by the formation of ice crystals, first in the intercellular spaces and then within the cells. Ice within the cells causes more injury by mechanical damage disrupting the structure of the protoplasm and plasma membrane. Freezing of water in intercellular spaces results in withdrawal of water from the cell sap, and increasing dehydration causes the cell to die.

In a study on freezing injury to fruits in Hungary, Szabo and colleagues (1995) found that apricot is the least cold hardy of stone fruit species grown in Hungary. In strawberry, flowers become more susceptible to freezing as development progresses (Ki and Warmund, 1992). In Belarus (Kozlovskaya and Myalik, 1998) the degree of damage within apple and pear seedling populations due to low temperature is variable. The hybrid seedlings of apple are more severely damaged by low temperatures than the other progenies. In the case of pear, the damage was more severe to the seedlings with a tap root system than to those with a branching root system.

FROST: DAMAGE AND CONTROL

Frost is a climatic hazard that causes serious damage to standing crops in temperate and subtropical climates. In Australia, crop losses due to frost are huge. In New South Wales (NSW) alone, a big portion of the state's fruit, vegetable, and grain crops is lost to low-temperature damage (Degan, 1989). The losses to the wheat crop vary between 5 and 35 percent due to heavy frost in late September and early October (Boer, Campbell, and Fletcher, 1993). Much distress can be avoided by properly understanding the characteristics of the frost, by using early warning information on frost, and by adopting frost protection measures. For this, the planning should begin before the crop is planted.

Frost is a weather hazard that occurs when the environmental temperature drops below the freezing point of water. It can be a white frost (also known as hoarfrost) or a black frost. White frost occurs when atmospheric moisture freezes in small crystals on solid surfaces. Black frost occurs when few or no ice crystals are formed because air in the lower atmosphere is too dry, but the damaging effect of the low temperatures on vegetation is the same as that of white frost.

Frost is formed through the physical processes of radiation and advective cooling. These are referred to as radiation frost and advection frost, respectively (Wickson, 1990).

- *Radiation frost* occurs when a clear sky and calm atmosphere (winds less than 8 kph) allow an inversion to develop, and temperatures near the surface drop below freezing. The thickness of the inversion layer varies from 10 to 50 m. The term inversion comes from atmospheric conditions being inverse to the normal daytime condition in which air temperature decreases with height. Plants can be successfully protected from radiation frost.
- *Advection frost* occurs when a cold air mass invades a relatively warm area suddenly. Under advection frost, winds may be above 7 kph and clouds may exist. The advected cold air mass may be 150 to 1,500 m deep. Plants can be protected from advection frost to a limited extent.

Frost Damage to Plants

Damage to plants from frost occurs because it results in freezing of the plant tissues. Freezing of plant tissues is a physical process triggered by ice-nucleating bacteria, the intensity and duration of the night temperature to which the plants are exposed, and the plant growth stage (Jamieson, 1986; Woodruff, Douglas, and French, 1997). Green plants contain mostly water, and on freezing, the water expands and ruptures the cell walls of the plant tissues. Because of the presence of chemicals in the sap, plant tissues freeze at temperatures lower than 0°C, the freezing temperature of water. When frozen water melts, it leaks away from the cells. The rupturing of the cells and leakage of water results in the death of tissues, giving a typical "burn" appearance to the plants.

Plants show different symptoms of frost injury, depending on the stage at which freezing occurs (Table 3.3). In the case of wheat, freezing stress can cause foliar injury and tiller death. Injury to developing foliage will not affect the crop yield because the plants can compensate. However, freezing injury during stem elongation can substantially reduce the final yield. Leaf injury can occur at any stage of development, and frozen leaves will appear dark in color. Slightly injured leaves will have yellow tips that should not be confused with the symptoms of nutrient stress (Youiang and Ellison, 1996). Injured stems appear discolored and often distorted near the nodes. Injury to young ears can cause the whole ear to die. At the booting stage, frost injury can damage the reproductive parts of some ears. The injury is easily detected after ear emergence because growth of floret and spikelet look stunted. During flowering, the reproductive parts of the plant may be damaged in some ears, and although they appear unaffected, they produce no grain.

Methods of Protection against Frost

Frost protection methods may be divided into passive and active forms (Powell and Himelrick., 1998; Mavi, 2000). Passive protection involves methods such as site selection and variety selection and several cultural practices such as brushing and soil surface preparation. These methods do not require expenditure of outside energy sources. Active protection systems replace radiant energy loss by using methods such as irrigation, heaters, and wind machines. Active methods require outside energy to operate. The proper choice of a protection method depends on many factors, such as site, crop, advantages and disadvantages of the protection methods, relative costs, and operating principles of the method.

TABLE 3.3. Frost damage to crops, vegetables, and fruits

Plants	Temperatures (°C) harmful to plants in the developmental phases		
	Germination	**Flowering**	**Fruiting**
Crops			
Wheat	–9 to –10	–1 to –2	–2 to –4
Oats	–8 to –9	–1 to –2	–2 to –4
Barley	–7 to –8	–1 to –2	–2 to –4
Lentils	–7 to –8	–2 to –3	–2 to –4
Peas	–7 to –8	–2 to –3	–2 to –4
Beans	–5 to –6	–2 to –3	–3 to –4
Sunflower	–5 to –6	–2 to –3	–2 to –3
Mustard	–4 to –6	–2 to –3	–3 to –4
Soybeans	–3 to –4	–2 to –3	–2 to –3
Maize	–2 to –3	–1 to –2	–2 to –3
Sudan grass	–2 to –3	–1 to –2	–2 to –3
Millets	–2 to –3	–1 to –2	–2 to –3
Sorghum	–2 to –3	–1 to –2	–2 to –3
Cotton	–1 to –1.5	–0.5 to –1	–2
Groundnuts	–0.5 to –1	–	–
Rice	–0.5 to –1	–0.5 to –1	–1
Vegetables			
Carrot	–6 to –7	–	–
Turnip	–6 to –7	–	–
Cabbage	–5 to –7	–2 to –3	–6 to –9
Potatoes	–2 to –3	–1 to –2	–1 to –2
Cucumbers	–0.5 to –1	–	–
Tomatoes	0 to –1	0 to –1	0 to –1
Fruits	*Bud closed but showing color*	*Full bloom*	*Small green fruit*
Almonds	–3 to –5	–2 to –4	–1.7
Apples	–3.5 to –4.5	–2.2 to –2.8	–1.7
Apricots	–4	–2.2	–0.6
Cherries	–2.2	–2.2	–1.1
Grapes	–1	–0.6	–0.6
Peaches	–4	–2.8	–1.1
Pears	–2.5 to –3.3	–2.2	–1.1
Plums	–4	–2.2	–1.1
Prunes	–3 to –5	–2.2 to –2.8	–1.1
Strawberry	–2	–2	–1.5
Walnuts	–1	–1.1	–1.1

Source: Adapted from Ventskevich, 1961; Rogers, 1970.

Site Selection

Many factors are involved to create pockets of very low temperatures. Before planning a crop or an orchard, the best method of frost protection is careful selection of the site. The site should be selected taking into account the climatic conditions prevailing in that location, its slope, and the soil characteristics. There is a possibility of cold air buildup in low paddocks or behind barriers such as fences, hedges, and wooded areas (Hutton, 1998).

Such paddocks are not the best locations for planting orchards and frost-sensitive crops. Removal or thinning of trees that create cold air dams is desirable. If a site has good cold air drainage, then it is likely to be a good production site as far as frost damage is concerned. Frost-sensitive fruit trees are usually planted on hillside slopes from which the cold air drains rapidly to the bottom of the valley. Such sites are usually 2 to 4°C warmer during radiational frost.

Frost-Resistant Cultivars

Planting frost-resistant cultivars and crop varieties is one approach to avoid frost damage to fruit trees and field crops. Oats are more tolerant to frost damage than barley, and barley is slightly more tolerant than wheat. The varieties could be those in which genetic resistance to freezing stress has been incorporated. Growers should refer to available extension publications on varieties that could withstand the low temperatures.

Optimizing Sowing Dates

The best and most cost-effective strategy to save field crops from frost is the choice of the optimum dates for crop plantings. As crops enter the flowering and grain-forming stage, their tolerance to frost is drastically reduced. If the sowing dates of crops are adjusted in such a way that these stages do not fall in the period of heavy frost, then its damaging action is avoided. In the case of wheat, it is necessary for anthesis to occur after the high-risk frost period is over. Results of experiments in NSW have shown that a wheat crop can be saved from frost damage to a great extent if the crop flowers in mid-September in areas around Trangie, in late September around Narrabri, and in early October around Tamworth. A late date of anthesis, however, needs to be balanced against the damage that can occur if grain filling takes place during the period of high temperatures or moisture stress. Each week's delay after these dates in anthesis can reduce yields dramatically (Boer, Campbell, and Fletcher, 1993).

Storing Heat in the Soil

Frost frequency and intensity is greater in orchards in which the soil is cultivated, dry, and covered with weeds or mulch as compared to orchards in which the soil is moist, compact, and weed free (Johns, 1986). This is because soil that is bare or weed free, compact, and moist stores more heat during the daytime than soil that is covered with shade and is dry. At night this heat is released to the lower layers of the air surrounding the crop plants and fruit trees, minimizing the damage from frost.

Standing weeds increase the incidence of frost in three ways: by shading the soil, which hampers the heat flow to the soil; by drying the soil; and by raising the cold radiating surface which comes close to the fruit level. Thick mulches also increase the incidence of frost through hampering the heat flow to the soil during the day and retarding the heat flow to the top of the straw during the night. A dry cultivated soil increases the incidence of frost, because cultivation creates more air pockets in the soil which act as insulating layers and hamper the flow of heat to the soil, lowering its heat storage during the day.

Therefore, keeping the soil moist with frequent light irrigation, maintaining it weed free, and making it compact with rollers is the best technique to minimize frost damage in orchards, vineyards, and wide-row crops.

Plant Cover

Planting large canopy trees with orchard plants provides some freeze protection. Date palms in California and pine trees in southern Alabama are used as canopy cover for citrus plantings (Perry, 1994)

"Brushing" is commonly used for protecting vegetable crops from frost damage. Shields of coarse brown paper are attached to arrowhead stems on the poleward side of the east-west rows of plants. The fields present a brushy look. During the day the shields act as windbreaks against cold wind, while at night they reduce radiation loss to the sky. Woven or spun-bonded polypropylene covers of varying thickness are among the latest forms of protection used on fruit crops. Depending on the material used, several degrees of protection are achieved. Copolymer white plastic has provided protection to nursery stock but is not used on fruit and vegetable crops. Light- and medium-weight covers provide excellent protection for low-growing crops such as strawberries.

Nutrition

Deciduous fruit plants, such as peach, that are not nutritionally sound, especially in regard to nitrogen, are more subject to frost damage. Fruit buds of such trees are less healthy and more easily damaged by frost. Using midsummer or postharvest application of nitrogen can induce vigor for strong fruit bud development and some delay in flowering in stone fruits such as peaches. However, tree fruits with low fertility requirements, such as apples and pears, do not normally require mid- to late-summer fertilization, whereas such applications do benefit blueberries (Perry, 1994).

Chemicals

Some inexpensive materials which could be stored easily until needed and are portable and easily applied to provide frost protection have been tested. The possibilities of using cryoprotectants, antitranspirants, and growth regulators are encouraging.

A number of materials that could change the freezing point of plant tissue, reduce the ice nucleating bacteria on the crop and thereby inhibit frost formation, or affect growth, i.e., delay dehardening, have been examined. Several products are advertised as frost protection materials; however, none of the commercially available materials has successfully withstood the scrutiny of scientific testing. Growers should be very careful about accepting the promotional claims of these materials (Ullio, 1986; Powell and Himelrick, 1998).

Growth regulator applications that could increase the cold hardiness of the buds and flowers, delay flowering, or both seem to hold the most promise at this time. Among the growth regulators tested, only the ethylene-releasing compound ethephon has shown promise (Gallasch, 1992; Powell and Himelrick, 1998). Ethephon increases winter fruit bud hardiness and delays flowering of peaches by four to seven days. It provides the same effects on cherries. In the United States, ethepon has been federally labeled for use on cherries, and it is on several state labels for use on peaches.

Irrigation

Irrigation is the oldest, most popular, and most effective method of protection from frost. Irrigation is done with sprinklers mounted above or below the crop canopy. Sprinkling the canopy with water releases the latent heat of fusion when water turns from liquid to ice. As long as ice is being formed, latent heat released by water efficiently compensates for the heat lost from the crops to the environment.

For most situations, sprinkler rotating once each minute and an application rate of 2 to 4 mm of water per hour is sufficient. A backup power source is essential, as power failure can be devastating. Once started, irrigation must continue until the morning sun hits the trees (Wickson, 1990). During the other seasons the sprinklers can be used for evaporative cooling, artificial chilling, delayed flowering, fruit drop prevention, sunburn injury, and color improvement of fruits (Spieler, 1994).

Heaters

Heating of orchards for protection against frost has been relied upon for centuries. The high cost of fuel has now provided an incentive to look at other methods. There are several advantages to using heaters. Most heaters are designed to burn oil and can be placed as freestanding units or connected by a pipeline network throughout the crop area. The advantage of connected heaters is the ability to control the rate of burning and shut all heaters down from a central pumping station simply by adjusting the pump pressure. A pipeline system can also be designed to use natural gas. Propane, liquid petroleum, and natural gas systems have been used for citrus.

Heaters provide protection by three mechanisms. The hot gases emitted from the top of the stack initiate convective mixing in the crop area and break the inversion. The bulk of a heater's energy is released in this form. The remaining energy is released by radiation from the hot metal stack. A relatively insignificant amount of heat is also conducted from the heater to the soil. Around the periphery, more heaters are required, because the ascending plumes of hot air allow an inflow of cold air.

Heaters provide the option of delaying protection measures if the temperature unexpectedly levels off or drops more slowly than predicted. The initial installation costs are lower than those of other systems, although the expensive fuels required increase operating costs. There is no added risk to the crop. Whatever heat is provided will be beneficial.

Wind Machines

The purpose behind using wind machines is to circulate warmer air down to the crop level. Wind machines are effective only under radiation frost conditions. They should be installed and operated after a thorough understanding of how frost affects a particular area or orchard (Lipman and Duddy, 1999). A typical wind machine with fans about 5 m in diameter and mounted on a 10 m steel tower can protect approximately four hectares of area, if the area is relatively flat and round. The fan is powered by an engine delivering 85 to 100 HP. Wind machines used in conjunction with heaters provide the best protection. When these two methods are combined, the required number of heaters per hectare is reduced by about half.

Wind machines provide noteworthy advantages in frost protection by minimizing labor requirements, reducing refuelling and storage of heating supplies, and requiring a low operational cost per hectare. Wind machines use only 5 to 10 percent of the energy per hour when compared to heaters. The original installation cost is quite similar to that for a pipeline heater system, making wind machines an attractive alternative to heaters for frost protection. They are also more environmentally friendly (except for noise) because they do not produce smoke or air pollution.

An overview of the advantages and limitations of the methods mentioned in this section is given in Table 3.4. Each grower must choose the proper method of frost protection for the particular site considered. Once the decision has been made, and if frost protection is to be practiced successfully, three guidelines apply to all systems:

1. Operation for protection against frost must be handled with the same care and attention as spraying, fertilizing, pruning, and other cultural practices.
2. Frost protection equipment must be used correctly with sound judgment and attention to detail and commitment.
3. Operation should not be delegated to someone with no direct interest in the result.

Frost Forecast

Season Ahead

The Southern Oscillation Index, particularly the SOI phase during autumn, provides a modest tool to determine the dates of the last frost and the number of frost days in eastern Australia with a lead time of three to four months (Stone, Nicholls, and Hammer, 1996).

The SOI phases representing either a consistently negative or a rapidly falling phase during late autumn indicate a greater chance of frosts with the last frost occurring late in the season. The reason for more frosts under these patterns is because these are associated with El Niño, resulting in less rainfall, more clear skies, and more radiational cooling of the earth's surface during winter and early spring.

TABLE 3.4. Frost protection alternatives—advantages and disadvantages

Protection method	Crops protected	Advantage	Disadvantage
Site selection	All fruit crops especially tree fruit	A location with good cold air drainage is a good production location	Good site may not be available
Resistant varieties	Grain and fruit crops	Resist frost action	Many varieties of crops already at the upper limit of frost resistance Very difficult to create genetic resistance to frost
Optimize sowing dates	Grain and fruit crops	Very effective	May not be practicable due to soil moisture and other unfavorable weather conditions
Improving soil heat storage	All fruit crops	Easy and inexpensive	None
Plant cover	Citrus and strawberries	Very effective	None
Nutrition	Stone fruit trees	Easy	None
Chemicals	Peaches and stone fruits	Stored easily	Very few scientifically tested so far
Irrigation	All fruit plants, vegetable crops	Operational cost lower than heaters Can be used for other cultural purposes, such as drought prevention	Installation costs relatively high Risk damage to crop if rates inadequate Ice buildup may cause limbs to break Overwatering can waterlog soil Does not provide protection in winds above 8 kmph
Heaters	All fruit plants, but mainly used in tree fruits	Installation cost lower than irrigation No risk if heating rates adequate	Fuel oil expensive
Wind machines	All fruit plants, but mainly used in tree fruits	Very effective	Not effective in winds above 5 kmph or advective freeze Expensive

A rapidly rising phase of SOI in late autumn is an indication of fewer frosts in the season and the occurrence of the last frost early in the season. In other words, the frost period will not be prolonged late into the season. The reason for this scenario is median to above-median rainfall and cloud cover for a greater number of days resulting in less radiational cooling of the earth's surface in the winter season.

Forecast for Next One to Two Days

The meteorological conditions for frost occurrence are clear skies, inversion of temperature, and wind speed less than 7 kph. These weather conditions are normally associated with high pressure systems during the winter season. If the sky is clear, the atmosphere is comparatively calm during a cold evening, and the weather map on television shows that the area of interest falls well within a high pressure system, frost is expected during the next one to two nights.

THERMOPERIODISM

The response of living organisms to regular changes in temperatures, either day/night or seasonal, is known as *thermoperiodism.* Thermoperiodism exerts effects on the seasonal biology of insects and the growth and development of plants. Effects on insects include rates of growth and development, determination of diapause and dormancy, and acclimatization to low temperatures. Effects on growth and development of plants vary from one species to another. Crops such as soybean, maize, tomato, potato, eucalyptus, and mango are classified as thermoperiodic, while wheat, oats, peas, and cucumber are classified as nonthermoperiodic.

In soybean, a cool day/night temperature combination of 18/14°C disrupts floral development, leading to physically malformed parts. Normal floral initiation and pod development occurs at 30/18°C and 30/22°C, respectively, while the greatest number of pods per plant is obtained at 26/14°C (Judith and Raper, 1981). Tomatoes grow faster when the temperature is 26°C by day and 17°C by night than at the constant temperature of 26°C or any intermediate temperature. For this reason, tomatoes do not grow well in warm countries except in those locations where the temperature falls appreciably at night.

Some maize cultivars respond to a daily temperature fluctuation. In experiments with three lines differing in earliness (early, mid, and late) grown in a phytotron from two-week-old seedlings raised in a 16 h photoperiod under three different day/night temperature regimes, there was a direct correlation between chlorophyll content and grain yield. It proved best for yield and energy savings to raise seedlings of early lines at a high diurnal temperature (day/night = 24/15°C), and midseason or late lines at a constant day and night temperature of 21°C (Stolyarenko et al., 1992).

Potato plants grown under the fluctuating temperature treatment develop normally, develop tubers, and have a fivefold or greater total dry weight compared to those under constant temperature. This suggests a thermoperiod could allow normal plant growth and tuberization in potato cultivars that are unable to develop effectively under continuous radiation (Tibbitts, Bennet, and Cao, 1990).

Two important aspects of the environment influencing induction of flowering in mango are photoperiodism and thermoperiodism. Studies in the Maharashtra region of India (Lad, Pujari, and Magdum, 1999) indicated that minimum temperature below 10°C and above the freezing point stimulated heavy flowering in mango. Furthermore, flowering occurred only in a single flush, compared to two to three flushes under normal environmental conditions.

Many crop seedlings will grow perfectly well at a constant temperature, but others, such as celery, germinate best at fluctuating temperatures. The emergence of carrot seedlings from soil is faster in fluctuating temperatures than at constant temperature. *Solanum elaeagnifolium* is a weed that produces solasodine (a steroidal alkaloid used for the production of corticosteroids). It has a strict requirement for alternating temperature to germinate. A constant day/night temperature prevents the germination process. Seeds become sensitive to alternating temperature five days after the start of imbibition. After that, three daily cycles of alternating temperature are required for 50 percent germination (Trione and Cony, 1990).

The ecological significance of this response to a diurnal alteration of temperature may be that it promotes germination of those seeds close to the soil surface. When such fluctuations do not occur, germination, especially of more deeply seated seeds, may remain suppressed.

Apart from daily fluctuations in temperature, seasonal fluctuations are important in the development of many plants. Annual plants do not need a cold period during their development, except for plants that germinate in autumn and flower in the spring or summer after a cold winter. An example is winter wheat. Peaches cannot flower at high temperatures, but the vegetative growth phase continues. They need a period of cold weather before flower buds can open. No flower primordia are laid down under conditions of continued high temperature.

TEMPERATURE AS A MEASURE OF PLANT GROWTH AND DEVELOPMENT

Growing Degree-Days

Growing degree-days (GDD), also called heat units, effective heat units, or growth units, are a simple means of relating plant growth, development, and maturity to air temperature. The concept is widely accepted as a basis for building phenology and population dynamic models. Degree-day units are often used in agronomy, essentially to estimate or predict the lengths of the different phases of development in crop plants (Bonhomme, 2000).

The GDD concept assumes a direct and linear relationship between growth and temperature. It starts with the assumption that the growth of a plant is dependent on the total amount of heat to which it is subjected during its lifetime. A degree-day, or a heat unit, is the departure from the mean daily temperature above the minimum threshold (base) temperature. This minimum threshold is the temperature below which no growth takes place. The threshold varies with different plants, and for the majority it ranges from 4. 5 to 12.5°C, with higher values for tropical plants and lower values for temperate plants.

Methods of Degree-Day Estimation

Many methods for estimating degree-days are available in the literature (Perry et al., 1997; Vittum, Dethier, and Lesser, 1995). The three most dependable and commonly used methods are the standard method, the maximum instead of mean method, and the reduced ceiling method. Numerous others have been proposed, a majority being a modification of one of these three. An exhaustive review of degree-day methods was reported by Zalom and colleagues (1993).

1. Standard degree-day method:

$$\text{GDD} = \sum[(\text{Tmax} + \text{Tmin})/2] - \text{Tbase} \tag{3.1}$$

   where (Tmax + Tmin)/2 is the average daily temperature and Tbase is the minimum threshold temperature for a crop.
2. Maximum instead of means method:

$$\text{GDD} = \sum(\text{Tmax} - \text{Tbase}) \tag{3.2}$$

3. Reduced ceiling method: where Tmax ≤ $T_{ceiling}$, then

$$GDD = \Sigma(Tmax - Tbase), \text{ or} \tag{3.3}$$

where Tmax > $T_{ceiling}$, then

$$GDD = \Sigma\left[\left(T_{ceiling} - \left(T\max - T_{ceiling}\right)\right) - Tbase\right] \tag{3.4}$$

If maximum temperature (Tmax) is greater than the ceiling temperature ($T_{ceiling}$), then set Tmax equal to $T_{ceiling}$ minus the difference between Tmax and $T_{ceiling}$.

Uses and Limitations of Growing Degree-Day Methods

The use of degree-days for calculating the temperature-dependent development of insects, birds, and plants is widely accepted as a basis for building phenology and population dynamics models. The simplicity of the degree-day method has made it widely popular in guiding agricultural operations and planning land use. Most applications of the growing degree-day concept are for the forecast of crop harvest dates, yield, and quality. It helps in forecasting labor needs for factories, and in reducing harvesting and factory costs. A potential area of application lies in estimating the likelihood of the successful growth of a crop in an area in which it has not been grown before. The growing degree-day concept can also be applied to the selection of one variety from several varieties of plants to be grown in a new area. Another application of the concept can be to change or modify the microclimate in such a way as to produce nearly optimum conditions at each point in the developmental cycle of an organism. The concept is also applied to plants other than crop plants and to the issues of growth and development of insects, plant pathogens, birds, and other animals.

Though the degree-day concept is simple and useful, it lacks theoretical soundness and has a number of weaknesses. A range of factors that influence the predictive capability of degree-day accumulations have been identified. Among these are the conditions that impact the physiological state of an organism (such as nutrition and behavior-based thermoregulation), error associated with the assumptions and approximation processes used in estimating developmental rates and thresholds, and the limitations of available weather data. In addition, it is emphasized that regardless of the calculation method, degree-days are never more than estimates of developmental time (Zalom et al., 1993; Perry et al., 1997; Roltsch et al., 1999; Bonhomme, 2000). Specific limitations identified are as follows:

- While using growing degree-days, the physiological and mathematical bases upon which they are founded are sometimes forgotten, resulting in questionable interpretations (McMaster and Wilhelm, 1997).
- Except for the modified equations, a lot of weightage is given to high temperature.
- No differentiation can be made among the different combinations of the seasons. For example, the combination of a warm spring and a cool summer cannot be differentiated from a cold spring and a hot summer.
- The daily range of temperature is not taken into consideration, and this point is often more significant than the mean daily temperature.
- No allowance is made for threshold temperature changes with the advancing stage of crop development.
- Net responses of plant growth and development are to the temperature of the plant parts themselves, and they may be quite different from temperatures measured in a Stevenson's screen. Though this difference at a particular time may be small, the cumulative effects through an entire growing period can be very large.
- The effects of topography, altitude, and latitude on crop growth cannot be taken into account.
- Wind, hail, insects, and diseases may influence the heat units, but these cannot be accounted for in this concept.
- Soil fertility may affect crop maturity. This cannot be explained in this concept.

In spite of these limitations, the degree-day or heat unit concept answers a number of questions in plant and insect phenology and growth.

Chapter 4

Climatological Methods for Managing Farm Water Resources

Almost all of the water available on the earth, 97 percent, occurs as saltwater in the oceans. Of the remaining 3 percent, 66 percent occurs as snow and ice in polar and mountainous regions, which leaves only about 1 percent of the global water as liquid freshwater. More than 98 percent of freshwater occurs as groundwater, while less than 2 percent occurs in rivers and lakes. Groundwater is formed by excess rainfall (total precipitation minus surface runoff and evapotranspiration) that infiltrates deeper into the ground and eventually percolates down to groundwater formations (aquifers). For temperate, humid climates, about 50 percent of precipitation ends up in the groundwater. For Mediterranean-type climates, this figure is 10 to 20 percent, and for dry climates it can be as little as 1 percent or even less. The global renewable water supply is about 7,000 m^3 per person per year (present population). The per capita minimum water requirement is estimated at 1,200 m^3 annually, of which 50 m^3 is for domestic use and 1,150 m^3 is for food production (Food and Agriculture Organization [FAO], 1994). In Western and industrialized countries, a renewable water supply of at least 2,000 m^3 per person per year is necessary for adequate living standards (Bouwerg, 2000). These figures suggest that enough water is available for at least three times the present world population. Hence, water shortages are due to imbalances between population and precipitation distributions.

WATER FOR CROP PRODUCTION

Rainfall contributes to an estimated 65 percent of global food production, while the remaining 35 percent of global food is produced with irrigation. In most parts of the world, rainfall is, for at least part of the year, insufficient to grow crops, and rainfed food production is heavily affected by annual variations in precipitation.

A major part of the developed global water resources is used for food production. In most countries, 60 to 80 percent of the total volume of developed water resources is used for agriculture and may reach well over 80 percent for countries in arid and semiarid regions (Smith, 2000).

Irrigation is an obvious option to increase and stabilize crop production. Major investments were made in irrigation during the latter half of the twentieth century by diverting surface water and extracting groundwater. The irrigated areas in the world, during the last three decades of twentieth century, increased by 25 percent (FAO, 1993). The expansion rate has slowed down substantially because a major part of the reliable surface waters have already been developed, while groundwater resources have become overexploited at an alarming rate.

With water resources becoming scarce, waters of inferior quality are increasingly used. Excessive use and poor management of such irrigation water has had, in some cases, detrimental effects on soil quality, causing whole areas to be taken out of production or requiring the construction of expensive drainage works. Defining strategies in planning and management of available water resources in the agricultural sector will become a national and global priority.

MAKING EFFECTIVE USE OF RAINFALL

An inadequate and variable water supply and extremes of temperatures are the two universal environmental risks in agricultural production. High temperatures in tropical climates limit the production of crops native to temperate latitudes, and low winter temperatures in high latitudes are a check on growing crops native to tropical areas. Inadequate and variable water supply, however, has a negative impact on crop production in every climatic region. The problem is more pronounced in tropical and sutropical semiarid and arid climates in which the water losses in evaporation and evapotranspiration are very high throughout the year. Management of water resources is a much greater and more universal problem than any other factor of the environment.

Not all rainfall that falls in a field is effectively used in crop growing, as part of it is lost by runoff, seepage, and evaporation. Only a portion of heavy and high-intensity rains can enter and be stored in the root zone, and therefore effectiveness of this type of rainfall is low. With a dry soil surface with no vegetation cover, rainfall up to 8 mm/day may all be lost by evaporation. A rainfall of 25 to 30 mm may be only 60 percent effective with a low percentage of vegetative cover. Frequent light rains intercepted by a plant canopy with full ground cover are close to 100 percent effective (FAO, 1977).

In most parts of the world crop production depends on rainfall. Knowledge of the probable dates of commencement and end of the rainy season and the duration of intermittent dry and wet spells can be very useful for planning various agronomic operations such as preparing a seedbed, manuring, sowing, weeding, harvesting, threshing, and drying. This results in minimizing risk to crops and in optimum utilization of limited resources including water, labor, fertilizer, herbicides, and insecticides. There are critical periods in the life history of each crop, from sowing to harvesting. With knowledge of frequency of occurrence of wet and dry spells, a farmer can adjust sowing periods in such a way that moisture-sensitive stages do not fall during dry spells. Under irrigated farming, irrigation can be planned using data regarding consecutive periods of rainfall to satisfy the demands for critical periods. Knowledge of wet and dry spells can also help a great deal in improving the efficiency of irrigation-water utilization.

Measurement of Effective Rainfall

Numerous studies have been done in many countries to identify rainfall patterns and characteristics which can be used for planning agricultural operations such as sowing dates, harvesting dates, and periods and frequency of irrigation. Many of these studies are based on statistical analysis of the historical rainfall records. To study these characteristics of rainfall, it is assumed that each year provides one observation for an event of characteristic interest, and the total observations are then analyzed, assuming that they are a simple random sample from a single distribution. An effective rain event has been defined in various ways for varied purposes.

1. The start of the rains in northern Nigeria is defined as the first ten-day period with more than 25 mm precipitation, provided that rainfall in the next ten days exceeded half the potential evapotranspiration (Kowal and Krabe, 1972).

2. Raman (1974), deciding on a criterion of rainfall favorable for commencement of sowing operations, considered two basic requirements that must be satisfied. First, a sustained rainspell, which more or less represents the transition from premonsoon to monsoon conditions, should be identified. Second, in the spell so chosen, the rain that falls should percolate into the soil down to a reasonable depth and also build a moisture profile after loss through evaporation. Keeping in view these requirements, Raman (1974) selected a criterion for rainfall occurrence favorable for the commencement of sowing operations as a spell of at least 25 mm of rain in a period of seven days, with 1 mm or more on any five of these seven days, assuming an evaporative loss of 18 mm at the end of five days in the spell. The weekly spell taken was compatible with the average life cycle of monsoon depression in the area. Based on this criterion, the dates of commencement

of the first spell were chosen for each year, their mean, median, standard deviation, and quartile range were calculated, and these were mapped. These values were used to study the spatial distribution of the dates of commencement of sowing rains in the black cotton soils of Maharashtra in India.

3. Dastane (1974) recommended two methods for estimation of effective rainfall. In the first method, a percentage of rainfall varying from 50 to 80 percent was assumed to be effective. In the second method, rainfall less than 6.25 mm or in excess of 75 mm on any day, or a rainfall in excess of 125 mm in 10 days, is considered to be ineffective.

4. The U.S. Department of Agriculture (USDA) Soil Conservation Service (SCS) method estimates the effective rainfall by the evaporation/precipitation ratio method (FAO, 1977). Tables are given in which relationships are shown between average monthly effective rainfall and mean monthly rainfall for different values of average monthly crop evapotranspiration values. For use in irrigation, a net depth of irrigation water that can be stored effectively in the root zone is assumed to be 75 mm. Correction factors are given for different depths that can be stored.

5. Benoit (1977) defined the start of the growing season in northern Nigeria as the date when rainfall exceeded evaporation and remained greater than zero for the remainder of the growing season, provided that a dry spell of five days or more did not begin in the week after this date. Based on this criterion, he determined the start of the growing season in northern Nigeria. The planting dates of millet in Nigeria are observed to coincide with the first occurrence of 20 mm of rain over a two-day period.

6. The India Meteorological Department uses a chart showing normal dates for the onset of the southwest monsoon over India, taking long-term averages of five-day accumulated rainfall at 180 stations (Ashok Raj, 1979). The period characterizing an abrupt rise in the normal rainfall curve was taken to define the onset of the monsoon. This chart assists in overall indication of the arrival and progress of the monsoon over the entire country. However, for agricultural planning over small areas, this chart has serious limitations. This criterion has no relationship to the buildup of a moisture reserve in the soil, which alone is vital for commencement of the sowing operation.

7. Ashok Raj (1979) proposed a method for forecasting rainfall characteristics, such as the onset of an effective monsoon, based on the following criteria:

a. The first day's rain in the seven-day spell, signifying the onset of an effective monsoon, should not be less than e mm, where e is the average daily evaporation.
b. The total rain during the seven-day spell should not be less than $5e$ + 10 mm.

c. At least four of these seven days should have rainfall, with not less than 2.5 mm of rain on each day.

Using these criteria, Ashok Raj determined the onset of an effective monsoon at various probability levels for several states of India.

8. Stern and Coe (1982) used a general definition for the start of rains with these criteria:

a. The event making the start of the season was not considered until after a stated date *D*.
b. An event *E* then indicates a potential start date, defined as the first occurrence of at least *x* mm rainfall totalled over *t* consecutive days.
c. The potential start could be a false start if an event *F* occurs afterward, where *F* was defined as a dry spell of *n* or more days in the next *m* days.

For determining the start of rains at Kano, Nigeria, *D* was taken as May 1, *x* as 20 mm in two consecutive days, and *F* as a ten-day spell in the next 30 days. By using frequency distribution, they determined the potential start and false starts at different probability levels.

In all the aforementioned models, workers defined the event signifying the start of rains as a particular amount of rainfall received over a period of days. However, they neglected the soil moisture characteristics, which decide the availability of water and workable condition of the soil. A potential start of rains must make the soil sufficiently moist to support the germination of seeds. Thus, while deciding the start of rains or the onset of monsoon, it is important to consider the soil's moisture characteristics.

9. Patwardhan and Nieber (1987) proposed a soil-water balance model based on the equation of conservation of water in the soil profile. The water balance of the entire soil profile is considered in terms of individual processes:

$$\mathrm{P + I - R - RN = ET + D + \Delta S} \tag{4.1}$$

where *P* is rainfall, *I* is irrigation, *R* is runoff, *RN* is rainfall interception, *ET* is evapotranspiration, *D* is deep drainage, and ΔS is the change in water content in the soil profile. All the measurements are in mm of water. Effective rainfall is defined in the model as being that portion of the rainfall that infiltrates into the soil and does not contribute to deep percolation. This is expressed as

$$EP = (P - RN - R) \qquad (4.2)$$

where *EP* is the effective rainfall. The time scale on which the effective rainfall can be defined can be as small as one day; however, there is no upper limit.

10. In Taiwan, Chin, Komamura, and Takasu (1987) developed a model for the estimation of effective rainfall in order to use rainfall more effectively. The basis of the model is the equilibrium equation of the water balance in a paddy field. An irrigation area of a farm pond in northwest Taiwan was chosen to test the model's accuracy because of its simple cropping and single-rotation irrigation block, where inflow and outflow could be easily measured. The average measured and computed values were in close agreement, and the effective rainfall rate for this area was 40 to 65 percent.

11. A modified water balance model was used to estimate effective rainfall for lowland paddy in Thailand (Mizutani et al., 1991). An interception component was included in the model. The relationship of interception to rainfall at three growth stages was established from field experiments and utilized in the model. Eight stations with records for 30 years were selected for analysis. Simulations were run with computed crop water requirement and various values of percolation rate, ponding depth, and irrigation interval to study their effects on effective rainfall, irrigation requirements, and types of irrigation practiced. A 150 mm ponding depth and a five-to-six-day irrigation interval provide the most efficient irrigation and effective use of rainfall for lowland rice.

12. Drainage lysimeters were used by Kanber and colleagues (1991) to determine the effective rainfall in the Cukurova region of Turkey. They concluded that the relationship between total and effective rainfall increased linearly. An equation was derived to estimate the total monthly effective rainfall. The study showed that 16 percent of the rainfall was lost by deep percolation and 84 percent was retained on plant surfaces or stored in the soil.

13. In Japan, Komamura (1992)assessed the lower limit of effective rainfall for a small rainfall event. It was concluded from the study that (a) the degree of interception varies with crop type and (b) the useful lower rainfall limit for increasing soil water content is a minimum of 2 to 3 mm.

14. Alqarawi, Aldoss, and Assaeed (1997, 1998) carried out studies to investigate the effect of amount of rainfall (100, 200, and 400 mm) and rainfall distribution (7 and 14 days between two rains) on seedling survival, establishment, and growth characteristics of three populations of *Hammada elegans* in different areas in Saudi Arabia. Water equivalent to the specified amounts of rainfall was evenly distributed every 7 or 14 days over a period of three months. Seedlings were then left to grow for another two months

without irrigation. The results showed that survival and establishment under 400 mm rainfall were significantly higher than the other two rainfall averages (47 percent and 11 percent, respectively). Survival percentage increased as the period between two rains was extended to 14 days, although not significantly. Establishment increased from 3 to 9 percent with extension of the period between two rains.

15. Mohan, Simhadrirao, and Arumugam (1996) proposed a model for determining effective rainfall for use in estimating irrigation requirements for lowland rice. The method assumes that a paddy field can store additional rainfall up to the paddy spillway. The water balance equation reflecting the storage at the end of a time period t is given as

$$S_t = S_{t-1} + I_t - ET_t + ER_t - P_t \quad (4.3)$$

where S_t is storage at the end of period t; S_{t-1} is storage at the beginning of the period t; I_t is irrigation applied during the period t; ET_t is actual evapotranspiration during the period t; ER_t is effective rainfall during the period t; and P_t is percolation loss during the period t.

Free board is the rainfall storage capacity in the field. It is the difference between the spillway height and the depth of water in the paddy field. The depth of water use by *ET* and percolation losses during the period are added with free board to obtain the available storage capacity. If the rainfall amount is greater than this capacity, the rainfall excess is taken as runoff. A field spillway height of 100 mm was adopted. The percolation losses were taken as 2 mm/day, according to the local data. The depth of the water layer was 50 mm throughout. Mohan, Simhadrirao, and Arumungam (1996) compared this method to a number of other methods, including the USDA (SCS) method, and found this to be more appropriate than all the other methods.

16. A numerical simulation model (E-RAIN) was used to estimate long-term average and extreme values of monthly and annual effective rainfall for both seepage (seep) and fully enclosed seepage (FES) irrigation systems in Florida (Smajstrla, Stanley, and Clark, 1997). The model calculates effective rainfall as the difference between rainfall and runoff.

$$\text{E-Rain} = \text{Rainfall} - \text{Runoff} \quad (4.4)$$

where E-Rain = effective rainfall (mm), Rainfall = rain depth (mm), and Runoff = runoff volume per unit land area (mm). Runoff is calculated as

$$\text{Runoff} = (\text{Rainfall} - 0.2\ \text{S})^2/(\text{Rainfall} + 0.8\ \text{S}) \quad (4.5)$$

where S = a watershed storage coefficient (mm). The model was used with 41 years of daily rainfall data at Bradenton, Florida, demonstrating that the average annual effective rainfall is 775 mm with FES and 577 mm with seep irrigation. The model also simulates probabilities of occurrence of effective rainfall extreme values. The researchers claimed that this model should be useful to water management districts that issue water use permits on a probability basis and to irrigation system designers and managers who require estimates of effective rainfall as a component of crop water use.

EVAPORATION AND EVAPOTRANSPIRATION

Evaporation

The change of the state of water from solid and liquid to vapor and its diffusion into the atmosphere is referred to as *evaporation*. It plays a major role in the redistribution of thermal energy between the earth and the atmosphere and is an essential part of the hydrological cycle.

The process of evaporation involves the supply of energy for the latent heat of vaporization and the transfer process. The transfer process is governed by turbulence. Evaporation is a continuous process as long as there is a supply of energy, availability of moisture, and vapor pressure gradient between the water surface and the atmosphere.

Water vapor diffuses into the atmosphere from different surfaces such as lakes, rivers, ponds, cloud droplets, rain drops, moist soil, animals, and plants, but there is no fundamental difference in the physics of the process. Evaporation also occurs directly from the solid state, that is, from snow and ice, provided an appropriate vapor pressure gradient exists.

Transpiration

Most of the water absorbed by plants is lost to the atmosphere. This loss of water from living plants is called transpiration. It can be stomatal, cuticular, or lenticular. Transpiration that takes place through stomata is called *stomatal transpiration*. The maximum stomatal transpiration takes place through leaves. Outside the epidermal cells of a leaf is a thin layer called the cuticle. Sometimes gaps or pores in the cuticle are present. Water loss through these gaps is called *cuticular transpiration*. Pores or gaps in roots or stems are called lenticules, and loss of water through lenticules is called *lenticular transpiration*. The rate of transpiration depends on both meteorological factors and crop characteristics.

Stomata open in light and close in the dark, and the opening of stomata during day leads to transpiration. Lowered humidity results in higher transpiration. An increased difference between atmospheric and leaf humidity leads to increased transpiration. Humidity or vapor pressure is a function of temperature. A decrease in temperature increases vapor pressure in the environment, reducing the saturation deficit. The reverse is the case at higher temperatures. It follows that at higher temperatures there will be an increase in transpiration. In windy conditions, fresh dry air will replace the saturated air around the plant, leading to increased transpiration.

If the root/shoot ratio is high, there will be more absorption and less transpiration and vice versa. With greater availability of water to plants, transpiration will rise, while under a water stress condition, transpiration is restricted.

Leaf characteristics also influence transpiration. If the leaf area is large, transpiration will be high. A thicker cuticle will result in lowered cuticular transpiration. The presence of epidermal hair on leaves restricts the loss of water vapor to the atmosphere.

Evaporation versus Transpiration

The fundamental difference between evaporation from a free water surface and transpiration from plants is that in transpiration a diffusive resistance occurs due to the internal leaf geometry, including the stomata. No such resistance exists in evaporation from a free water surface. Because the stomata closes at night, the rate of transpiration drops to 5 to 10 percent of that occurring during the day, but the rate of evaporation remains relatively high because of the availability of energy stored at night.

Evapotranspiration and Potential Evapotranspiration

Over a land surface covered with vegetation, evaporation involves the following processes:

1. movement of water within the soil toward the soil surface or toward the active root system of the plants;
2. movement of water into the roots and then throughout the plant tissues to leaf surfaces;
3. change of water into vapor at the soil surface or at the stomata of plants;
4. change of rain water or snow from the outer surface of plants into vapor; and
5. the physical removal of water vapor from the boundary layer.

The overall process that involves these activities is termed *evapotranspiration* (ET).

Evapotranspiration is the combined loss of water from vegetation—both as evaporation from soil and transpiration from plants. Both processes are basically the same and involve a change of state from liquid to vapor. When water is adequately available at a site of transformation (i.e., soil or plant surfaces) the rate of evapotranspiration is primarily controlled by meteorological factors, including solar radiation, wind, temperature, and the evaporating power of the atmosphere.

The dependence of evapotranspiration on meteorological factors at a given place has led to the concept of potential evapotranspiration (PET). It is the upper limit of evapotranspiration. The concept assumes that there is an ample supply of water at the site of evaporation and that the rate is governed by the evaporating capacity of the atmosphere. However, the aerodynamic properties and stomatal behavior of the crop may modify the effect of meteorological factors on evapotranspiration. Potential evapotranspiration is therefore defined as the rate of evapotranspiration from an extensive surface of 8 to 15 cm green grass cover of uniform height, actively growing, completely shading the ground, and not short of water (Doorenbos and Pruitt, 1977; Smith, 2000).

When empirical methods of determining potential evapotranspiration are calibrated under conditions of unlimited water supply, they provide reasonably quantitative estimates. This is due to the conservativeness of potential evapotranspiration and because the variance from average values of potential evapotranspiration is correlated with variances of many climatic variables from their means. Empirical formulae, in general, correlate evapotranspiration with air temperatures, incident solar radiation, wind, atmospheric humidity, or a combination of these.

The measurement of evapotranspiration under normal conditions is of great importance in the estimation and management of present and future water resources and for solving many theoretical problems in the field of hydrology and meteorology. In planning irrigation, evapotranspiration data are used as a basis for estimating the acreage of various crops or combination of crops that can be irrigated with a given water supply or as a basis for estimating the amount of water that will be required to irrigate a given area. There has been a tremendous increase in the use of evapotranspiration data in scheduling irrigation. Evapotranspiration data are also used as a basis for evaluating the overall efficiency of irrigation in the field. As an agroclimatic index it has been widely used to assess the effect of the water supply on both the growth and yield of the crops.

Measurement of Evaporation and Evapotranspiration

There are several simple devices and empirical methods of estimating evaporation. Small containers of different kinds can measure evaporation quite accurately. However, for practical purposes, the measurement of evaporation from the surface of large water bodies, crop fields, bare soil, or catchment basins has greater significance. The relationship between the size of the evaporating surface and the rate of water loss is illustrated in Figure 4.1. The rate of evaporation is fairly independent of the size of the measuring pan under high humidity conditions. However, when the air is dry the size of the pan greatly influences the rate of evaporation. Therefore, to make use of measurements taken from these pans and the other bodies, a relationship between them needs to be established. There are five main types of evaporimeters or pans used for measuring evaporation. These are pans placed above the ground, pans sunk in the soil, floating pans, lysimeters, and Piche evaporimeters.

1. *Pans placed above the ground:* The U.S. Weather Bureau Class A pan is widely used in most countries of the world. The Class A evaporation pan

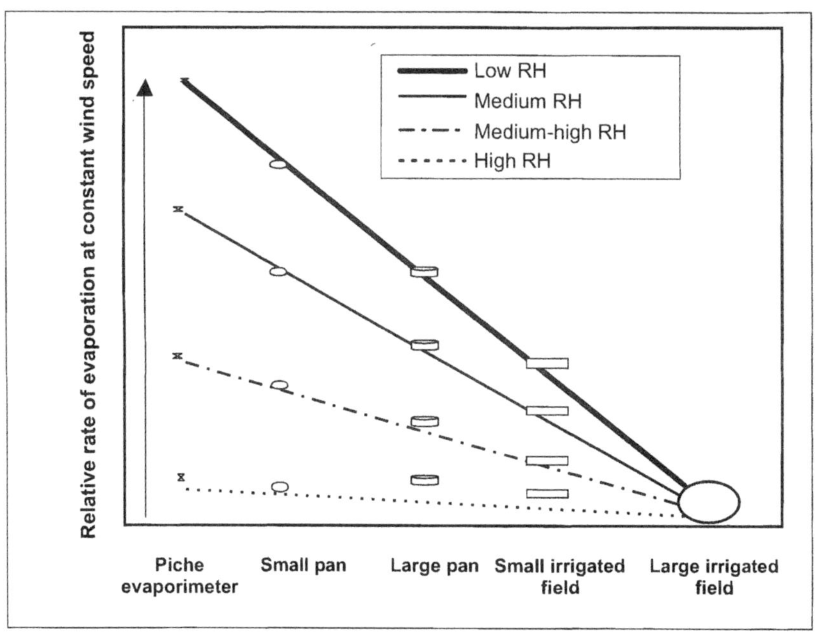

FIGURE 4.1. Size of evaporating surface and rate of evaporation

is circular, 121 cm in diameter, and 25.5 cm in depth. It is made of galvanized iron (22 gauge). The pan is mounted on a wooden frame platform with its bottom 15 cm above ground level. The pan must be level. It is filled with water below the rim, and water level should not drop to more than 7.5 cm below the rim (FAO, 1977). The major drawback of this pan is that the sensible heat flux from the sides and bottom results in increased evaporation, and it gives inflated values of evaporation.

2. *Sunken pans:* Many countries use sunken pans for measuring evaporation. The U.S. Bureau of Plant Industry and the British Institute of Water Engineers use pans of different dimensions in which the water surface is kept close to the earth's surface. The most common is the Sunken Colorado pan. It is 92 cm square and 46 cm deep. It is made of glavanized iron, set in the ground with the rim 5 cm above the ground (FAO,1977). The water inside the pan is maintained at or slightly below ground level. Sunken pans suffer from several operational difficulties including cleaning and heat leakage.

3. *Floating pans:* These pans are made to float in water bodies with suitable rafts. Water loss from these pans is similar to the water loss from the surrounding water surface. The installation and operation of these pans in water bodies are costly. Moreover, their operation becomes difficult when the wind is strong.

4. *Lysimeters:* Lysimetry is defined as the calculation of the vertical output fluxes using the volume and concentration of leached water over a period of time from a defined volume of soil (Muller, 1995). Lysimeters are tanks, filled with soil and buried in the ground, to measure the loss of water from the soil. They are commonly used for measuring evapotranspiration from a crop. However, they can also be used to measure the evaporation from a bare soil. Lysimeters are of the drainage and weighing types, with the latter the most commonly used.

The weighing lysimeter can measure evaporation and evapotranspiration for very short intervals of time. In addition to the measurement of evaporation and evapotranspiration, weighing lysimeters can give information such as the diurnal patterns of evaporation, variations in energy partitioning, and the relationships between transpiration and soil moisture tension. The biggest drawback of lysimeters is the high cost of their installation and their immobility.

5. *Piche evaporimeters:* A Piche evaporimeter consists of an inverted graduated tube filled with water and a filter paper clamped over its mouth. The instrument is kept in a Stevenson's screen. The Piche evaporimeter is not very reliable. It overestimates the effects of wind and underestimates the effects of solar radiation.

Empirical Methods

There is no end to the list of empirical methods that have been proposed for measuring evapotranspiration. The methods enumerated in this section are only a sample of that population. These are in common use within the irrigation profession. A brief description of each of these methods for computing reference crop evapotranspiration (in mm/day) is given.

1. *Hargreaves method:* This method (Hargreaves and Samani, 1985) estimates grass-related reference evapotranspiration. According to this method,

$$ET_0 = 0.0023R_a.\sqrt{TR}.(T + 17.8) \tag{4.6}$$

where ET_0 is reference crop evapotranspiration in mm/day; R_a is extra terrestrial radiation in equivalent evaporation in mm/day; TR is temperature range in °C, and T is mean daily air temperature in °C.

Because this is basically a temperature-based method, it is less accurate. However, local calibration of this method gives reasonably accurate ET_0 estimates. It requires only the measurements of maximum and minimum air temperatures. The method is recommended for ET estimates over ten days or longer periods (Smith, 1992).

2. *Ritchie method:* This method, as quoted in Meyer, Smith, and Shell (1995), is principally based on the radiant energy concept. It can be expressed as

$$E_{eq} = R_s(0.00488 - 0.00437\alpha) * (T_d + 29) \tag{4.7}$$

$$E_0 = 1.1E_{eq} \tag{4.8}$$

where E_{eq} is equilibrium evapotranspiration (mm/day); α is albedo, equal to 0.23; T_d is adjusted mean daily temperature, defined as (0.6 T_{max} + 0.4 T_{min}); and E_0 is daily potential evaporation (mm/day).

3. *Class A pan method (FAO-24 Pan):* Doorenbos and Pruitt (1977) provided a simple proportional relationship to estimate the ET, from U.S. Class A pan evaporation as

$$ET_0 = K_p.E_{pan} \tag{4.9}$$

where K_p is the pan coefficient, which depends on the pan environment in relation to nearby surfaces, obstructions, and the climate itself. K_p values can be obtained from FAO-24 Table-18 (Doorenbos and Pruitt, 1977).

4. *Penman-Monteith method* (as quoted by Chiew et al., 1995):

$$ET_0 = \frac{0.408\Delta(R_n - G) + \gamma\frac{900}{T+273}U(e_a - e_d)}{\Delta + \gamma(1+0.34U)} \quad (4.10)$$

5. *FAO-24 Penman method* (as quoted by Chiew et al., 1995):

$$ET_0 = c\left[0.408\frac{\Delta}{\Delta+\gamma}(R_n - G) + 2.7\frac{\gamma}{\Delta+\gamma}(1+0.864U)(e_a - e_d)\right] \quad (4.11)$$

c depends on shortwave radiation, maximum relative humidity, daytime wind speed, and ratio of daytime to nighttime wind.

6. *FAO-24 Radiation method* (as quoted by Chiew et al., 1995):

$$ET_0 = c(0.408WR_s) \quad (4.12)$$

W depends on temperature and altitude; c depends on mean relative humidity and daytime wind speed.

7. *FAO-24 Blaney-Criddle method* (as quoted by Chiew et al., 1995):

$$ET_0 = c[p(0.46T + 8)] \quad (4.13)$$

Explanations of the symbols used and where not explained along with methods 4 to 7 are as follows:

p is daily percentage of total annual daytime hours and depends only on the latitude and time of year.
c is a correction factor and depends on minimum relative humidity, sunshine hours, and daytime wind speed. It can be calculated with the procedure oulined in FAO-24.
R_n is net radiation at crop surface (MJ m^{-2}/day).
R_s is shortwave radiation (MJ m^{-2}/day).
G is soil heat flux (MJ m^{-2}/day).
T is average daily temperature (°C).
U is wind speed at 2 m above ground surface ($m \cdot s^{-1}$).
e_a is saturation vapor pressure at air temperature (kPa).
e_d is actual air vapor pressure (kPa).
Δ is slope of saturation vapor pressure/temperature curve (kPa/°C).
γ is psychrometric constant (kPa/°C).

8. *Computerized crop water use simulations:* Computer programs have been developed for the estimation of reference crop evapotranspiration from

climatic data and allow the development of standardized information and criteria for planning and management of rainfed and irrigated agriculture. The FAO CROPWAT program (Smith, 1992) incorporates procedures for reference crop evapotranspiration and crop water requirements and allows the simulation of crop water use under various climate, crop, and soil conditions.

9. *ET estimates from National Oceanic and Atmospheric Administration (NOAA) imageries:* Di Bella, Rebella, and Paruelo (2000) used multiple regression analysis to relate evapotranspiration computed from a water balance technique data obtained from NOAA satellite imagery. This approach, based on only remotely sensed data, provided a reliable estimate of ET over the Pampas region of Argentina. The approach is useful to estimate evapotranspiration on a regional scale and not at a particular point.

As stated at the beginning of this section, there is no dearth of methods available in the literature that are proposed to measure evapotranspiration. Numerous studies have been conducted at locations in different parts of the world with a wide range of climatic conditions to compare the relative performance of various methods of ET estimation (Jensen, Burman, and Allen, 1990; McKenny and Rosenberg, 1993; Chiew et al., 1995; Kashyap and Panda, 2001). There are some common conclusions from these studies.

- Combination methods (based on a number of parameters) generally provide more accurate ET estimates because they are based on physical laws and rational relationships.
- Depending on the climatological situation of a specific site, a locally calibrated, limited data input, simple ET estimation method may produce better results than a data extensive, complicated method.
- Availability of climatic data alone should not be the sole criterion in selecting a method since some of the data needed can be estimated from other variables with sufficient accuracy to permit using one of the better ET estimating methods.
- Penman estimates are consistently 20 to 40 percent higher than the Penman-Monteith estimates. Given that Penman-Monteith is the current standard method recommended by FAO, ET values calculated using FAO-24 Penman should therefore be used with caution.
- The FAO-24 radiation, FAO-24 Blaney-Criddle, and Penman-Monteith give similar monthly ET estimates. The Blaney-Criddle method, which uses only temperature data and some long-term average climate information, is adequate for applications in which only monthly estimates of ET are required. The radiation method gives daily ET estimates similar to Penman-Monteith. Unlike Penman-Monteith, that also requires wind data, the FAO-24 radiation method estimates ET from temperature and sunshine hours, climate variables that are rela-

tively conservative in spatial dimensions. The FAO-24 radiation method can thus be used as a surrogate for Penman-Monteith to estimate daily ET for areas where wind data are not available.

- The use of a pan method for estimating ET is controversial. Some researchers do not favor the use of this method, as extreme care is required in the operation of a pan as compared to any other climatic instrument. On the other hand, others favor the use of this method due to the availability of long-term evaporation records and the ease of use.
- There is a satisfactory correlation between Class A pan data and Penman-Monteith evaporation totals over three or more days. However, pan data are useful only if an accurate pan coefficient is used to relate the pan data to Penman-Monteith ET. The pan coefficient is very much dependent on local conditions and should be determined by comparing the pan data with the Penman-Monteith ET estimates.

WATER USE AND LOSS IN IRRIGATION

Surface irrigation and sprinkler irrigation are the main systems of irrigation practiced in the world. Although surface irrigation is the oldest and most extensively used method of applying water to crops, the use of sprinklers is emerging fast. The latest addition is the microirrigation system, but its use is currently confined to intensive horticulture (Periera, 1999).

Sprinklers and microirrigation systems are definitely better and much more efficient than surface irrigation systems (Pitt et al., 1996; Lamaddalena, 1997; Ramalan and Hill, 2000). However, the cost involved in installation and maintenance will keep their adoption restricted, and surface irrigation will continue to be the main irrigation system (Heermann, 1996).

Most surface-irrigated areas are supplied with water from a canal system. In general, supply rules are rigid, and often the time interval between successive deliveries is too long. Irrigators tend to compensate for this by applying all the water they are entitled to use. This leads to substantial water wastage through evaporation and percolation. Excessive applications also result in reduced crop growth, leaching of plant nutrients, and waterlogging.

According to an estimate (FAO, 1994), on average, only 45 percent of the water is used by the crop, with an estimated 15 percent lost in the water conveyance system, 15 percent in field channels, and 25 percent in inefficient field applications (Figure 4.2). Based on certain assumptions, T. Cummins (personal communication) estimated that in the Murray-Darling Basin of Australia, average water application efficiency in dairying is 0.4, 0.6 in rice, 0.75 in cotton, and 0.6 in horticulture.

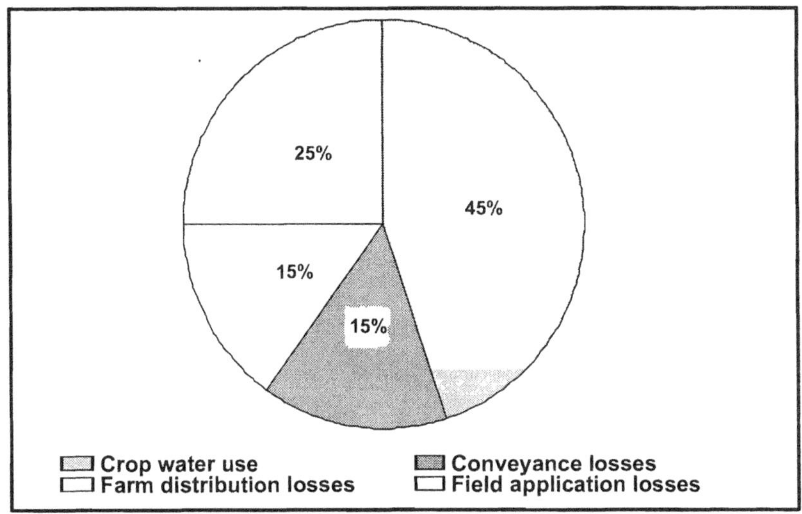

FIGURE 4.2. Water use and losses in irrigation (*Source:* Reprinted from *Agricultural and Forest Meteorology,* 103, M. Smith, The application of climatic data for planning and management of sustainable rainfed and irrigated crop production, pp. 99-108, 2000, with permission from Elsevier Science.)

Considerable space exists for more accurate and efficient crop water application by improved field irrigation methods and better crop water management techniques through the introduction of irrigation scheduling and water-supply control (Goussard, 1996; Liu et al, 1997). Irrigation scheduling is not yet utilized by the majority of farmers, and only limited irrigation scheduling information is utilized worldwide by irrigation system managers, extension officers, or farm advisers. It is well recognized that the adoption of appropriate irrigation scheduling practices could lead to increased yields and greater profit for farmers, significant water savings, reduced environmental impacts of irrigation, and improved sustainability of irrigated agriculture (Malano, Turral, and Wood, 1996; Smith et al., 1996). This concept is illustrated in Figure 4.3.

CLIMATOLOGICAL INFORMATION IN IMPROVING WATER-USE EFFICIENCY (WUE)

Climatological information plays a major role in evolving strategies for improving water-use efficiency. The prerequisites for water-supply control

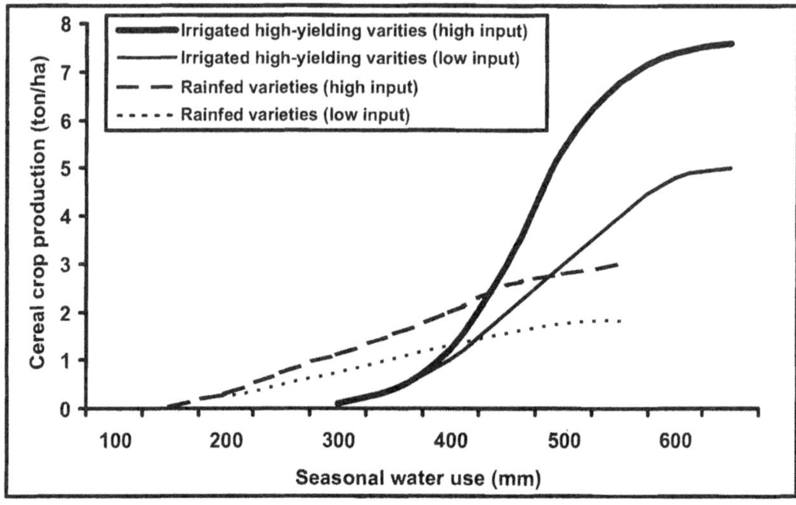

FIGURE 4.3. Water use efficiency of irrigated and rainfed cereal crops (*Source:* Reprinted from *Agricultural and Forest Meteorology,* 103, M. Smith, The application of climatic data for planning and management of sustainable rainfed and irrigated crop production, pp. 99-108, 2000, with permission from Elsevier Science.)

and introduction of irrigation scheduling strategies are analysis of the climatic conditions of the region and use of weather forecast information. Stochastically determined variability of rainfall and evapotranspiration is required for simulation of expected yield improvements and options for water storage (Smith, 2000).

Climatological Models for Irrigation Scheduling

Computerized procedures are available (Joshi, Murthy, and Shah, 1995; Smith, 2000) that greatly facilitate the estimation of crop water requirements from climatic data. The FAO CROPWAT program incorporates procedures for reference evapotranspiration and crop water requirements and allows the simulation of crop water use under various climate, crop, and soil conditions (Smith, 1992). As a decision support tool, there are several functions of the CROPWAT. These are (1) the calculation of reference evapotranspiration according to the FAO Penman-Monteith formula; (2) crop water requirements using crop coefficients and crop growth periods; (3) effective rainfall and irrigation requirements; (4) irrigation supply scheme to a given cropping pattern; and (5) daily water use computation.

The water balance procedure in CROPWAT allows the development of irrigation schedules and evaluation of irrigation practices in terms of water-use efficiency and the impact of water stress on crop yield. The system also allows the assessment of impact of rainfall, dry spells, and drought on crop production (Smith, 2000).

Use of Short-Range Weather Forecasts

A short-range weather forecast (up to a lead time of five days) would be of significant value to farmers, particularly in surface irrigation. This value comes from such advantages as:

- Better ability to manage waterlogging, particularly in surface irrigation: This could mean the difference between scheduling an irrigation event or not. By chosing not to irrigate, farmers may reduce both waterlogging and irrigation-associated costs.
- Better ability to manage soil moisture and plant stress: In viticulture and horticulture this could have significant implications for the quality of the crop, with significant financial implications as well.
- Better time management (when to apply) of sprays for disease and pest control: Again, such information could have important financial and environmental rewards.
- Water-use savings and the associated cost savings: These include cost of water; labor; fuel; access to the crop, especially at harvest; soil compaction; less water table accessions; management time, etc.

Nonetheless, more predictive information on likely weather adds another string to the bow for improving water-use efficiency. There are a number of reasons for this. A forecast information is a race with time, which makes it more likely to be acted upon. Farmers are already accustomed to using weather forecasts. Use of a new service would take time as farmers learn, from experience, how to use the information and the degree of trust they can put in it. Adoption is also likely to be high because it is also driven by the other potential gains in farm management and profitability.

Water-use savings with seasonable predictions can also allow farmers to determine their need for storage and the associated economic costs.

Checking Water Losses from Soil

Considerable potential exists to optimize the use of water for crop production, but strategies for more efficient usage are different for rainfed and irrigated agriculture (Smith, 2000).

Rainfed crop production is subject to frequent fluctuations in precipitation. Failing rains result in droughts and yield deficits, while excessive rains cause flooding and crop losses. Yields and water-use efficiency will therefore remain low even in periods with ample water supply or increased fertility levels. Crop water use needs to be optimized through more effective use and conservation of rainwater.

Extensive literature is available on technologies that can help conserve soil moisture and reduce evaporation and seepage from soil (Bos et al., 1994; Ventrella et al., 1996; Jalota and Prihar, 1998; Kazinja, 1999; Boldt et al., 1999; Mando and Stroosnijder, 1999). Most of the technologies have proved successful in achieving their objectives.

Use of Petrochemicals

Petrochemicals have shown promising results in reducing water movement and evaporation from bare soil. In Saudi Arabia, petrochemical soil conditioner (Hydrogrow 400) has been tested for reducing water losses from a sandy soil. The application of the conditioner resulted in an increase of stored water under saturated conditions due to the minimization of both gravitational and evaporation losses. An application rate of 0.75 percent of Hydrogrow 400 was found to be the optimum for water conservation and the maintenance of an adequate supply of water for plant growth. This rate reduced water movement under saturated conditions by 79.2 percent and loss of water from evaporation by 30 percent (Sabrah, 1994).

Agronomic Practices

Strip cropping, contour plowing, and terracing are used for reducing runoff, resulting in significantly increased soil moisture content. The protection provided by vegetation is also a major factor in runoff control. Plants intercept part of the rainfall and reduce the velocity of raindrops. They also slow down the movement of water on the soil's surface. Mulches of straw or crop residues, shredded bark, and wood chips break the impact of the raindrops and markedly improve infiltration and check evaporation (Smith, 1992).

The effect of tillage methods on crop growth and yields is to a large degree attributable to an increased soil moisture reservoir. This is achieved by creating soil conditions that favor root growth and penetration and improved infiltration and conservation of water. Tillage can be effective in reducing surface runoff if it is carried out according to soil conservation practices. By sacrificing a crop, moisture is conserved from one season to the next so that the combined precipitation of two seasons is sufficient for one crop.

Integrated Watershed Development

An integrated watershed development approach was successfully used in peninsular India (Raoa Mohan Rama et al., 1996). Measures adopted were (1) diversion drains and staggered contour trenches in nonarable land, (2) terraces of trapezoidal cross section with a graded channel on the upstream side (locally termed a graded bund) and stone checks in arable lands and rockfill dams, and (3) archweir (a curved barrier) and earthen embankment across a gully. Hydrological analysis revealed that integrated measures consistently improved the groundwater regime. Surface runoff from the treated forest and agricultural catchments were only 27.4 and 57.4 percent, respectively, of the untreated agricultural catchment, reflecting high infiltration of rainwater due to enhanced opportunity time. Consequently, water levels in the open wells rose by 0.5 to 1.0 m, thereby increasing the area irrigated by the wells by 172 percent when compared to the pre-project period, which in turn improved crop yields by 70 percent.

Temporal and Spatial Management Concept

A concept of temporal and spatial management of soil water (TSMSW) as a means to ensure effective use of soil water was developed by Jin and colleagues (1999) in North China. Four aspects were studied: readjusting crop structures and rotations to fit changes in soil water; increasing soil water resources; reducing soil water evaporation; and managing soil water to meet temporal and spatial crop water demand. Field experiments showed that temporal and spatial management of soil water can significantly increase water-use efficiency. For cotton, adopting an integration of microtopography and plastic mulch increased WUE from 0.49 to 0.76 to 0.86 $kg \cdot m^{-3}$; stalk mulch with manure for winter wheat reached 2.41 $kg \cdot m^{-3}$; and straw mulch with deep furrows (microtopography) for summer maize increased it from 2.06 to 2.34 $kg \cdot m^{-3}$.

Water Spreading

Water spreading schemes are applicable at specific sites to (1) assist in the control of erosion of susceptible soils and (2) increase the infiltration of water into the soil following rainfall (Wheeler, 1994). The additional soil moisture increases the yield of crops and pasture and may contribute to a general improvement in the soil condition.

An effective way to increase the benefit of high-intensity rainfall on pasture and crops is the construction of small spreading banks. Banks are usually from 100 to 120 m long and store a depth of water up to 400 mm. A gap of approximately 10 m between adjacent banks is left to allow for access during and after construction and to allow a passage for outflows when they occur. These outflows lead into the next bank downstream until they reach a natural stream or gully. Provision is made for water to be distributed downfield via sill boards, pipes, or open sections in the bank. Runoff from adjacent areas such as roads and rocky ridges or in nearby gullies is sometimes diverted to water-spreading schemes to supplement local rainfall.

Deficit Irrigation

Two levels of deficit irrigation strategies (irrigation limited to a certain growing period) for maize crop were evaluated by Boldt and colleagues (1999) in Nebraska. The first limited irrigation period started when the crop had accumulated 560 growing-degree days and ended when 1,220 GDDs were accumulated. On average, this is a five-week irrigation season. The second limited irrigation period started when 720 GDDs had accumulated and ended when 1,110 GDDs were accumulated, representing approximately a 3.5-week irrigation season.

The five-week irrigation season resulted in little yield reduction; however, applied water was reduced by 19 percent compared to irrigating for maximum yield. Limiting the irrigation season to 3.5 weeks decreased applied water depths further but had a more noticeable impact on grain yields. The yield was reduced by about 15 percent as compared to the maximum yields. Applied water decreased by 39 percent.

Partial Root-Zone Irrigation

The Commonwealth Scientific and Industrial Research Organisation (CSIRO) in Australia conducted successful field trials (Anonymous, 2000) into partial root-zone drying (PRD) to make horticultural crops more water efficient. This system involves the irrigation of only one side of the root zone. This causes a biochemical change in the roots that in turn leads to a reduced water loss from plant foliage. Fruit yield remains unaffected. Trials were done with drip-irrigated citrus fruit trees and grapevines. Results suggest that using the PRD system could provide farmers with water allocation savings of up to 50 percent.

REDUCING WATER LOSSES FROM RESERVOIRS

Evaporation

The evaporation rate from reservoirs is influenced by energy source, vapor pressure, air and water temperature, wind, reservoir volume, reservoir area, and water quality. Measures to reduce evaporation from reservoirs include designing deep reservoirs with minimal surface areas; concentrating storage in one central reservoir in preference to several; avoiding the creation of shallow areas; planting trees as windbreaks; preventing growth of water plants; designing the reservoir so that the wind blows along the length rather than the width of the reservoir; covering the reservoir; and using a reflective layer on the surface (Harrosh, 1992).

Aquacaps

The "aquacap" could be something that arid countries would be interested in. The Royal Melbourne Institute of Technology developed (Payten, 1999) a way of effectively halting evaporation with the invention of the aquacap. Aquacaps essentially are plastic domes that sit on rings, and when they are slung together over a water reserve they can curtail water loss by up to 70 percent. The lightweight aquacap is constructed of polyvinyl chloride (PVC), wire, bubble film, and polypropylene but is extremely hardy—actually improving its performance when weather conditions worsen. Aquacaps can achieve over 70 percent gross evaporation reduction, which is quite substantial. In areas with higher-than-average evaporation rates, it appears that these modules are very effective and perform at an even better rate.

The greatest benefit of aquacaps is in the farm dam situation, and, depending on the economics, they could be suited to larger storages. Benefits of aquacaps may reach further than minimizing evaporative water loss. An observation made through the trials (Payten, 1999) was that the modules significantly reduced incidents of algal bloom. Aquacaps prevent direct sunlight from hitting the water and also reduce stratification (in which water temperatures rise near the surface), both of which are factors in algal growth. The domes also appear to minimize wave action, which helps alleviate erosion on the banks, and can assist with the high salinity that evaporation can encourage.

Aquacaps are not yet commercially available, but mass production and interest by farmers all over Australia who suffer from water shortages due to notoriously irregular rainfalls and river flows may see the modules become financially realistic.

Underground Dams

Trials are being conducted across southern Australia (Collis, 2000) to evaluate the use of underground dams as alternative storages of winter water. Dams could collect water, which is then pumped into an aquifer and stored until summer when it is drawn back to the dam. This system has many benefits, including improved water quality, as long-term underground storage kills microorganisms in the water. It also greatly reduces water evaporation losses.

Seepage

A traditional small earthen pond is a good method for storing incident rainfall. There are, however, two significant problems. First, these small farm ponds are normally shallow which means their high surface area (in relation to their volume) fosters excessive evaporation. More important, however, they are usually constructed in soils that are permeable and not conducive to holding water for a long time. The dual loss of evaporation from the top and seepage from the bottom and sides is the main reason why the ponds often go dry at the time when water is most needed to keep crops and livestock alive. Various methods of reducing water loss have been tried. These can be classified roughly into chemical, physical, and biological approaches.

- *Chemical:* Where clays are of an appropriate type, certain sodium salts can reduce seepage in earthen ponds. Sodium ions cause clay to swell and clay particles to become dispersed (as opposed to coagulated), thereby reducing or plugging water-conducting pores in the soil. Sodium chloride, tetrasodium pyrophosphate, sodium hexametaphosphate, and sodium carbonate have been tested under field conditions. Sodium carbonate performed the best, both for reducing initial seepage and for subsequently maintaining low seepage rates. Narayana and Kamra (1980) recommend that mixing sodium carbonate with locally available soil (1 percent by weight) and applying the mixture reduces seepage losses by sedimentation.
- *Physical:* Soil in ponds can be physically compacted to reduce seepage losses. This is done with either manual or tractor-mounted compactors. The amount of compaction achieved depends on the load applied and the wetness of the soil. The soil's physical and chemical properties are also important. Sometimes merely walking cattle or buffalo over the area will help.

- *Biological:* Layers of organic materials that are rich in colloids (compounds that swell in water) can reduce percolation losses. One form of this is the so-called "bioplastic," a sandwich made of successive layers of soil, manure (from pigs, cattle, or other livestock), vegetative materials, and soil. This creates an underground barrier to seepage. Kale and colleagues (1986) obtained a seepage reduction of approximately 9 percent by using a mixture of cow dung, paddy husk, and soil.

Ahmad, Aslam, and Shafiq (1996) evaluated chemical, physical, and biological methods for reducing seepage in small ponds created in a permeable, calcareous silt-loam soil. The chemical method involved treating the top 10 cm of soil with sodium carbonate ($NaCO_3$). The physical method involved compacting the soil, and the biological method (the bioplastic sandwich) consisted of successive layers of soil, manure, and vegetative material. The chemical treatment proved less efficient than the other two methods. Compared to the untreated soil, the physical and biological methods reduced the mean cumulative seepage rates (measured 350 days after the initial wetting) by 72 percent and 67 percent, respectively. Both of these treatments seem to be cost effective and can be easily applied to farm ponds.

Chapter 5

Drought Monitoring and Planning for Mitigation

Drought is a climatic hazard that occurs in almost every region of the world. It causes physical suffering, economic losses, and degradation of the environment. A drought is a creeping phenomenon, and it is very difficult to determine when a dry spell becomes a drought or when a severe drought becomes an exceptional drought. It is slower and less dramatic than other natural disasters, but its effects are long lasting and widespread.

The cost and misery suffered from a drought are more than typhoons, earthquakes, and all other sudden climatic hazards. A drought results in less water in the soil, streams, and reservoirs, less water for livestock and wildlife, and poor crops and pastures. A chain of indirect effects follows which may include depressed farm income, closure of farm-supporting industries, and reduced hydroelectric power. A drought often induces malnutrition, disease, famine, population migration, and a chain of consequences for farm families (Stehlik, Gray, and Lawrence, 1999). The costs associated with drought are wide-ranging—economical, social, and environmental (National Drought Mitigation Center, 1996b,e).

The economic cost may include losses from crop, dairy, livestock, fishery, and timber production. Economic development, recreational business, and manufacturing are slowed, unemployment increases, and prices of essential commodities soar. Social costs of a drought may encompass food shortages, malnutrition, conflict between water users, water and garbage sanitation problems, increased poverty, decreased living standards and reduced quality of life, social unrest, and population migration from rural areas to urban centers. People experience shock, anger, and denial (Cheryl, 2000). The environmental cost may be in the form of damage to wildlife, wind erosion, higher concentrations of salt and pollutants in water, and decay of vegetation.

DEFINITION OF DROUGHT

The definition of drought is not very simple, and the question "What is drought?" continues to pose a problem (Sivakumar, 1991). This is because drought could mean different things to different people, and there are probably as many definitions of drought as there are users of water. In general, a drought is when a shortfall in precipitation creates a shortage of water, whether it is for crops, utilities, municipal water supply, recreation, wildlife, or other purposes. According to a WMO definition (Bogardi et al., 1994), "drought is a sustained, extended deficiency in precipitation."

Operational definitions of drought vary from place to place and are crucial to identify the beginning and intensity of drought. There are three main types of drought: meteorological, agricultural, and hydrological (National Drought Mitigation Center, 1996c).

1. *Meteorological drought* is an expression of rainfall departure from normal over some period of time. Meteorological drought definitions are usually region specific and are based on a thorough understanding of the climatology of the region.
2. *Agricultural drought* occurs when there is not enough soil moisture to meet the needs of crops at a particular time.
3. *Hydrological drought* refers to deficiencies in surface and subsurface water supplies. It is measured as stream flow and as lake, reservoir, and groundwater levels.

Some economists look at drought in socioeconomic terms. According to them, a socioeconomic drought is when physical water shortages start to affect supply and demand of goods.

METEOROLOGICAL INDICATORS OF DROUGHT

Drought conditions are basically due to a deficit of water supply in time and/or space. The deficit may be in precipitation, stream flow, or accumulated water in storage reservoirs, ground aquifers, and soil moisture reserves. In describing a drought situation, it is important to understand its duration, spatial extent, severity, initiation, and termination. Depending on the areal extent, a drought can be referred to as a point drought, small-area drought, or a continental drought. The point and small-area drought frequency are very high but are not major sources of concern at the national scale, unless they continue for a prolonged period. When the areal extent of the drought assumes a wide dimension, its assessment and mitigation measures become state and national concerns.

Over time, a number of drought assessment methods have been proposed. Some methods are based on qualitative observations, some on scientific criteria, and others on actual field surveys. However, to date, no comprehensive assessment method is available that has universal appeal. Different countries use different criteria to define and assess the drought situation. It is beyond the scope of this book to enumerate each and every indicator of drought that has been proposed and referred to in the literature. Some of these are very simple and old but still widely used. Others are more comprehensive, having sound scientific bases and holding good promise for application. The National Drought Mitigation Center (Hayes, 1996) has done a detailed comparative evaluation of the most widely used indices and those proposed during the recent past. Another evaluation was performed by Quiring and Papakryiakou (2003).

Percent of Normal

The percent of normal precipitation is one of the simplest measurements of drought for a location. It is calculated by dividing actual precipitation by the normal (considered to be a 30 or more years mean) and multiplying by 100. The percent of normal is calculated for a variety of time scales. Usually the time scales range from a single month, to a group of months representing a particular season, to an annual climatic year.

Analyses using the percent of normal are very effective when used for a single region or a single season. However, it is also easily misunderstood and gives different indications of conditions depending on the location and season.

One of the disadvantages of using the percent of normal precipitation is that the mean, or average, precipitation is often not the same as the median precipitation, which is the value exceeded by 50 percent of the precipitation occurrences in a long-term climatic record, largely because precipitation on monthly or seasonal scales does not have a normal distribution. Use of the percent of normal comparison implies a normal distribution in which the mean and median are considered to be the same. Because of the variety in precipitation records over time and locations, there is no way to determine the frequency of the departures from normal. Therefore, the rarity of an occurring drought is not known and cannot be compared to a different location.

The India Meteorological Department defines drought on the basis of rainfall deficiency during the southwest monsoon season on the basis of the percent of normal rainfall (Murty and Takeuchi, 1996). It employs two measures, the first describing rainfall conditions and the second representing drought severity. Rainfall conditions (based on the average rainfall of the last 70 to 100 years) are described as rainfall thresholds (Table 5.1), with rainfall expressed on a weekly or monthly basis. The intensity of drought is described as drought thresholds (Table 5.2).

A drought-prone area is defined as one in which the probability of drought in a given year is greater than 20 percent. A chronic drought-prone area is defined as one in which the probability of drought in a given year is greater than 40 percent. A drought year is defined as when less than 75 percent of the normal rainfall is received.

The National Institute of Hydrology, India, while analyzing the drought of 1987 (Murty and Takeuchi, 1996) proposed indices describing rainfall deficits, low flows in streams, and a fall in the water table. The drought conditions were classified in terms of runoff as shown in Table 5.3.

TABLE 5.1. Rainfall thresholds

Class	Range
Scanty	–50% or less than the normal
Deficient	–20% to –50% of the normal
Normal	+19% to –19% of the normal
Excess	+20% or more than the normal

TABLE 5.2. Drought thresholds

Class	Range
Moderate drought	Seasonal rainfall –26% to –50% of the normal
Severe drought	Seasonal rainfall < –50% of the normal

TABLE 5.3. Hydrological classification of drought

Drought class	Departure in runoff volume from normal (%)
Severe drought	50 and above
Moderate drought	25 to 50
No drought	Less than 25

In the Philippines, percent of normal index is used to assess the drought situation. A drought warning is issued when less than 40 percent of normal rainfall is received within three consecutive months. In Thailand, a generalized monsoon rainfall index, based on percent of normal rainfall, is also used to assess the impact of rainfall on crop conditions (Murty and Takeuchi, 1996).

Deciles

To avoid some of the weaknesses within the "percent of normal" approach, Gibbs and Maher (1967) developed the technique of ranking rainfall values in deciles as an indicator of drought. The rainfall occurrences over a long-term precipitation record are divided into sections for each ten percent of the distribution. Each of the sections is called a "decile." The first decile is the rainfall amount not exceeded by the lowest 10 percent of the precipitation occurrences. The second decile is the precipitation amount not exceeded by the lowest 20 percent of occurrences. These deciles continue until the rainfall amount identified by the tenth decile is the largest precipitation amount within the long-term record. By definition, the fifth decile is the median, and it is the precipitation amount not exceeded by 50 percent of the occurrences over the period of record. The deciles are grouped into five classifications, as shown in Table 5.4. The Australian Bureau of Meteorology prepares and displays tables and maps of precipitation deciles for the previous one, three, six, and twelve months across Australia.

The decile method was selected as the meteorological measurement of drought in Australia because it is relatively simple to calculate and requires less data and fewer assumptions than the Palmer Drought Severity Index. In this system, a drought is an exceptional event if it occurs only once in 20 to 25 years (deciles 1 and 2 records) and has lasted longer than 12 months. This uniformity in drought classifications, unlike a system based on the percent of normal precipitation, has been more useful to Australian authorities in

TABLE 5.4. Decile ranges and moisture thresholds

Decile range	Percent values	Classification
Deciles 1-2	Lowest 20% values	Much below normal
Deciles 3-4	Next 20% values	Below normal
Deciles 5-6	Middle 20% values	Near normal
Deciles 7-8	Next highest 20% values	Above normal
Deciles 9-10	Highest 20% values	Much above normal

determining appropriate drought responses. The disadvantage of the decile system is that a long climatological record is needed to calculate the deciles accurately.

Dependable Rains (DR)

Dependable rains (DR) is defined as the amount of rainfall that occurs in four of every five years (statistically, not consecutively). The index has been applied to the African continent (Le Houerou, Popov, and See, 1993). Dependable rains have potential for use in agricultural planning outside of Africa as well, especially in comparatively dry regions. The concept is, however, not a very good drought-monitoring index.

National Rainfall Index (RI)

The National Rainfall Index compares precipitation patterns and abnormalities on a continental scale. It was utilized to characterize precipitation patterns across Africa (Gommes and Petrassi, 1994). The index is calculated for each country by taking a national annual precipitation average weighted according to the long-term precipitation averages of all the individual stations. The country-size scale is designed to correlate with other countrywide statistics, especially agricultural production.

The RI allows comparisons to be made between years and between countries. RI is well correlated with national crop yields in Africa. Because it is weighted by annual rainfall, those stations in wetter areas of a country have a greater influence on the RI than stations in naturally drier areas. In many countries, especially in Africa, the wetter stations are also located in more agriculturally productive regions. RI has, therefore, a natural bias toward agriculture, and it is a useful tool where country-scale crop production is correlated with rainfall.

RI is independent of absolute amounts of rainfall, which may be localized, and allows general comparisons to be made regarding an entire country. The long-term record makes available a frequency distribution of RI values, which allows historical comparisons to be made, an analysis not possible with the percent of normal. Even if the record is not complete for an individual station, the RI can still be calculated without that station.

The RI may be less useful when looking at overall drought conditions and the hydrological, environmental, and social impacts resulting from drought.

Palmer Drought Severity Index (PDSI)

The Palmer Drought Severity Index measures abnormalities in the moisture supply (Table 5.5). The index developed by Palmer (Palmer, 1965) is based on the supply-and-demand concept of the water balance equation, taking into account several other factors in addition to precipitation deficit at specific locations. The objective of the Palmer Drought Severity Index was to provide a measurement of moisture conditions that were "standardized," so that comparisons using the index could be made between locations and between months.

The PDSI is essentially a meteorological drought index and is based on precipitation and temperature data and the locally available water content (AWC) of the soil (Karl and Knight, 1985). From the inputs, all the basic terms of the water balance equation can be determined, including evapotranspiration, soil recharge, runoff, and moisture loss from the surface layer.

The Palmer Index has been widely used for a variety of applications across the United States. It is most effective in measuring impacts sensitive to soil moisture conditions, such as agriculture. It has also been useful as a drought-monitoring tool and has been used as an indicator on which to base the start or end of drought contingency plans. The index is popular because it provides decision makers with (1) a measurement of the abnormality of recent weather for a region; (2) an opportunity to place current conditions in a historical perspective; and (3) spatial and temporal representations of historical droughts.

TABLE 5.5. Palmer Drought Severity Index classifications for dry and wet periods

Index value	Classification
–4.00 or less	Extreme drought
–3.00 to –3.99	Severe drought
–2.00 to –2.99	Moderate drought
–1.00 to –1.99	Mild drought
–0.50 to –0.99	Incipient dry spell
0.49 to –0.49	Near normal
0.50 to 0.99	Incipient wet spell
1.00 to 1.99	Slightly wet
2.00 to 2.99	Moderately wet
3.00 to 3.99	Very wet
4.00 or more	Extremely wet

Along with its merits, the Palmer Index also has drawbacks (Alley, 1984; Karl and Knight, 1985):

1. The values quantifying the intensity of a drought and signaling the beginning and end of a drought or wet spell are arbitrarily selected.
2. The Palmer Index is sensitive to the AWC of a soil type. Applying the index for a climate division may be too general.
3. The soil layers within the water balance computations are simplified and may not accurately represent a location.
4. Snowfall, snow cover, and frozen ground are not included in the index.
5. All precipitation is treated as rain, so the timing of PDSI values may be inaccurate in the winter and spring months in regions where snow occurs.

Bhalme and Mooley Drought Index (BMDI)

The BMDI was developed by Bhalme and Mooley in 1980 (Bogardi et al., 1994) and is a simplified version of the Palmer Index. The calculations of BMDI need only precipitation data, but its performance, according to the authors, is comparable to that of PDSI.

The index expresses situations that vary from extreme drought to extreme wet (Table 5.6). BMDI = <–4 for extreme historical drought and proportionally increases to higher values. For normal conditions, BMDI = 0, and for extreme wet, BMDI = >4.

The simplicity of the calculations is the major merit of this index. The index has performed well under Indian and Hungarian climatic conditions. The performance has been equally good in the Great Plains of North America.

TABLE 5.6. Bhalme and Mooley Drought Index based drought categories

Index value	Character of the weather
Greater than 4	Extremely wet
4 to 3	Very wet
3 to 2	Moderately wet
2 to 1	Slightly wet
1 to –1	Near normal
–1 to –2	Mild drought
– 2 to –3	Moderate drought
– 3 to –4	Severe drought
Less than – 4	Extreme drought

Surface Water Supply Index (SWSI)

To overcome the limitations of the Palmer Index, Shafer and Dezman (1982) designed the Surface Water Supply Index (SWSI) to be an indicator of surface water conditions. They described the index as "mountain water dependent," in which mountain snowpack is a major component. The intention was to use the index as a complement to the Palmer Index in Colorado.

The SWSI incorporates both hydrological and climatological features into a single index value resembling the Palmer Index for each major river basin in a state. These values would be standardized to allow comparisons between basins. The inputs required are snowpack, stream flow, precipitation, and reservoir storage. Because it is dependent on the season, the SWSI is computed with only the snowpack, precipitation, and reservoir storage in the winter. During the summer months, stream flow replaces snowpack as a component within the SWSI equation. The procedure to determine the SWSI for a particular basin is as follows:

1. Monthly data are collected and summed for all the precipitation stations, reservoirs, and snowpack/stream flow measuring stations over the basin.
2. Each summed component is normalized using a frequency analysis gathered from a long-term data set.
3. Each component has a weight assigned to it depending on its typical contribution to the surface water within that basin, and these weighted components are summed together to determine a SWSI value representing the entire basin.
4. The SWSI is centered on zero and has a range between –4.2 and +4.2.

One of its advantages is that it is simple to calculate and gives a representative measurement of surface water supplies across the region/state. The SWSI has been used to trigger the activation and deactivation of a drought plan in Colorado.

Several characteristics of the SWSI create limitations in its application. The discontinuance of any station means that new stations need to be added to the system and new frequency distributions need to be determined for that component. Additional changes in the water management within a basin, such as flow diversions or new reservoirs, mean that the entire SWSI algorithm for that basin needs to be redeveloped to account for changes in the weight of each component. Thus, it is difficult to maintain a homogeneous time series of the index. Extreme events also cause a problem. If the events are beyond the historical time series, the index will need to be reevaluated to include these events within the frequency distribution of a basin component.

Standardized Precipitation Index (SPI)

The Standardized Precipitation Index (SPI) is based on the fact that a deficit of precipitation has different impacts on the groundwater, reservoir storage, soil moisture, snowpack, and stream flow (McKee, Doesken, and Kleist, 1993). The SPI quantifies the precipitation deficit for multiple time scales (3, 6, 12, 24, and 48 months). These time scales reflect the impact of drought on the availability of the different water resources. Soil moisture conditions respond to precipitation anomalies on a relatively short scale, while groundwater, stream flow, and reservoir storage reflect the longer-term precipitation anomalies.

SPI is calculated by taking the difference of the precipitation from the mean for a particular time scale and then dividing by the standard deviation. Because precipitation is not normally distributed for time scales shorter than 12 months, an adjustment is made which allows the SPI to become normally distributed. Thus, the mean SPI for a time scale and location are zeros and the standard deviation is one. This is an advantage, because the SPI is normalized so that wetter and drier climates can be represented in the same way.

A classification system is used to define drought intensities resulting from the SPI (Table 5.7). A drought event occurs any time the SPI is continuously negative and reaches intensity when the SPI is –1.0 or less. The event ends when the SPI becomes positive. Therefore, each drought event has a duration defined by its beginning and end and its intensity for each month that the event continues. An accumulated magnitude of drought can also be measured. It is called the drought magnitude (DM) and is the positive sum

TABLE 5.7. Standardized Precipitation Index

SPI value	Moisture category
2.0 and above	Extremely wet
1.5 to 1.99	Very wet
1.0 to 1.49	Moderately wet
0 to –0.99	Near normal
–1.00 to –1.49	Moderately dry
–1.50 to –1.99	Severely dry
–2.0 or less	Extremely dry

of the SPI for all the months within a drought event. This standardization allows the SPI to determine the rarity of a current drought.

The SPI has been used operationally to monitor conditions across Colorado during 1994 and 1995 (McKee, Doesken, and Kleist, 1995). The potential exists for the SPI to provide near-real-time drought monitoring for an entire country. The number of applications using the SPI around the world are increasing, because the index has the advantages of being easily calculated, having modest data requirements, and being independent of the magnitude of mean rainfall, and hence comparable over a range of climatic zones. It does, however, assume the data are normally distributed, which can introduce complications for shorter time periods (Agnew, 2000; Hayes, 2000).

Crop Moisture Index (CMI)

The Crop Moisture Index (CMI) was developed by Palmer in 1968 and uses a meteorological approach to monitor week-to-week crop conditions from procedures he used to calculate the PDSI (Palmer, 1968; McKee, Doesken, and Kleist, 1995). Whereas the PDSI monitors long-term meteorological wet and dry spells, the CMI was designed to evaluate short-term moisture conditions across major crop-producing regions. It is based on the mean temperature and total precipitation for each week and the CMI value from the previous week (Table 5.8). The CMI responds rapidly to changing conditions. It is weighted by location and time, so maps, which commonly display the weekly CMI across a state or a region, can be used to compare moisture conditions at different locations.

The Crop Moisture Index is designed to monitor short-term moisture conditions impacting a developing crop, so it is not a good tool for long-term drought monitoring. The CMI's rapid response to changing short-term conditions may provide misleading information about long-term conditions. The CMI typically begins and ends each growing season near zero. This limitation prevents the CMI from being used to monitor moisture conditions outside the general growing season, especially in drought situations that extend over a year or more.

DROUGHT MONITORING IN AUSTRALIA

Drought is not rare in Australia. There may be few countries in the world where drought occurrence is more frequent than in Australia. Every state has developed its own procedures to identify and monitor the drought situation.

TABLE 5.8. Crop Moisture Index (CMI)

CMI values when index increased or did not change from previous week		CMI values when index decreased from previous week	
3.0 and above	Excessively wet, some fields flooded	3.0 and above	Some drying, but still excessively wet
2.0 to 2.99	Too wet, some standing water	2.0 to 2.99	More dry weather needed, work delayed
1.0 to 1.99	Prospects above normal, some fields too wet	1.0 to 1.99	Favorable, except still too wet in spots
0 to 0.99	Moisture adequate for present needs	0 to 0.99	Favorable for normal growth and field work
0 to –0.9	Prospects improved, but rain still needed	0 to – 0.9	Topsoil moisture short, germination slow
–1.0 to –1.99	Some improvement, but still too dry	–1.0 to –1.99	Abnormally dry, prospects deteriorating
–2.0 to –2.99	Drought eased, but still serious	–2.0 to –2.99	Too dry, yield prospects reduced
–3.0 to –3.99	Drought continues, rain urgently needed	–3.0 to –3.99	Potential yields severely cut by drought
–4.0 and below	Not enough rain, still extremely dry	–4.0 and below	Extremely dry, most crops ruined

NSW Agriculture has been monitoring the monthly status of drought in New South Wales. Based on this assessment, the government has provided various forms of assistance to drought-affected people. To complete this procedure, Rural Land Protection Boards (RLPBs) supply the relevant data.

The RLPBs supply the information in a standard format. The information given is on meteorological conditions; agronomic conditions; stock numbers (change from normal); livestock condition; agistment of stock (change from normal); hand feeding (change from normal); water supply; environmental conditions; other drought-related factors; and the overall recommendation of the board. Except for meteorological conditions and other drought-related factors, the information on all the other factors is on a graded scale. The information on meteorological conditions covers data on rainfall and evaporation for as many stations as are available in the board's boundaries. In addition, data are also sought on wind, frost, and temperature.

All the data received from the board are compiled, analyzed, and mapped to demarcate the areas that meet the drought criteria. The meteorological data obtained from the Bureau of Meteorology are analyzed separately to assess the drought situation on the basis of decile ranking of the current season's rainfall. The map showing the spatial extent of the drought, along with recommendations pertaining to the drought, is then sent to the minister of agriculture.

In Queensland, procedures adopted to identify properties affected by drought are different from those used to identify areas affected by drought. Furthermore, conditions of drought considered for livestock-dominated enterprises are different from those considered for agricultural, horticultural, and sugar enterprises (Queensland Department of Primary Industries, 1995; Rural Industry Business Services, 1997).

Official drought declaration in Queensland is made under extreme drought conditions. Events of an extreme nature under the Queensland drought policy, based on historical records, occur once in every 10 to 15 years. Such events are usually associated with an extreme lack of effective rains over two or more consecutive seasons. When drought becomes widespread in a shire, local drought committees make an assessment of the seasonal conditions in terms of

- rainfall,
- availability of pasture and water,
- condition of the stock,
- whether drought mortalities of stock are occurring,
- the extent of movement of stock to force sales or slaughter and to agistment,
- quality of fodder introduced,
- assessment of agricultural and horticultural industries,
- number of individual property declarations that have been issued, and
- whether other abnormal weather factors have affected the situation.

In addition, field officers of the Queensland Department of Primary Industries (QDPI) are required to hold consultations with fellow officers of the department, local drought committees, and other knowledgeable persons concerning conditions. The local drought committee makes a formal recommendation and submission to the Natural Disaster Relief Section (NDRS) through the stock inspector (coordinator) for processing the submission.

The Natural Disaster Relief Section analyzes the monthly rainfall records of the last 12 months, and these are compared to the historical records of the area to identify those areas experiencing an extreme event (one in ten to fifteen years). Provided all criteria have been met, the NDRS makes a recommendation to the minister for primary industries who, in consultation with the treasurer, declares an area to be drought affected.

DROUGHT EXCEPTIONAL CIRCUMSTANCES

Drought conditions of some magnitude are present almost every year in some part of Australia because of its vast size and semiarid to arid climate. Such occurrences are a part of normal life and are not of major concern at the national level. Sustained droughts, usually lasting one to two years, possibly for three years, and extending across large tracts of the country have created great disasters. These are of relatively less frequent occurrence, and each of them has different spatial, duration, and intensity characteristics. When drought conditions are so intense and protracted that they are beyond those that can reasonably be factored into normal risk management strategies, they are termed drought exceptional circumstances (Lembit, 1995). In practice, this is a drought of such rarity and severity that it occurs no more than once in every 20 to 25 years and is more than 12 months in duration (Clark et al., 2000; Dixon, 1995).

Assessment of Drought Exceptional Circumstances

The National Drought Policy has laid out a process with a framework for the determination of drought exceptional circumstances and a set of six core criteria to be taken into account by both the commonwealth and the states in consideration of drought exceptional circumstances declarations (Lembit, 1995). The six core criteria are

1. meteorological conditions,
2. agronomic and stock conditions,
3. water supplies,
4. environmental impacts,
5. farm income levels, and
6. scale of the event.

Drought exceptional circumstances are indicated when the combined impact on farmers of the core criteria is a rare and severe occurrence. Meteorological conditions are the threshold or primary condition for exceptional circumstances but should be assessed in terms of "effective rainfall." The threshold conditions would involve a "rare and severe event"; rare being one in twenty years, and severe being either more than twelve months or at least three consecutive failed seasons depending on the nature of the production system being considered (Queensland Department of Primary Industries, 1995).

Assessment will go further if the criterion of meteorological conditions is satisfied. The remaining criteria should collectively indicate drought exceptional circumstances. The criteria are used together to form an overall judgment on exceptional drought circumstances. A similar process must be followed for the revocation of drought exceptional circumstances.

White (1997) has given a summary of the indices that are presently used for the assessment of drought exceptional circumstances by the commonwealth and state and territory governments in Australia. Although none of the major indices is superior to the rest in all circumstances, some indices are better suited than others are for certain uses. Some of the methods used or with potential for use are summarized here.

Rainfall Analysis

Rainfall is the main criterion for assessing drought exceptional circumstances. Analysis based on rainfall data averaged over a meteorological division has a drawback of not necessarily coinciding with the area of interest with respect to a drought event. Analysis of individual rainfall stations selected to represent the region of interest is more useful, as it could give a quick indication of the most affected areas and where boundaries might lie. Several indices measure how much rainfall for a given period of time has deviated from a historically established normal.

Statistical techniques for the analysis of drought events based on historical rainfall records of individual stations using commonly available spreadsheet packages operating on desktop computers are suggested (Bedo, 1997). These analyses complement more sophisticated approaches available only to specialists. A series of Microsoft Excel macros, which analyze rainfall in several ways to test the meteorological criteria, are used. These macros provide three techniques of rainfall analysis to identify an exceptional circumstances event.

1. For visual checking of individual monthly rainfall values and confirming patterns in the past, decile and percentile values for every month of the historical record are calculated and presented in a tabular form.
2. The cumulative rainfall anomaly is calculated and plotted for any period within the historical records of a rainfall station. For example, anomalies can be calculated for periods restricted to agriculturally important seasons. Plotting the cumulative rainfall anomaly for the en-

tire historical record provides useful indications of past exceptional events.

3. Analysis of rainfall for seasonal periods appropriate to different agricultural regions and farming systems is possible. The monthly periods can be selected to correspond to winter or summer rainfall climates or split to represent autumn and spring within a calendar year or a cool and warm season extending over two calendar years. Rainfall totals are calculated for the selected seasons over the historical record, and the percentile (or decile) ranking is determined. The macro then scans the result table and marks those seasons that qualify under the current criterion of three consecutive seasons at or below a critical percentile. The critical percentile value is adjusted so that the number of events over the historical record occurs about one in 20 to 25 years.

Statistical Models

Stephens (1997) proposed a Drought Exceptional Circumstances Index (DECI) as a criterion for defining exceptional drought in cropping areas. The index is based on long-term rainfall records of stations spread across the wheat belt, representing the major agrometeorological zones. For each region, he defined a cropping year, which covered the essential period of soil moisture accumulation and crop growth. This began on October 1 in the previous year and ended when rainfall stopped contributing to wheat yield. A five-step approach is used to derive the index:

1. Long-term wheat yields were calculated assuming no change in technology for all years of available rainfall data. He used a yield forecasting model (STIN) based on a "moisture stress index" (Stephens, 1996).
2. Growing season rainfall was added and ranked as percentiles. Abnormal years were discarded.
3. Relative winter wheat yields were combined with relative mean soil moisture to form a Yearly Productivity Index (YPI). This index ranges between 0 and 1.
4. Individual yearly yields must be in the lowest 30 percent of values for conditions to be exceptional. For a two-year interval, both yields must be below this cutoff point, whereas three- or four-year intervals should allow for one year with yields in the 30 to 40 percent range, before conditions fall back into exceptional circumstances again.
5. Individual years that qualify were ranked in ascending order of severity on the basis of four-, three-, and two-year mean yields to identify

> the worst years. Drought duration and severity were integrated with a summation of yield (YPI) deviations below a 40 percent value. This summation is called the Drought Exceptional Circumstance Index.

This method is claimed to be more responsive to the ground situation than many other empirical, agroclimatic, and simulation models (Brook, 1996).

Simulation Models

The primary argument for simulation of system performance is that meteorological conditions alone do not easily capture the true state of the agricultural system. Rainfall at one time of the year can be carried over under fallows to be used at other times of the year. Failure of planting rains at a critical time may downgrade otherwise average seasonal rainfall conditions in terms of production potential. A crop-soil management system simulation model has the potential to integrate the meteorological and agricultural dimensions of the production system.

Pasture simulation models: McKeon (1997) has advocated the use of the National Drought Alert Strategic Information System for assessing the events of drought exceptional circumstances. The National Drought Alert Strategic Information System is a good combination of rainfall analysis, seasonal climate forecasts, satellite and terrestrial monitoring, and simulation models of meaningful biological processes.

The core simulation model used in the National Drought Alert Strategic System is GRASP (GRASs Production), which has been thoroughly validated in Queensland (Carter and Brook, 1996). GRASP produces estimates of pasture growth, biomass in green and dead pools, green cover, soil moisture, animal liveweight gain, and pasture utilization on a daily basis and can be run forward up to 180 days into the future. When pasture production is combined with stock estimates, calculations of the degree of pasture utilization can be made and displayed as maps of feed availability and land condition, with a resolution of a quarter to half a shire. These maps form a core product of the strategic information system for assessing drought exceptional circumstances.

The National Drought Alert Strategic Information System produces percentile views of meaningful biological and agricultural variables. So it is possible to construct a percentile view of grassland production and condition that is more aligned with the actual extent and severity of drought than are rainfall percentile maps. A particular month's or season's grass biomass can be compared to the last 30 years or 100 years of biomass that would have existed at that location.

Smith and McKeon (1997) used the simulation models for assessing the historical frequency of drought events on rangelands. They analyzed the results in terms of various measures that could be used to identify an exceptional circumstances drought event on the basis of those occurring once in 20 years in the long term.

Another pasture simulation model that has been advocated to identify drought is GrassGro (Donnelly and Freer, 1997). GrassGro models pasture and animal production which in turn identify severe drought conditions. With the GrassGro model, simulation results are tabulated for monthly rainfall, weight of green herbage available, and weight of supplementary feed required to maintain the stock. In the pasturelands of southern Australia, severe droughts identified with GrassGro are more realistic than those identified with rainfall alone.

Crop simulation models: Keating, Meinke, and Dimes (1997) explored the potential role for crop simulation models, such as APSIM (Agricultural Production System Simulation Model), to assist in the objective assessment of drought. The study concluded that the dynamic simulation of agricultural systems has much to offer to the objective identification of drought exceptional circumstances. This does not mean that other, simpler methods which relate crop performance to weather could not achieve similar results. It should be possible to combine the various models on the strengths and weaknesses of the alternative approaches.

Remote Sensing

Satellite data can be directly related to land cover (vegetation and soil) status or functioning. Remotely sensed data are unsurpassed in supporting the formulation of drought indicators because they are actual observations of landscape status and its performance. Obtaining a time series of remotely sensed images allows information to be extracted regarding the location and duration of below-average biomass and below-average soil moisture (Smith, 1996). Normalized Difference Vegetation Index (NDVI) data are used to monitor vegetation health and to fine tune regional differences. Remotely sensed data (visible, thermal, etc.), geographic information system (GIS) data layers (soils, geology, etc.), and point-based measurements (climate, biomass, etc.) all have space and time dimensions and can be integrated for a better appreciation of the environment. This information can then be combined with other necessary information, such as agronomic, economic and social data, which allow drought exceptional circumstances to be determined objectively (McVicar, 1997).

Several federal and state departments and organizations use remotely sensed data to assess the seasonal quality and quantity of agricultural production for temporal comparisons of environmental conditions for their specific uses (Graetz, 1997; Cridland, 1997; McVicar et al., 1997). The topic is further discussed in Chapter 7.

OVERVIEW OF DROUGHT ASSESSMENT METHODS

Rainfall Data Analysis

Drought is a consequence of rainfall deficiency in relation to potential water loss through evaporation. Analysis of rainfall is therefore the primary basis of identification of drought. However, a number of pitfalls have to be kept in mind while depending on rainfall data for assessing drought.

Rainfall measurements are at points that are often widely separated. Maps of rainfall deficiency or surplus drawn on the basis of point measurements are frequently far from the reality of the conditions some distance from these points. Rainfall occurring at one time of the year can be carried over under fallow to be used at other times of the year, making average monthly rainfall values irrelevant. Results from various studies suggest that ranking the year according to rainfall may be quite different to ranking the year from simulated pasture growth. Failure of rain at the optimum time may downgrade the otherwise average rainfall conditions in terms of production potential. Average rainfall conditions mask the influence of rainfall intensity and rainy-spell duration on the actual performance of crops and pastures.

An examination of temporal rainfall records in Australia at some locations has shown a tendency toward higher rainfall in the second half of this century. If this observation turns out to be true, then the emerging pattern is likely to be strengthened further under the global warming scenario and will have a significant impact on the identification and temporal comparisons of droughts on the basis of severity and impacts. By the present definition, there may not be any drought exceptional circumstances event in the near future to be compared with those that occurred in first half of the twentieth century.

Simulation Models

Simulation models hold great promise in objective identification of drought and drought exceptional circumstances because they hold the potential to integrate the climate and farming dimensions of production systems. At the same time, simulation models are not capable now, nor will they be in the near future, of replacing other measures of drought identification.

Any minor wrong information about plant characteristics or soil parameters included in running the simulation can greatly influence the output of the model. Efforts to overcome this shortcoming need to be balanced against the expected gains over the simpler models.

The majority of the models can easily identify a major drought event, but they differ considerably in highlighting marginal events. Some models have a tendency to amplify minor events, while at other times major events are presented in a suppressed form. The differing results can create a confusing situation.

Modeling living systems is very complex. Models available at present are capable of giving outputs only approximating reality.

Remote Sensing

Satellite-derived images are useful in broad-scale assessment of greenness of ground cover, especially the vegetation response to a rainfall event. Another major value of remote sensing information is for spatial and temporal validation of the simulation models. However, remote sensing has inherent limitations in providing a total solution to drought monitoring.

Images from satellites give poor information on biomass, and tree cover confounds the signals. Interpretations of the images do not consider the effects of vegetation structure and cover. Remotely sensed data products currently available contain a lot of noise from the instruments onboard and the atmosphere. This noise is of sufficient magnitude to disqualify remotely sensed products for use in drought analysis, and if used unhindered, the state-of-the-art products may not stand up to the test of law courts.

Data availability from remote sensing satellites is not guaranteed because the majority of the countries in the world purchase data from foreign sources. Furthermore, a failure of systems onboard a satellite at a critical time during the drought season may render the previous satellite information completely useless. The remotely sensed data are currently available for less than two decades. This is too short a period to identify an exceptional circumstance event on a temporal scale. No future projection of the intensity and magnitude of the event is possible through remote sensing measurements. These limitations of remotely sensed information suggest that it can not be used exclusively in defining a drought or drought exceptional circumstances. It is a supplement to other measures of the event.

This leads to the conclusion that a combination of methods—rainfall analysis, crop and pasture simulation, and monitoring the health of vegetation through remote sensing—supported with field surveys is the most realistic approach to assess the extent and intensity of drought.

MEETING THE CHALLENGE: A DROUGHT MITIGATION PLAN

Effective drought mitigation should be based on a comprehensive view of drought, because drought is not simply a deficiency of rainfall but is a more complex phenomenon that influences the whole society. Strategies to minimize the impact of drought at a farm scale are different from those needed at the state or national level. Strategies normally adopted at the farm level are based on local experience. Some of these are discussed in Chapter 10. Combating drought at the national or state level is a three-stage process. The first stage is monitoring the drought development in terms of spread and intensity as realistically as possible. In the second stage, the monitored information is used as an early warning system. Activation of a readily available drought mitigation plan is the third step of the process (Anonymous, 2000; National Drought Mitigation Center, 1996d).

Three groups of people are the key players in tackling a drought situation. In the first group are climatologists and others who monitor how much water is available now and in the foreseeable future. The second group includes natural resource managers and others who determine how the lack of water is affecting various interests, such as agriculture, municipal supplies, and recreation. The third group of people is comprised of high-level decision makers who have the authority to act on information they receive about water availability and the drought's effects. The major challenge in successful drought planning is bringing together all these groups on a platform to communicate effectively with one another.

In the United States, a systematic plan is suggested for drought management. The plan is referred as "10 Steps to Drought Preparedness" (National Drought Mitigation Center, 1996a). Some salient points of this plan described in this section can serve as a model and could be adopted by other countries/regions with modifications and alterations as deemed necessary (Wilhite, 2001).

Drought Task Force

Creating a task force is the first step of the drought mitigation plan. The task force has two purposes. First, during plan development, it will supervise and coordinate the development of the plan. Second, after the plan is implemented and activated during times of drought, the task force will assume the role of policy coordinator, reviewing and recommending alternative policy options.

The task force includes representatives from the most relevant agencies within government and from universities. The composition of the task force recognizes the multidisciplinary nature of drought and its impacts. It may also include a representative of the media in an advisory capacity to ensure public awareness of drought severity and the actions implemented by government.

Drought Policy and the Plan's Purpose and Objectives

The drought task force develops a drought policy that specifies the general purpose for the drought plan. State officials consider many questions as they define the purpose of the plan. These include the purpose and role of state government in drought mitigation efforts; the scope of the plan; the most drought-prone areas of the state; and the most vulnerable sectors of the state's economy. It also includes the role of the plan in resolving conflict between water users during periods of shortage; the resources (human and economic) that the state is willing to commit to the planning process; the legal and social implications of the plan; and the principal environmental concerns caused by drought. Answers to these and other questions help to determine the objectives of drought policy.

Resolving Conflicts between Water Users

Political, social, and economic values are bound to clash as competition for scarce water resources intensifies during a drought. To lessen conflict and develop satisfactory solutions, it is essential that the views of citizens, the public, and environmental interest groups be considered early in the drought planning process. In fact, these groups are likely to impede progress in the development of plans if they are not included in the process. Local groups could be set up to bring neighbors together to discuss their water-use problems and seek cooperative solutions.

Inventory of Resources and Constraints

The drought task force should prepare an inventory of resources and constraints that might enhance or inhibit fulfilment of the objectives of the planning process. Resources include natural resources, human expertise, infrastructure, and capital available to the government.

The most obvious natural resource of importance is water and its location, accessibility, and quality. Biological resources refer to the quantity and quality of grasslands, rangelands, forests, and wildlife. Human resources include the labor needed to develop water sources, lay pipeline, haul water and hay, process citizen complaints, provide technical assistance, and direct citizens to available services.

Financial and legal constraints are likely to emerge during a drought. Financial constraints include costs such as hauling water or hay and new program or data collection costs. These costs must be weighed against the losses that may result from not having a drought plan. Legal constraints include water rights, methods available to control usage, the kinds of public trust laws in existence, requirements for contingency plans for water suppliers, and the emergency and other powers of state agencies during water shortages.

Drought Mitigation Procedures

A drought plan has three primary organizational tasks: monitoring, impact assessment, and response and mitigation. Each task is assigned to a separate group or a committee, but the groups need to work together well, with established communication channels.

The monitoring committee includes representatives from agencies with responsibilities for forecasting and monitoring the principal meteorological, hydrological, and agricultural indicators. The monitoring committee meets regularly, beginning in advance of the peak demand season. Following each meeting, reports are prepared and disseminated to the state's drought task force, relevant state and federal agencies, and the media. The committee ensures that accurate and frequent news bulletins are issued to the public to explain changing conditions and complex problems.

Drought impacts cut across economic sectors. An impact assessment committee represents those economic sectors most likely to be affected by drought. The impact committee considers both direct and indirect losses as drought effects ripple through the economy. It is responsible for determining impacts by drawing information from all available reliable sources. Unfortunately, the quantification of drought impacts is very complicated, and some impacts may be so subtle that detection is very difficult. Working groups composed of specialists in each impact sector are created for this purpose.

The drought task force, or a similar group of senior-level officials, acts on the information and recommendations of the impact assessment committee and evaluates the state and federal programs available to assist agricultural producers, municipalities, and others during times of emergency. During periods of severe drought, the committee makes recommendations to the government about specific actions that need to be taken.

Integration of Science and Policy

The policymaker's understanding of the scientific issues and technical constraints involved in addressing problems associated with drought is often negligible. Likewise, scientists generally have a poor understanding of existing policy constraints for responding to the impacts of drought. Communication and understanding between the science and policy communities are poorly developed and must be enhanced if the planning process is to be successful. Direct and extensive contact is required between the two groups to distinguish what is feasible from what is desirable for a broad range of science and policy issues.

Publicity

The drought plan is unveiled and presented to the public in a way that gives maximum visibility to the program and credit to the agencies and organizations that have a role in its operation. For purposes of gaining publicity and attention, it is a a good idea to announce and implement the plan just before the most drought-sensitive season. The cooperation of the media is essential to publicizing the plan. A representative of the media on the drought task force is a valuable resource in carrying out the publicity.

Education on Drought

The drought task force initiates an information program aimed at educating the general population about drought and drought management and what individuals can do to conserve water in the short run. Educational programs are long term in design, concentrating on achieving a better understanding of water conservation issues for all age groups and economic sectors. Without such programs, governmental and public interest in water conservation vanish as soon as the drought is over.

Drought Plan Evaluation

Periodic evaluation and updating of the drought plan is essential to keep the plan responsive to the state's needs. To maximize the effectiveness of the system, two modes of evaluation are in place: (1) An ongoing or operational evaluation keeps track of how social changes such as new technology, new research, new laws, and changes in political leadership may affect the drought plan. (2) A postdrought evaluation of the plan teaches lessons from past successes and mistakes. Postdrought evaluation documents the assessment and response actions of government, nongovernmental organizations, and others and implements recommendations for improving the system. Attention is focused not only on those situations in which coping mechanisms failed but also on the areas in which the success achieved has been exemplary. Evaluations of previous responses to severe drought are a good planning aid.

Research Needs and Institutional Gaps

Research needs and institutional gaps become apparent during drought planning and plan evaluation. The drought task force compiles those deficiencies and makes recommendations on how to remedy them to the relevant state agencies.

DESERTIFICATION

The appearance of desertlike conditions that were nonexistent previously in an area is termed *desertification.* More specifically, desertification may be defined as land degradation in arid, semiarid, and subhumid areas resulting from climatic variation and human activities (Hare, 1993). Desertification is a widespread and discrete process of land degradation. With desertification, the fraction of bare soil increases, and vegetation is reduced to small patches. With more bare soil, fine mineral and organic material is rapidly removed by wind. Gully and sheet erosion by occasional heavy rainfall tends to accumulate the eroded material on the low-lying areas or the valley floors.

More than 250 million people in 110 countries are directly affected by desertification, while more than 1 million people are threatened by it. Six million hectares each year are affected worldwide by desertification, causing famine, death of livestock, and the loss of cultivated land (di Castri, 1990; Horstmann, 2001). According to another estimate, approximately 70 percent of the susceptible drylands are undergoing various forms of land degradation (Ayoub, 1998). The West African Sahelian region is one example of increasing desertification and its impact. In this region, original field data show that forest species richness and tree density have declined in the last half of the twentieth century. Average forest species richness of areas of

4 km in northwest Senegal fell from 64 ± 2 species in 1945 to 43 ± 2 species in 1993. Densities of trees of height greater than or equal to 3 m declined from 10 ± 0.3 trees/ha in 1954 to 7.8 ± 0.3 trees/ha in 1989. Arid Sahel species have expanded in the north. The changes also decreased human-carrying capacity to below actual population densities. For example, the carrying capacity for firewood from shrubs in 1993 was of 13 people/km^2 compared to the rural population density of 45 people/km^2 (Gonzalez, 2001).

Significant climatic anomalies, both in time and space, may result in lengthening dry periods, higher temperatures, and strong winds resulting in the permanent loss of vegetation from an area (Abdel-Samie, Gad, and Abdel-Rahman, 2000; Oba, Post, and Stenseth, 2001). Likewise, increasing human pressure results in extension of cultivated areas and overuse of dryland natural resources. Details of the desertification processes vary from region to region. However, the common processes are water and wind erosion; overgrazing by livestock; deforestation for more firewood and building; bush and forest fires; alkalization; and waterlogging (Kerley and Whitford, 2000). Desertification is further accelerated by disadvantageous social factors (Zhang and Bian, 2000). None of these processes is capable of affecting the natural ecosystems seriously. The entire structure of these systems is adapted to seasonal distribution of rainfall. Desertification increases very rapidly when human misuse of land combines with the occurrence of drought. The documented impacts of desertification foreshadow possible future effects of climate change. Under that situation, the natural mechanism of repair and renewal cannot cope with the added stresses.

Livestock overgrazing causes the semiarid grasslands to shift in community structure toward the shrublands, with associated changes in the structure and functioning of faunal communities (Kerley and Whitford, 2000). Vegetation cover, plant height, herbage yield, and root weight decrease, and the composition of the grasses changes. The surface is exposed to wind and soil-holding ability is decreased, soil erosion increases, and sands begin to accumulate on the leeward slope. This process is sped up by the heavy tread of animals.

There is a sequential degradation of soil properties in the process of desertification. The concentration of fine soil particles (clay and silt) decreases, but the concentration of sand particles increases with the deterioration of the ecosystem. Bulk density is at its maximum in a degraded land ecosystem, and consequently, pore space is at its minimum. Water holding and field capacity of soil decreases with the degradation of the ecosystem and exhibits a positive correlation with the clay content of soil. Eventually, desertification causes progressively drier soil conditions. The concentration of organic carbon, total nitrogen, and available phosphorus significantly decreases, whereas concentration of calcium, magnesium, potassium, and sodium significantly increases with the deterioration of ecosystem. The reduction of concentration of organic carbon, total nitrogen, and available phosphorus and increase of concentration of sodium, potassium, and calcium are related to the reduction of the amount of soil macroaggregates. Electrical conductivity and pH of soil increases, suggesting that soil salinity develops with desertification (Pandey, Parmar, and Tanna, 1999).

Irrigation expansion results in ever-increasing water withdrawal from rivers. The main desertification processes associated with irrigation are a decline in the groundwater level, increased mineralization and chemical pollution of watercourses, and soil salinization. Many of the salt-sensitive plant species stop growing in such soils. (Saiko and Zonn, 2000).

Regular monitoring of climate and associated hydrological and ecological processes is the first and most essential part of any program for checking desertification. Because desertification is a discrete process, its tackling requires continued monitoring and research. We need to know how the climate systems in tropical and subtropical areas work. In many arid zones of the world, the network of weather stations is still very poor and climate records are not complete (Sivakumar, Stigter, and Rijks, 2000; Stigter, Sivkumar, and Rijks, 2000). More weather stations and more weather records are needed in fragile areas. Using satellite remote sensing facilities, vegetation changes should be observed regularly (Wang, 1990). Another important action for combating desertification is the control on land use (Olulumazo, 2000). A land-use control is also a microclimate control. Good land-use practices lead to enhanced biological productivity and maintain healthy ecosystems. Apart from control measures, a political and managerial will is needed to control the processes of desertification.

Realizing the consequences of the impacts of desertification, the United Nations held Conventions to Combat Desertification in mid-1990s. The UN conventions recommended that efforts should move from global assessment to national and local level assessments, and focus should shift from physical parameters and be directed more to people issues (Ayoub, 1998). Acting on these recommendations, a majority of countries have taken initiatives for the halting of desertification (Wang, Zheng, and Yang, 2000; Hoven, 2001).

Chapter 6

Climate, Crop Pests, and Parasites of Animals

ROLE OF WEATHER AND CLIMATE

Weather is the most important factor that determines the geographical distribution and periodic abundance of crop insect pests and parasites of animals. Weather controls the development rate, survival, fitness, and level of activity of individual insects; the phenology, distribution, size, and continuity of insect populations; migration and their establishment; and the initiation of insect outbreaks (Pedgley, 1990; Drake and Farrow, 1988). Weather influence can be immediate, cumulative, direct, indirect, time lagged, exported, or imported. Indirect effects arise through host quality and parasite populations. A time-lagged influence is one that occurs at a later stage as a consequence of both past and current weather. Imported/exported influences arise because insects are highly mobile, and outbreaks may be initiated by windborne migrations (Drake, 1994; Baker et al., 1990).

Among the weather elements, temperature, humidity, and wind play the major roles in insect life. Other elements of lesser importance are solar radiation and photoperiod. In interpreting the role of weather in an insect's life, we must remember that all weather elements are interrelated, so the role played by any individual element is not simple to understand and explain.

Temperature

Each species of insect has a range of temperature within which it can survive. This range is referred as the tolerable zone (Atkins, 1978). Within the tolerable zone, there are different optimal temperature ranges for a variety of vital functions. Exposure to a temperature toward the upper or lower limit of the tolerable zone will usually result in death if it persists for a long enough time. At extremes of the tolerable zone, death will occur after a short duration of exposure.

The actual temperatures that limit the tolerable zone vary from species to species, but extremes exist that apply to all insects. Most insects have an upper temperature tolerance between 40 and 50°C, and no known insect can survive temperatures in excess of 63°C (Gerozisis and Hadlington, 1995). Some insects can adapt physiologically to survive several months of hot, dry weather in a dormant state called summer diapause. The absolute minimum temperature tolerated by any insect is not well defined but is almost certainly below –30°C. Some insects cannot survive for long if the temperature drops below the lower threshold for development. Other species can become dormant at low temperatures. Activity and development cease but begin again as soon as the temperature exceeds the activity or developmental thresholds. Yet others usually overwinter in a particular stage that is physiologically adapted and therefore can survive periods of extreme cold. Several insects inhabiting temperate or even arctic regions are able to survive by the process of supercooling, in which tissues are able to withstand the freezing of their fluid for extended periods without damage. The mean supercooling point for larvae of elm bark beetle *Scolytus laevis* (Coleoptera: Scolytidae) reached as low as –29°C in midwinter (Hansen and Somme, 1994).

The duration of the time of exposure to extremes of temperature is also important to the survival of insects. The pecan aphid, *Monelliopsis pecanis* (Hemiptera: Aphididae), could survive fairly well even at temperatures near freezing if exposed for only one hour, but many deaths occurred if they were forced to spend five hours at these low temperatures. At very high temperatures, survival is reduced even after a one-hour exposure (Kaakeh and Dutcher, 1993).

Insects are able to function faster and more efficiently at higher temperatures. They can feed, develop, reproduce, and disperse when the climate is warm, though they may live for a shorter time (Drake, 1994). Higher temperatures are not always favorable to insects, usually reducing their life span. All other factors remaining normal, insects live for shorter times at higher temperatures. An example is that of the parasitic wasp *Meteorus trachynotus* (Hymenoptera: Braconidae). Variation of temperature from 15 to 30°C reduces its adult life span from 40 days to a mere 10 days or so (Thireau and Regniere, 1995).

Low temperature is an advantage under certain conditions. For example, there are lower energy demands at low temperatures (Hunter, 1993). If resources such as food are in short supply, insects can survive longer without starving. Under extremely low temperatures some insects can remain in suspended animation until warmer conditions return.

In temperate regions, where the seasonal variation in temperature is often large, development starts slowly in early spring, progresses more rapidly as the season advances, and perhaps slows or is suspended in the heat of midsummer. Temperature is less limiting in the tropics and subtropics, but arrested development is commonly used to survive a dry season. Survival through unfavorable seasons is usually possible only in particular developmental stages.

Temperature also impacts insect activity. As the temperature rises, insects move more rapidly. Short-distance flight activity in relation to temperature has been documented for a number of insect species in Australia (Drake, 1994). Using radar observations, it has been established that daytime flights of *Chortoicetes terminifera* adults occur mainly when the sun is not obscured. The locusts often become concentrated in rising air in the walls of a warm convection funnel (Reid, Wardhaugh, and Roffey, 1979). Adults of *Lucilia cuprina* are most active at temperatures between 16 and 27°C and move further when the temperature is high. Flight and general activity of *Musca vetustissima* are limited by temperatures below 10 to 12°C (Hughes, 1981). Sterile *Bactrocera tryoni,* mass-reared in warm insectaries, are uncompetitive with wild flies (and ineffective as control agents) unless preconditioned to field temperatures before release (Fay and Meats, 1987). Field-cage observations indicate that *Eudocima salaminia* ceases feeding at temperatures below 20°C, and laboratory studies have shown that at this temperature the flight capacity of *Epiphyas postvittana* is at its maximum (Gu and Danthanarayana, 1992). Females of the latter species fly longest at a relative humidity of 60 percent. Egg laying by the jarrah leaf miner, a pest of the native eucalypt forests of southwestern Australia, is most intense at 15 to 20°C (Mazanec, 1989).

Major meteorological factors affecting migration are the vertical profiles of temperature, wind velocity, and the presence of convergence. Temperature may determine the time of flight, height of flight, and thus the speed and direction of the transporting wind, as well as flight duration. *Chortoicetes terminifera* adults initiate their long-distance nocturnal flights only if the temperature at dusk (the time of takeoff) exceeds 20°C and rain has fallen on them at certain stages during their development (Hunter, 1993; Hunter, McCulloch, and Wright, 1981). The flight duration and dispersal distance of *E. postvittana* adults is much greater when the larvae are reared at high (25-28°C) rather than low (15°C) temperatures (Gu and Danthanarayana, 1992). Flight duration may sometimes be determined by the migrants' supplies of physiological fuel, which could depend on food abundance and quality (and thus in turn on weather conditions) at the development stage. Direct influences leading to the termination of migration include falling temperatures below the threshold for flight, the onset or cessation of thermal convection, and an encounter with rainfall that can cause flying insects to be washed down.

Temperature also affects behavior of insects. Insects may remain totally inactive at both high and low temperatures or move actively along a temperature gradient until a preferred zone is encountered. The influence of temperature on dispersal, mating, and reproduction is of great importance. If conditions are adverse for dispersal, local populations neither increase because of influx, nor decline because of exodus. If temperatures are not suitable for mating for several days, some adults may die without leaving offspring and others may become less fertile due to age.

Moisture

The moisture content in the habitat of an insect directly determines whether or not an individual survives. Moisture also has indirect effects on insect populations through its influence on plant growth. All forms of environmental moisture (atmospheric humidity, rain, snow, hail, dew, soil moisture, and surface water) influence the water balance of insects.

The humidity in an insect's habitat may have some indirect effects. Some parasites do not search for hosts or oviposit in them if the relative humidity is either low or high. The susceptibility of insects to fungal, bacterial, and viral diseases also changes with environmental moisture. Moist conditions seem to facilitate the spread of some insect pathogens and may also affect their survival and virulence.

Rainfall can act as a direct cause of mortality. Insect eggs and small larvae can be permanently washed from their host plants by heavy rain. Rain may also saturate the soil and drown insects that are unable to escape. Many insects cease feeding during periods of precipitation and may seek refuges in which to pass a rainy period. Small parasitoids have difficulty moving around in wet conditions. A prolonged rainy spell, particularly when the temperature is suitable for development, may lengthen the time required to complete development or cause mortality by starvation.

Heavy or excessive rain can cause high mortality, either directly through knockdown, saturation, or flooding, or by providing conditions favorable for disease. Heavy rain washes aphids off of their host plants, and both beetles and bugs may be killed by violent thunderstorms.

Insect abundance varies with seasonal variations in rainfall. Some species are more abundant in the dry season, whereas others proliferate only during the rainy season. Lack of rain can cause desiccation and death of insects. In Australia, the onset of hot, dry conditions in summer reduces populations of the aphid vectors of a variety of plant-virus diseases (Drake, 1994). In the eucalyptus forests of northern Australia, the sap feeder (Hemiptera: Psylloidea) is much more common in the late dry season than at any other time of year, whereas defoliator grasshoppers (Orthoptera) were most abundant during the rainy season. It is suggested that sap feeders receive nutrients from sap produced by the regrowth of trees in response to fires which sweep through these forests in the late dry season. The defoliating grasshoppers, on the other hand, benefit from the relatively luxuriant production of new leaves during the rains (Fensham, 1994).

Rain may not have an influence at the time it falls but may promote insect performance some months later. This phenomenon is well illustrated by the seasonal outbreaks of African armyworm, *Spodoptera exempta* (Lepidoptera: Noctuidae) which can reach enormous numbers to become serious pests of cereals and pasture. In Kenya, the number of outbreaks was negatively correlated with rainfall in the six to eight months preceding the start of the armyworm season. The high correlation between rain and later armyworm outbreaks has been used to construct a prediction model for Kenya, providing an accurate forecast of the likelihood of armyworm outbreaks (Haggis, 1996).

Rain also plays a role in altering a host's susceptibility to windborne insects and disease vectors. Rain-drenched or moisture-stressed crops and stock may be particularly vulnerable to insects or the pathogens they carry (Risch, 1987). Sheep are especially susceptible to strike by *L. cuprina* during periods of frequent rainfall, which increase the incidence of predisposing factors such as fleece rot and nematode infestation (Wardhaugh and Morton 1990; Waller, Mahon, and Wardhaugh, 1993). Epizootics of Akabane disease, which causes calving losses in cattle, occur when winds carry the vector *(Culicoides brevitarsis)* outside its normal range. In the region where the vector overwinters, the disease is endemic, and cows usually develop immunity before becoming pregnant (Murray, 1987). Jarrah trees stressed by low rainfall during the previous winter are particularly susceptible to attack by *Perthida glyphopa* (Mazanec, 1989).

Wind

Wind is an important factor of the environment of insects, and it influences insect populations in a number of ways. It is a vital component of broad weather patterns, giving rise to fronts and convergence zones. Low pressure systems and anticyclones in temperate regions determine migration trajectories of insects, while trade winds and monsoons determine the trajectories in tropical and subtropical areas. Wind causes insect displacement and therefore affects population changes by influencing the numbers moving into or out of an area. It can carry them considerable distances away to new habitats and regions. Many insects and pathogens appear to undertake enormous migrations covering hundreds if not thousands of kilometers on occasions. They perform this feat by exploiting the wind as an external source of energy (Shields and Testa, 1999; Byrne, 1999).

In Australia, warm northerly or northwesterly winds emanating from the Intertropical Convergence Zone have introduced Japanese encephalitis virus. Outbreaks of Akabane disease and bluetongue infection have been linked to long-range windborne dispersal (Mackenzie, Lindsay, and Daniel, 2000). These winds bring *C. terminifera, Nysius vinitor,* various noctuid moths, and *Musca vetustissima* into the cropping regions of southeastern Australia during spring and summer, often in large numbers. In southwestern Australia, similar movements of insect populations occur along with the northeasterly winds. Movements also occur in other directions, and these may be important in reestablishing populations in the inland regions from which the major invasions originate. The spread or reestablishment of disease infection by windborne migration of pathogens and insect vectors has been recorded for crops and cattle (Drake, 1994; Limpert, Godet, and Müller, 1999; Aylor, 1999).

In southeast and east Asian countries, most migrations of rice insect pest populations are determined by the direction and extent of wind (Rutter, Mills, and Rosenberg, 1998). Out of nearly 2,600 trajectories drawn upwind from 15 catching sites, only 5 percent of the trajectories failed to locate a possible source, and over 90 percent were completed in 40 hours or less. Nearly 80 percent of the trajectories were constructed in the prevailing winter monsoon and trade winds, resulting in a southward displacement of insects toward overwintering areas. Tropical cyclones in autumn produced trajectories that differed in both direction and extent from those in the prevailing winds, supporting the suggestion that the contraction of the distribution areas of rice pests at that time of year is the product of a series of movements in different directions. The results suggest migrations continue throughout the year in the tropics and subtropics and indicate this may be one way the capacity for long-distance migration is maintained in some rice pest populations.

In atmospheric convergence zones, some insects rise until they reach their flight ceiling and subsequently land to feed and reproduce. Normally, the same convergence zones are the harbingers of rain. The semiarid tropical regions of the world are particularly affected by such weather patterns. Rain associated with these systems results in luxuriant growth of vegetation on which large densities of new-generation insect larvae feed. Pests such as the desert locust, *Schistocerca gregaria* (Orthoptera: Acrididae), and the armyworm, *Spodoptera exempta* (Lepidoptera: Noctuidae), are two classic examples of this phenomenon that is particularly noted in Africa.

A risk for insects who rely on wind to aid in dispersal is the chance of going too far from their destination. Wind velocity to a certain extent helps movement and host finding. Strong wind could kill the insects by carrying individuals to unsuitable areas, completely out of their habitat range. Therefore, some species will fly only in winds of a certain velocity. Most insects will not take flight when the speed of the wind exceeds their normal flight speed, simply because they will lose control over the direction of movement.

Some insects, especially aphids, rely on daytime thermals (convective currents) to carry them aloft. However, many insects ascend under their own power at dusk and migrate above the nocturnal boundary layer in the fast-moving, stable airflows found at altitudes of 100 to 2,000 m (Drake and Farrow, 1988).

Like adult insects, larvae can also be carried considerable distances, especially if they are attempting to escape because of limited food supply. In Europe and North America, female adult gypsy moths, *Lymantria dispar* (Lepidoptera: Lymantriidae), are wingless (except in the Asian strain), and the insect population disperses mainly as first instar larvae which use "ballooning" as a means of colonizing new areas (Diss et al., 1996).

An important property of wind is its ability to convey chemical messages to insects from point sources. These messages can include information about the distance and location of a specialized food plant or of a suitable and receptive mate. A chemical plume formed in the shape of a tongue in a laminar, wind-driven system provides an unbroken guide to the location of the source. Insects can then fly along a concentration gradient of a chemical signal to precisely locate their prize.

Light

Light is not a true climatic factor, but it is interrelated with solar radiation and temperature. As such, it is often included as a component of weather. Photoperiodism (variations in the photoperiod in a 24-hour day-night cycle) exercises a great deal of control over processes directly related to survival of insects. Light intensity greatly influences insect behavior. Many species are active during the hours of full sunlight but remain quiescent at night. Some insects are active during the faint light of dawn and dusk.

Day length can also be used as a signal or trigger by insects to enter diapause during potentially harsh conditions such as summer heat, winter cold, and drought. In some cases, the day length experienced by an insect larva provides information about the progression of the seasons. The ability to vary growth and development rates enables the insect to achieve efficient timing relative to favorable conditions (Leimar, 1996). In the case of *Kytorhinus sharpianus* (Coleoptera: Bruchidae), a wild bean weevil from Japan, the duration of the various stages in the life cycle from egg to adult vary according to the photoperiod at a constant temperature. The whole cycle can be accomplished between 75 and 80 days, when the insect receives 15 or 16 hours of daylight per day-night cycle. This period increases dramatically as the hours of daylight shorten to 14 and then to 12. With only 12 hours of daylight, the pupal stage is never reached until longer hours of light return.

Hill and Gatehouse (1992) studied the influence of daylight on many insect pests and suggested the migratory capacity of insects may be influenced more by photoperiod during development than by temperature.

SOME IMPORTANT INSECT PESTS OF CROP PLANTS

Aphids

Weather factors, especially temperature and rainfall, play a dominant role in the population dynamics of aphids in all the climatic regions of the world where crop production is possible. Aphids are highly sensitive to temperature changes. Field observations, climate chamber experiments, and computer simulations confirm this fact. Skirvin, Perry, and Harrington (1996) used a model to assess the population dynamics of aphids at various temperatures. The model predicted that an increase in temperature leads to a greater number of aphids in the absence of the predator. However, the presence of predators reduces the number of aphids predicted. Temperatures below 20°C and above 25°C limit the buildup, while an increase from 20 to 22°C enhanced the intrinsic rate of increase of aphid populations (Freier and Triltsch, 1996). A maximum temperature of 45°C in the postrainy season has been observed to be lethal for the sugarcane aphid species *(Melanaphis sacchari)* in sorghum (Waghmare et al., 1995).

Aphids in the tropics show remarkable adaptability to climate, regulating their population by suitably adjusting their life history strategy development, reproduction, survival, and dispersal. In the northeastern states of India, aphid species, both oligophagous *(Ceratovacuna silvestrii)* and polyphagous *(Toxoptera aurantii),* show a shift in the abundance of their populations in space and time in response to seasonal variation in temperature, which brings about changes in their host's quality. In contrast, the monophagous species *(Cinara atrotibialis)* escapes the hot summer of plains and uplands by occurring exclusively in the milder temperatures of the hills. The three species performed optimally at 20°C with respect to development, reproduction, and adult survival (Agarwala and Bhattacharya, 1994).

Studies on aphids in a cotton crop revealed that a drop in mean temperature to below 25°C when the cotton is in the elongation stage could cause a sharp increase in the aphid population (Chattopadhyay et al., 1996). Minimum temperature, evening humidity, sunshine hours, and rainfall influenced aphid incidence in both the 45 and 52 standard meteorological weeks. If there were substantial increases in minimum temperature and evening humidity and an appreciable decrease in sunshine hours with occasional rain, aphid infestation was observed at high levels in these weeks. Rainfall was found to be the predominant variable controlling the aphid population in the boll formation stages. When the crop was in the maturity stage, a decrease in maximum temperature and rainfall affected the aphid population in the fourth standard meteorological week. Between the two, rainfall was found to be the predominant meteorological variable. Even a decrease in maximum temperature in the second week alone increased aphid infestation in the fourth week. Cloudiness also plays an important role in an aphid population during the first generation.

In Mediterranean climatic environments, several weather conditions are highly related to the population dynamics of cereal aphids in winter (Pons, Comas, and Albajes, 1993). In northeastern Spain, population dynamics of *Rhopalosiphum padi* L. and *Sitobion avenae* F., the species found most frequently, were affected not only by low temperatures causing aphid mortality but also by other factors. Dry and cold weather, through the effect they have on host plant phenology and quality, reduce aphid developmental rates.

In the pastures growing in a dryland farming rotation system in the 300 to 450 mm rainfall region of southwestern Australia (Mediterranean climate), aphids were not found on plants in hot, dry summers but were present from April until November, when the weather was mild and humid (Ridsdill, Scott, and Nieto, 1998). However, late summer/early autumn rainfall is the key factor in determining the aphid population in lupins. This rainfall can be used to forecast an aphid population as it maintains weeds on which aphids build up before they move into crops. Little or no rain at this time results in very little green plant material to support aphids, and hence aphid arrival in crops is late. Thackray (1999) has built a computer model that incorporates data on climate, aphid population, and yield losses at a specific site to forecast aphid arrival in lupins in Western Australia.

Studies in Britain, representing temperate climates of the world, suggest temperatures have an important role in the flight phenology of aphids. The first migratory period appeared to be more strongly correlated with winter temperature than summer temperature for most of the prevalent aphid species. Warm winters will probably lead to advances in the first migratory period and large intervals between the adjacent migratory periods. The study indicates climate change will lead to more frequent and severe attacks of many aphid species and the virus diseases they transmit (Zhou et al., 1996).

Armyworms

Economic infestations of armyworms in many parts of the world have revealed that precipitation is the primary factor influencing pest populations. However, temperature and the availability of weather transport systems were the most important climatic factors governing pest abundance (Pair and Westbrook, 1995; Rose, Dewhurst, and Page, 1995; Tucker, 1994; Stewart, Layton, and Williams, 1996).

Findings from studies in Africa (Rose, Dewhurst, and Page, 1995) show that the onset of the first outbreaks of an epidemic is caused by oviposition at high density by *S. exempta* concentrated by wind convergence at storm outflows. The sources of these insects seem to be low-density populations, which survive from one season to the next at sites receiving unseasonal rainfall. Some areas in Tanzania and Kenya are particularly prone to early outbreaks that are potentially critical for the initiation of a subsequent spread of outbreaks downwind throughout eastern Africa. These areas have low and erratic rainfall and are near the first rising land inland from the coast. Below-average rainfall prior to the development of outbreaks increases the probability of their occurrence. Their subsequent spread is enhanced by storms downwind, which concentrate insects in flight, and by sunshine during larval development. Persistent wet weather reduces the spread of outbreaks. Seasons with many outbreaks often had rainstorms separated by dry periods during the rainy seasons.

Grasshoppers

Temperature sensitivity determines the geographical distribution of grasshoppers, with generalist species widespread and thermally specialized species restricted to warmer habitats. For all the species that are thermal specialists, variation in their sensitivity to temperature is a good predictor of their distribution. The developmental and reproductive responses to different rearing temperatures of grasshoppers, examined in a laboratory experiment, showed that growth and development rates increase with temperature for each species. Nymphal development, adult mass and size, and egg production rate also increases with temperature (Willott and Hassall, 1998).

Examination of life history variations among many populations of the grasshopper *Chorthippus brunneus,* from around the British Isles, revealed a relationship between grasshopper life histories and the climates of their ancestral sites (Telfer and Hassall, 1999). The grasshoppers from cooler sites are heavier at hatching, and those from northern sites grow faster and develop through fewer instars, attaining adulthood earlier at the expense of adult size. Adults are larger in warmer, sunnier, or more southerly locations.

A study of rangeland grasshopper population dynamics in Wyoming (Lockwood and Lockwood, 1991) has indicated that the climate in December to January has predictive value with regard to population dynamics the following spring. However, when temperature and precipitation during hatching and early development of grasshoppers (April to May) are the control variables, forecast of the observed population is better.

Prolonged drought conditions suppress the population of grasshoppers. Kemp and Cigliano (1994) monitored the abundance of the rangeland grasshopper (Acrididae) species at various sites in Montana during the period 1986 to 1992, which included an extreme drought year (1988). Significant post-1988 reductions were observed in rangeland acridid species abundance in the eastern and south-central regions of Montana, where drought intensity had been increasing during the previous 20 years. In the north-central region, which also experienced the 1988 drought but showed no long-term drought trend, a postdrought reduction in overall acridid species abundance was not observed.

The black cone-headed grasshopper lives in hot, arid environments. The sexual dimorphism of the species suggests the larger females may have an advantage in water storage over the males. Both sexes were able to depress their internal temperatures below the higher temperatures of their environment by evaporative cooling. The males lost proportionately more water by evaporation, produced drier feces, and may have been more constrained by water availability. The females appeared to be more profligate with their water reserves, which supports the theory that large body mass may be an advantage to an insect in the desert (Prange and Pinshow, 1994).

Locusts

Locusts are able to travel long distances and colonize new habitats. Therefore, their distribution is variable in time and space and can occur within a large area. The pest is feared for both its destructive capacity and its constant threat to the region. It is capable of sudden appearances and severe devastation to standing crops. The most affected part of the world is the continent of Africa, where in some years the infested area may cover several million hectares. Distribution and sequence of rainfall is the principal determinant of locust population increase over several generations and of the concentration of widespread populations into highly mobile and destructive swarms.

Survival and populations are greatest with an increased frequency of sufficient rainfall, where rainwater is enhanced by runoff and flooding, and where vegetation and soils provide suitable habitats. However, excessive rainfall affects the first three nymphal stages, limiting its development (Hunter, 1981; Montealegre, Boshell, and Leon, 1998). Locusts move in swarms from one area to another where rain has fallen. Wind direction and speed at 850 hpa (about 1.5 km above the earth's surface) and convergence zones determine the paths of the locusts' movements.

In the Indian subcontinent, monsoon rainfall and, to a certain extent, winter-spring rainfall play a role in the resurgence, establishment, and termination of locust plague upsurges (Chandra, 1993).

Temperature affects the rate of development, body size, and adult color (Gregg, 1983). Capinera and Horton (1989) and Nikitenko (1995) suggest grasshoppers and locusts in the cooler regions of North America and Europe are favored by warm, dry summer conditions, whereas in warmer areas they appear to require spring and summer moisture. Locusts avoid thermal extremes by taking refuge in appropriate sheltering sites, loss of water by flying during favorable climatic conditions, and cannibalism (El Bashir, 1996).

Cotton Bollworms

The effect of weather conditions on various species of bollworms has been investigated in many cotton-growing regions of the world in laboratory as well as in field. The combined effect of weather factors, maximum temperature, minimum temperature, rainfall, and relative humidity on the population density of cotton bollworms is very high (El Sadaany et al., 1999).

Laboratory studies have demonstrated the temperature-dependent development of larvae and pupae of pink bollworm, *Pectinophora gossypiella* (Saund). No eggs hatched at less than 10° or more than 37.5°C. Mortality of larvae and pupae also increased at terperatures greater than 37.5°C. The development rate of all stages of the pest increased with temperature. Development of larvae was successful at all temperatures between 15° and 35°C. Larval period and adult longevity decreased as relative humidity increased (Wu, Chen, and Li, 1993; Gergis et al., 1990).

The most important weather factor for the abundance of cotton bollworm is the amount and distribution of rainfall. The incidence of cotton bollworms (Lepidoptera) in southern China is significantly affected by July and August rainfall. Populations of the pests were sparse when total precipitation was greater than 500 mm and dense when total precipitation was less than 400 mm, showing a very significant negative correlation. Continuous rain produced more severe damaging effects on the pupal stage (Li et al., 1996).

In the cotton fields of northwestern India, significant relationships are observed between the buildup of *Heliothis* spp. and pink bollworms *(Pectinophora gossypiella)* and mean air temperature and relative humidity. The optimum temperature and humidity range for the buildup of *Heliothis* during the growing season is observed to be 20 to 24°C and 46 to 60 percent, respectively. In the case of *P. gossypiella* it was 22 to 23°C and 52 to 72 percent, respectively (Bishnoi et al., 1996). In central India, the optimal conditions for emergence was observed to be 26.7 to 31.4°C maximum temperature and 62.2 to 77.7 percent relative humidity. Minimum temperature showed a significant correlation only with the emergence of *P. gossypiella* (Gupta, Gupta, and Shrivastava, 1996).

Fruit Fly

Queensland fruit fly *Bactrocera tryoni* (Froggatt) (QFF) is one of Australia's most costly horticultural pests, with major potential impacts that have local, regional, and policy dimensions. Its range extends from northern Queensland to eastern Victoria, and populations occur in many inland towns of Queensland and NSW as far west as Broken Hill. The distribution of fruit fly (including QFF) is primarily determined by climate (Bateman, 1972; Yonow and Sutherst, 1998).

The abundance of fruit fly is greater in regions where the daily maximum temperature does not exceed 38°C during summer. Immature adults are unlikely to survive when the maximum temperature exceeds 40°C for four continuous days. However, adults are mobile and can seek cooler habitats in the field and survive very hot weather periods when the temperatures suggest that they should not (Meats, 1981). Conversely, the lowest monthly mean of 2°C is only marginally favorable for winter survival.

Moisture appears to be the primary determinant of the number of fruit flies, and correlations between summer rainfall and fruit fly populations are significant. Populations reach extremely high numbers in wet years and decline in dry years. Other factors being favorable, a relatively high population would survive with a mean monthly rainfall of 48 mm and more. Mavi and Dominiak (2000) observed a highly significant correlation between the availability of moisture, as measured by summer rainfall, and the peak numbers achieved each year. Rainfall in excess of 170 mm in November, January, and February resulted in high fruit fly populations, and less than 170 mm resulted in low populations. Meats (1981) reported that more than 48 mm rainfall per month in summer resulted in at least three generations.

Studies conducted in southwestern NSW confirm (Mavi and Dominiak, 2000) that infestation has been severe in years when mild winters were followed by humid summers. Infestation has been almost negligible in years when severe winters were followed by comparatively less humid summers.

The vulnerability of Australian horticulture to the QFF under climate change was studied by Sutherst, Collyer, and Yonow (2000). The study revealed that climatic warming to the extent of 1°C reduces the severity of cold season stress in southern parts of Australia and thus increases the suitability of southern states for both population growth and survival over the winter period. Under this scenario, damage costs will increase to the extent that horticulture may become hardly economical.

CLIMATE AND PARASITES OF ANIMALS

Worms, flies, lice, and ticks are the major parasites of cattle, sheep, pigs, and poultry. Diseases caused or carried by parasites constitute a major obstacle to the development of profitable livestock enterprises. Some of these parasites also have significance for human health.

Parasites and many parasitic diseases are influenced by climate. Diseases spread by insects are encouraged when climatic conditions, temperature and excessive moisture in particular, favor the propagation of the vector. Many parasites and parasitic diseases in cattle and sheep reach their peak incidence in warm, wet summers and are relatively rare in dry seasons. Internal parasites are similarly influenced by climate. The direction of prevailing winds is of importance in many disease outbreaks, particularly in relation to the contamination of pasture and drinking water by fumes from factories and mines and the spread of diseases carried by insects (Blood and Radostits, 1989).

HELMINTH PARASITES

The helminth parasites of sheep and cattle are classified into nematodes, trematodes, and cestodes. In everyday language these are roundworms, flukes, and tapeworms.

Climate is the major factor determining the development of free-living parasites, and it affects different species of parasites differently. It also affects the survival of the parasites; for example, as temperature varies from warm to cold, and levels of soil and plant moisture vary from moist to dry, populations of the free-living stages of parasites fluctuate. The differential effects of climate for different species of nematodes explains why haemonchosis is the common parasitic disease in summer-rainfall areas and ostertagiosis is the common one in winter-rainfall areas (Cole, 1986).

Most helminth parasites undergo a period of development outside the host before becoming infective for another host. During this time in the outside world, the rate of their development and chances of survival are influenced by climatic and other environmental factors.

The rate of development of the extra-host stages tends to rise with increasing temperature. Moisture is required for survival, and extreme desiccation is usually lethal. In the wetter parts of the tropics, extra-host stages are seldom exposed for long to the destructive effects of desiccation, and temperatures are commonly optimal throughout the year. Under these conditions, the survival rate of parasitic forms outside the host is high; they develop rapidly to the infective stage and large populations are established. Even where prolonged dry seasons alternate with very short wet seasons, the extra-host stages of many endoparasites can take full advantage of the warm, wet conditions associated with the latter, while the adult stages which subsequently develop within the host are well protected and safely carried over to the ensuing dry season.

Nematodes

Roundworms or nematodes generally cause major trouble to livestock in tropical areas. The level of parasitism in grazing animals depends to a large extent on the numbers of free-living stages on the pasture. Climate and pasture management affect the survival of these free-living stages of gastrointestinal nematodes. Roundworms or nematodes require oxygen, warmth, and moisture for their development and completion of their full life cycles. They are susceptible to desiccation. In areas where the dry season is prolonged and severe, pastures are often free of infection long before the beginning of the rains. In such areas worm infection is carried over from one wet season to the next by infected carrier animals.

In Australia, the most important parasite in the summer-rainfall region is *Haemonchus contortus,* while in the winter-rainfall region, the most important are *Ostertagia* spp., *Trichostrongylus* spp., and *Chabertia ovina* (Cole, 1986).

Gastrointestinal Parasites in Sheep

The relationship between climate and gastrointestinal parasites is well established. Temperature is negatively correlated with the level of pasture infectivity (except for *Trichostrongylus*), and rainfall is positively correlated with pasture infectivity. The number of larvae of gastrointestinal nematodes recovered per kilogram of grass in the grazing lands of Mexico in the dry season was significantly lower than during the rainy season (Hernandez, Prats, and Ruiz, 1992).

Examination of the pattern of pasture contamination and the influence of some climatic factors on the development of ovine trichostrongyles in dry pastures of central Spain revealed two peaks of pasture contamination, from midwinter to early spring, and from midautumn to early winter (Romero, Valcarcel, and Vazquez, 1997).

In the Western Australian Mediterranean climate of hot, dry summers and cool, wet winters, the relative prevalence of *Trichostrongylus vitrinus, T. colubriformis,* and *T. rugatus* in sheep are closely correlated with the weather conditions of the region. The prevalence of *T. colubriformis* was positively correlated with the mean autumn, winter, and spring temperatures. There were suggestions of an association between the amount of rainfall of a locality and prevalence of *T. colubriformis*. The prevalence of *T. rugatus* was not correlated with the temperature of any season but was negatively correlated with the mean annual rainfall and length of growing season of a locality (De Chaneet and Dunsmore, 1988).

Flukes

Fluke or trematode infections are found in all species of domestic animals, including poultry, but cattle and sheep are the principal victims. Flukes are prevalent in most animal and sheep grazing areas of the world.

Fluke commonly use a water snail as an intermediate host. Infection with this species is associated with stock grazing in land that may be infested by migrating snails. The level of infestation with liver fluke is often heaviest following rain, flooding, and irrigation, though in areas of limited infestation, such as banks of streams or dams, it is heaviest in dry times when animals graze these areas more closely.

ARTHROPOD PARASITES

Arthropod parasites, such as flies, mosquitoes, and midges, assume economic significance by impairing productivity or carrying pathogens. Their greater importance is because they transmit infectious diseases.

Flies

Biting and bloodsucking flies affect cattle by sucking blood and irritating the stock. Because the larvae and pupae of these flies need water for their metamorphic life, they are more plentiful during the rainy season, as compared to the dry season (Johnson, 1987). Horse flies, belonging to the family Tabanidae, are restless feeders and may bite a number of animals in a short space of time. Thus, they are efficient mechanical transmitters of a variety of viral, bacterial, and protozoan diseases. The bloodsucking stable fly, *Stomoxys calcitrans,* is very common around milking sheds and on farms where horses are kept (Smeal, 1995). It transmits disease mechanically. The common housefly, *Musca domestica,* transmits a variety of bacterial and viral infections as a result of its feeding on feces and on the food of humans and other animals.

Blowfly strike *(Lucilia cuprina)* in sheep is primarily governed by climatic conditions. Flywaves that cause colossal mortalities of sheep occur mainly in spring and autumn when prolonged rainy spells keep the sheep wet to the skin.

The tsetse fly is notoriously important because of its role in the transmission of trypanosomiasis, or "nagana," of cattle and other animals in Africa. It also causes sleeping sickness in humans. Vast tracts of tropical Africa are infested with tsetse flies and are literally closed to successful animal husbandry because of trypanosomiasis.

There are many different species of tsetse fly, and various groups have particular biological requirements. In general, riverine species require the cool, moist environment associated with tree-lined riverbanks. On the other hand, savanna species inhabit dry, open parkland. The distribution of tsetse, as revealed by various surveys, is strongly influenced by environmental conditions such as climate and vegetation cover (Robinson, Rogers, and Williams, 1997a,b).

Studies of the behavior of *Glossina pallidipes* and *G. morsitans* have been conducted in Zimbabwe (Torr and Hargrove, 1999). Attributes of samples of tsetse from refuges, odor-baited traps, targets, and mobile baits were compared. The results suggested that during the hot season, refuges were significantly cooler than the surrounding riverine woodland during the day, and tsetse experienced temperatures 2°C cooler than the daily mean in a Stevenson's screen located in woodland. Compared to the catches from traps, refuges had higher proportions of tsetse because temperatures in the refuges do not exceed the lethal level of 40°C. Tsetse populations declined by 90 percent during the hot season. This decline in numbers is not due to direct mortality effects of temperature on adults but may be due, in part, to a doubling in the rates of reproductive abnormality during the hot season and an increase in adult mortality related to a temperature-dependent decrease in the pupal period.

Mosquitoes

Mosquitoes are common and widespread throughout the world. They are important as vectors of Rift Valley fever (hepatitis enzootica) of sheep and other animals, African horse sickness, malaria, lymphatic filariasis, yellow fever, western equine encephalitis, and *Dirofilaria immitis* (a dog heartworm) (Speight, Hunter, and Watt, 1999). In semiarid tropical and subtropical regions, severe plagues usually occur after spring and summer rains. A close relationship exists between El Niño and seasonal moquito populations and mosquito-associated diseases (Bouma and van der Kaay, 1996).

Midges

Midges are tiny flies of the family Ceratopogonidae, the most important genera being *Culicoides* spp. They suck blood and, apart from causing annoyance and worry, can act as vectors for a number of arboviruses, including those that cause ephemeral fever, bluetongue, Akabane disease of cattle, African horse sickness, and epizootic hemorrhagic disease of deer.

Midges pass through the egg, larval, and pupal stages, and the immature stages are aquatic. The number of generations per year depends on temperature. In temperate regions there is usually a single generation, while in the tropics three or four may be completed. These flies are plentiful in the warmer months and are most active at dusk and in the early morning. In climates such as that of northern Queensland, the cycle continues in all the seasons and adult midges are active throughtout the year (Smeal, 1995). Because of their small size they are capable of being carried long distances by wind (Arundel and Sutherland, 1988).

Midge-Related Diseases

Bluetongue. Ecosystems in which bluetongue virus circulates between equine or ruminant hosts and *Culicoides* have been described in many parts of the world. Orbiviruses transmitted by *Culicoides* midges are found in areas with varying climates, the main factor being warmth for all or part of the year. Moisture is provided through rain or, in semiarid areas, through irrigation. Spread of the virus is through movement or migration of hosts and through unaided flight or carriage on the wind of infected *Culicoides* midges (Sellers and Walton, 1992).

The incidence, seasonality, and geographical distribution of bluetongue virus infection in cattle herds in Queensland, Australia, has been exhaustively investigated (Ward, 1996a,b; Ward and Johnson, 1996). Cases of seroconversion, which mostly occurs in autumn and winter, were associated with summer and autumn temperature and rainfall. In the far north, most cases were associated with temperature and rainfall in the summer months. Elsewhere, most cases were associated with autumn temperature and rainfall. It was suggested (Ward, 1996b) that two ecological cycles of infection of cattle exist, supporting a hypothesis of differential transmission by vector species.

In Australia, subclinical infection of cattle with bluetongue occurs mostly in autumn, and the median month of seroconversion is May. The prevalence of infection is suppressed approximately fourfold in a series of dry, cool, autumn seasons, as compared to other combinations of seasons. Occurrence of dry, cool, autumn seasons at least once every four years or less keeps a good check on bluetongue virus infection. The association between the El Niño/ Southern Oscillation Index and cases of seroconversion to bluetongue viruses also indicates that more cases occur during months in which the SOI is positive, compared to months in which it is negative. Studies suggest drought conditions in Australia may affect the endemic stability of bluetongue virus infection. Instability in the system could lead to cyclical epidemics of infection (Ward and Johnson, 1996).

Horse sickness. The incidence of African horse sickness with climatic conditions has been studied by Baylis, Mellor, and Meiswinkel (1999) and Baylis and colleagues (1998) in Morocco and South Africa. There is evidence of an association between African horse sickness and El Niño/Southern Oscillation (ENSO). Baylis, Mellor, and Meiswinkel (1999) suggested that the association is mediated by the combination of rainfall and drought brought to South Africa by ENSO. The combination of heavy rain followed by drought is thought to affect disease transmission, with breeding sites of the insect vector *Culicoides imicola* being altered, or, during drought, the animal reservoir for the virus (zebra [*Equus burchellii*]) may congregate near the few remaining sources of water where they are in contact with and infect more vector midges. High temperatures during droughts increase vector population growth rates and favor disease transmission.

Lice

Lice are small obligate parasites that are highly host specific. They are dorsoventrally flattened, wingless, and have tarsi adapted for clinging to hair or feathers. Lice infestations are common in cattle, sheep, goats, and horses in late winter and early spring.

According to Arundel and Sutherland (1988), *Bovicola ovis* is the most common and important louse on sheep, causing considerable irritation that results in poor wool growth and damaged wool. Infested sheep rub against objects such as fences and fence posts and bite and chew their fleeces. The fleeces therefore appear deranged and have a pulled and ragged appearance, particularly the areas on the sides behind the shoulder, which sheep can reach with their mouths.

Solar radiation, temperature, and rainfall all have profound effects on lice numbers. Lice is an important parasite of sheep and cannot survive off sheep for more than a few hours under extremes of temperature and light. Laboratory studies have, however, shown that lice held away from sheep could survive for 11 days at 25°C. In shearing sheds in winter and early spring, lice could survive for up to 14 and 16 days, respectively (Crawford, James, and Maddocks, 2001). Shearing has the most dramatic effect in that it physically removes the lice with wool and exposes the remaining lice to extreme weather conditions. Lice numbers drop to the lowest after one to two months of shearing. Saturation of the fleece by heavy and prolonged rain also kills many lice. When temperatures fall from late autumn to early winter, there is an increase in lice numbers, and they decline over summer. However, this pattern is influenced strongly by the timing of shearing.

Ticks

Ticks occur on a global scale but they transmit many more serious viral and protozoan diseases in tropical and subtropical areas than in other regions. Climate and vegetation are the major factors affecting tick distribution (Estrada and Genchi, 2001). When weather conditions are favorable, ticks lie in wait (on grass or rocks) or move in active search of a host. When conditions are unfavorable the ticks return to shelter (under a stone, in litter, under vegetation). Depending on the climatic characteristics of a season and region, ticks may be active during the day (morning or evening). Ticks are active at night in dry plains where strong sunlight prevents any diurnal movement on the ground.

Each species of tick has a particular threshold temperature below which a diapause occurs in all instars. Various climatic factors such as sunlight, temperature, rainfall, and wind patterns condition the presence or absence of a tick species (Shah and Ralph, 1989).

Development and survival of free-living stages of tick are related to temperature, while the duration of survival is influenced mainly by rainfall and consequent relative humidity (Pegram and Banda, 1990). Appropriate relative humidity, rather than wet conditions, is essential for the development and survival of eggs and pupae and the survival of unfed hatched ticks. Each species of tick is adapted to a particular relative humidity range which varies with the instar and its size. The requirements range from extremely humid to very low relative humidity. Immature stages have very specific requirements, while adults can protect themselves better against evaporation because of their larger size and thicker tegument. Immatures adjust their humidity requirements by locating in holes in the ground, cracks in rocks, under litter, at the base of the vegetation layer, and in other sheltered places. Some adults are remarkably drought resistant and can survive for several months or years in a semidesert environment.

In temperate climates, the temperature determines the distribution of ticks. Abrupt or slow changes of temperature can modify the life cycle within a few days or weeks. In northern Eurasia, central Eurasia, and adjoining Mediterranean regions, the greater frequency of activity is mostly in summer.

In tropical climates, the dominant factor is rainfall (Pegram et al., 1989). The start and end of the rainy season influence the different phases of the life cycle. Parasitism is reduced during the dry months (March to June north of the equator) and increases sharply within days following the first major winter rainfall. The population remains stable for a few weeks, then slowly diminishes. At the end of the rainy season there is a marked decrease, with a progressive fall to almost zero in the dry season. In such regions the more rapid tick development pattern is determined by the dry season, and the life cycle takes one year.

Climatic uniformity and the absence of an unfavorable season in equatorial regions allow tick development throughout the year (Smeal, 1995). There is no annual cycle determined by a diapause. Generations overlap or follow one another in a pattern depending on the species.

Every year during the spring and summer, *Ixodes holocyclus,* a paralysis tick also known as Australian paralysis tick or dog tick, is at its most dangerous along the eastern seaboard of Australia, from around the lakes district in Victoria to the northernmost tip of the country. The female tick produces neurotoxins in its saliva which is injected into the host animal while feeding. These toxins affect nervous tissue causing paralysis in cattle, sheep, and goats (Smeal, 1995).

Mites

Mite-Related Diseases

Sarcoptic mange (barn itch). Sarcoptic mange occurs in all species of animals, causing a severe itching dermatitis. The causative mite, *Sarcoptes scabiei,* is usually considered to have a number of subspecies, each specific to a particular host. Animals in poor condition or underfed appear to be most susceptible. The disease is most active in cold, wet weather and spreads slowly during the summer months. The females form shallow burrows in the horny layer of the skin in which to lay their eggs. The larval and nymphal stages may remain in the tunnels or emerge onto the skin. The normal exfoliation of the skin eventually exposes the tunnels, and any of the life cycle stages may transmit by contact to other animals (Blood and Radostits, 1989).

Among domestic species, pigs are most commonly affected, but it is an important disease of cattle and camels and also occurs in sheep. It is a notifiable disease in most countries and is important because of its severity.

Sheep itch mite. The itch mite *(Psorergates ovis)* has been recorded as a parasite of sheep in Australia, New Zealand, South Africa, and the United States. The life cycle, comprising eggs, larvae, three nymphal stages, and adults, takes four to five weeks and is completed entirely on the sheep. Only the adults are mobile, and they effect the spread of the disease by direct contact between recently shorn sheep, when contact is close and prolonged. All stages occur in the superficial layers of the skin and cause skin irritation, leading to rubbing and biting of the affected areas and raggedness of the fleece. Merino sheep are most commonly affected. The highest incidence is observed in this breed, particularly in areas where the winter is cold and wet. There is a marked seasonal fluctuation in the numbers of mites. The numbers are very low in summer, rise in the autumn, and peak in the spring (Blood and Radostits, 1989).

Chapter 7

Remote-Sensing Applications in Agrometeorology

SPATIAL INFORMATION AND THE ENVIRONMENT

Spatial information is geographically located data, that is, data that can be related to a position on the earth's surface. It can be obtained from many different sources: satellite imagery, aerial photographs, field-recorded surveys, and weather station reports. The great power of location is that it naturally integrates data that lie close together in space but may be otherwise unrelated.

"Spatial information" is an umbrella term covering information about the environment and its many aspects: agricultural activities, properties, infrastructure such as roads and administrative boundaries, weather and climatic data, and so on. Information is collected in many ways. Some is acquired through on-ground measurements and sampling, but often the most efficient or only method is through remote sensing, for example, aerial photography, satellite imaging, or airborne electromagnetic surveys.

At a basic level, spatial information relates to points, lines, and polygons or areas. Some examples are as follows:

Point—the properties of a soil profile from a sample taken at a single location

Line—a fence line or an irrigation channel

Polygon—a weed patch in a paddock, a paddock property, water catchment, province, or state

Spatially located information can be stored, manipulated, analyzed, and mapped using a computer-based geographic information system (GIS).

The information in this chapter gives an explanation of the principles of remote sensing, geographic information systems, and the global positioning system (GPS), as well as examples of how remote sensing is being used in agrometeorology.

REMOTE SENSING

Remote sensing is defined as the science of obtaining and interpreting information from a distance, using sensors that are not in physical contact with the object being observed. Animals (including people) use remote sensing via a variety of body components to obtain information about their environment. The eyes detect electromagnetic energy in the form of visible light. The ears detect acoustic (sound) energy, and the nose contains sensitive chemical receptors that respond to minute amounts of airborne chemicals given off by the materials in our surroundings. Some research suggests migrating birds can sense variations in the earth's magnetic field, which helps explain their remarkable navigational ability.

At its broadest, the science of remote sensing includes aerial, satellite, and spacecraft observations of the surfaces and atmospheres of the planets in our solar system, although the earth is obviously the most frequent target of study. The term *remote sensing* is customarily restricted to methods that detect and measure electromagnetic energy, including visible light, which has interacted with surface materials and the atmosphere.

Remote sensing of the earth is used for many purposes, including the production and updating of planimetric maps, weather forecasting, and gathering military intelligence. The focus in this chapter is on remote sensing of agriculture and the associated environment and resources of the earth's surface. It explores the physical concepts that underlie the acquisition and interpretation of remotely sensed images, the characteristics of images from different types of sensors, and common methods of processing images to enhance their information content. For additional information on remote sensing refer to <http://www.microimages.com> for a useful tutorial on remote sensing of the environment.

The Electromagnetic Spectrum

The science of remote sensing began with aerial photography, using visible light from the sun as the energy source. Visible light, however, makes up only a small part of the *electromagnetic spectrum,* a continuum that ranges from high-energy, short-wavelength gamma rays to lower energy, long-wavelength radio waves. The portion of the electromagnetic spectrum that is useful in remote sensing of the earth's surface is illustrated in Figure 7.1. (See color section at end of chapter.) The earth is naturally illuminated by electromagnetic radiation from the sun. The peak solar energy is in the wavelength range of visible light (between 0.4 and 0.7 μm), and the visual systems of most animals are sensitive to these wavelengths. Although visible light includes the entire range of colors seen in a rainbow, a cruder subdivision into blue, green, and red wavelength regions is sufficient in many remote-sensing studies. Other substantial fractions of incoming solar energy are in the form of invisible ultraviolet and infrared radiation. Only tiny amounts of solar radiation extend into the microwave region of the spectrum. Imaging radar systems that are used in remote sensing generate and broadcast microwaves, then measure the portion of the signal that has returned to the sensor from the earth's surface. The nature and laws of the electromagnetic spectrum are discussed in Chapter 2.

Interaction Processes in Remote Sensing

The sensors in remote-sensing systems measure electromagnetic radiation (EMR) that has interacted with the earth's surface. Interactions with matter can change the direction, intensity, wavelength content, and polarization of EMR. The nature of these changes is dependent on the chemical makeup and physical structure of the material exposed to the EMR. Changes in EMR resulting from its interactions with the earth's surface therefore provide major clues to the characteristics of the surface materials. EMR that is *transmitted* passes through a material (or through the boundary between two materials). Materials can also *absorb* EMR. Usually absorption is wavelength specific: that is, more energy is absorbed at some wavelengths than at others. EMR that is absorbed is transformed into heat energy, which raises the material's temperature. Some of that heat energy may then be emitted as EMR at a wavelength dependent on the material's temperature. The lower the temperature, the longer the wavelength of the emitted radiation. As a result of solar heating, the earth's surface emits energy in the form of longer-wavelength infrared radiation. For this reason, the portion of the infrared spectrum with wavelengths greater than 3 μm is commonly called the *thermal infrared* region. EMR encountering a boundary such as the earth's surface can also be reflected. If the surface is smooth at a scale comparable to the wavelength of the incident energy, *specular reflection* occurs in which most of the energy is reflected in a single direction, at an angle equal to the angle of incidence. Rougher surfaces cause *scattering,* or *diffuse reflection,* in all directions (Figure 7.2).

To understand how different interaction processes impact the acquisition of aerial and satellite images, let us analyze the reflected solar radiation that is measured at a satellite sensor. As sunlight initially enters the atmosphere, it encounters gas molecules, suspended dust particles, and aerosols. These materials scatter a portion of the incoming radiation in all directions, with shorter wavelengths experiencing the strongest effect. An example is the preferential scattering of blue light in comparison to green and red light, which accounts for the blue color of the daytime sky. Clouds appear opaque because of intense scattering of visible light by tiny water droplets. Although most of the remaining light is transmitted to the surface, some atmospheric gases are very effective at absorbing particular wavelengths. The absorption of dangerous ultraviolet radiation by ozone is a well-known example.

As a result of these effects, the illumination reaching the surface is a combination of highly filtered solar radiation transmitted directly to the ground and more diffuse light scattered from all parts of the sky, which helps illuminate shadowed areas. As this modified solar radiation reaches the ground, it may encounter soil, rock surfaces, vegetation, or other materials that absorb, transmit, and reflect the radiation. The amount of energy absorbed, transmitted, and reflected varies in wavelength for each material in a characteristic way, creating a spectral signature. The selective reflectance of different wavelengths of visible light determines what we perceive as a material's *color.* Most of the radiation not absorbed is diffusely reflected (scattered) back up into the atmosphere, some of it in the direction of the satellite. This upwelling radiation undergoes a further round of scattering and absorption as it passes through the atmosphere before finally being detected and measured by the sensor. If the sensor is capable of detecting thermal infrared radiation, it will also pick up radiation emitted by surface objects as a result of solar heating.

Atmospheric Effects

Scattering and absorption of EMR by the atmosphere have significant effects that impact sensor design as well as the processing and interpretation of images. When the concentration of scattering agents is high, scattering produces the visual effect we call haze. Haze increases the overall brightness of a scene and reduces the contrast between different ground materials. A hazy atmosphere scatters some light upward, so a portion of the radiation recorded by a remote sensor, called *path radiance,* is the result of this scattering process. Because the amount of scattering varies with wavelength, so does the contribution of path radiance to remotely sensed images. The path radiance effect is greatest for the shortest wavelengths, falling off rapidly with increasing wavelength. When images are captured over several wavelength ranges, the differential path radiance effect complicates comparison of brightness values at the different wavelengths. As detailed in Chapter 2, the atmospheric components that are effective absorbers of solar radiation are water vapor, carbon dioxide, and ozone. Each of these gases tends to absorb energy in specific wavelength ranges. Some wavelengths are almost completely absorbed (Figure 7.3). Consequently, most broadband remote sensors have been designed to detect radiation in the "atmospheric windows," those wavelength ranges for which absorption is minimal and, conversely, transmission is high.

REMOTE SENSORS AND INSTRUMENTS

All remote-sensing systems designed to monitor the earth's surface rely on energy that is either diffusely reflected by or emitted from surface features. Current remote-sensing systems fall into three categories on the basis of the source of the EMR and the relevant interactions of that energy with the surface.

Reflected Solar Radiation Sensors

These sensor systems detect solar radiation that has been diffusely reflected (scattered) upward from surface features. The wavelength ranges that provide useful information include the ultraviolet, visible, near-infrared, and middle-infrared ranges. Reflected solar-sensing systems discriminate materials that have differing patterns of wavelength-specific absorption, which relate to the chemical makeup and physical structure of the material. Because they depend on sunlight as a source, these systems can provide useful images only during daylight hours. Changing atmospheric conditions and changes in illumination with time of day and season can pose interpretive problems. Cloud cover is a particular problem. Reflected solar remote-sensing systems are the most common type used to monitor earth resources.

Thermal Infrared Sensors

Sensors that can detect the thermal infrared radiation emitted by surface features can reveal information about the thermal properties of these materials. Because the temperature of surface features changes during the day, thermal infrared sensing systems are sensitive to the time of day at which the images are acquired.

Imaging Radar Sensors

Rather than relying on a natural source, these systems "illuminate" the surface with broadcast microwave radiation, then measure the energy that is diffusely reflected back to the sensor. The returning energy provides information about the surface roughness and water content of surface materials and the shape of the land surface. Long-wavelength microwaves suffer little scattering in the atmosphere, even penetrating thick cloud cover. Imaging radar is therefore particularly useful in cloud-prone tropical regions.

Remote-sensing instruments are either active or passive, and they are further divided into scanning and pointing instruments, with scanning instruments being the most commonly used. Also, remote-sensing instruments are typically multispectral, that is, they detect multiple wavelengths of radiation.

Active Instruments

A remote-sensing instrument that transmits its own electromagnetic radiation to detect an object or to scan an area for observation and receives the reflected or backscattered radiation is called an active instrument. Examples are radars, scatterometers, and lidars.

- *Radar (radio detection and ranging):* Radar uses a transmitter operating at either radio or microwave frequencies to emit electromagnetic radiation and a directional antenna or receiver to measure the reflection or backscattering of radiation from distant objects. Distance to the object can be determined because electromagnetic radiation propagates at the speed of light.
- *Scatterometer:* A scatterometer is radar that measures the backscattering coefficient of the surface of the viewed object. The backscattering coefficient can be used to define surface characteristics such as surface roughness, moisture content, and dielectric properties. Over ocean surfaces, measurements of the backscattering coefficient in the microwave spectral region can be used to derive maps of surface wind speed and direction.
- *Lidar (light detection and ranging):* A lidar uses a laser (light amplification by stimulated emission of radiation) to transmit a light pulse and a receiver with sensitive detectors to measure the backscattered or reflected light. Distance to the object is determined by recording the time between the transmitted and backscattered pulses and using the speed of light to calculate the distance traveled. Lidars can determine the profile of aerosols, clouds, and other constituents in the atmosphere.
- *Laser altimeter:* A laser altimeter uses a lidar to measure the altitude of the instrument platform by measuring the distance to the surface below. By independently knowing the location of the platform, the topography of the underlying surface can be determined.

Passive Instruments

Passive instruments sense only radiation emitted by the object being viewed or reflected by the object from a source other than the instrument. Reflected sunlight is the most common external source of radiation sensed by passive instruments. Various types of passive instruments are used, including radiometers and spectrometers.

- *Radiometer:* This instrument quantitatively measures the intensity of electromagnetic radiation in some bands of wavelengths in the spectrum. Usually a radiometer is further identified by the portion of the spectrum it covers, for example, visible, infrared, or microwave.
- *Imaging radiometer:* A radiometer includes a scanning capability to provide a two-dimensional array of pixels from which an image may be produced. An array of detectors can perform scanning mechanically or electronically.
- *Spectroradiometer (spectrometer):* A spectrometer has the capability for measuring radiation in many wavelength bands (i.e., multispectral), often with bands of relatively high spectral resolution designed for the remote sensing of specific parameters such as sea surface temperature, cloud characteristics, ocean color, vegetation, and trace chemical species in the atmosphere.
- *Imaging spectrometer:* Hyperspectral images are produced by instruments called *imaging spectrometers.* The development of these complex sensors has involved the convergence of two related but distinct technologies: *remote imaging* of the earth and planetary surfaces and spectroscopy.

Spectroscopy is the study of light that is emitted by or reflected from materials and its variation in energy with wavelength. As applied to the field of optical remote sensing, spectroscopy deals with the spectrum of sunlight that is diffusely reflected (scattered) by materials at the earth's surface. Spectrometers (or spectroradiometers) are used to make ground-based or laboratory measurements of the light reflected from test materials. An optical dispersing element such as a grating or prism in the spectrometer splits this light into many narrow adjacent wavelength bands, and a separate detector measures the energy in each band (Figure 7.4). By using hundreds or even thousands of detectors, spectrometers can make spectral measurements of bands as narrow as 0.01 μm over a wide wavelength range, at least 0.4 to 2.4 μm (visible through middle infrared wavelength ranges). Remote sensors are designed to focus and measure the light reflected from many adjacent areas on the earth's surface. In many sensors, sequential measurements of small areas are made in a consistent geometric pattern as the sensor platform moves, and subsequent processing is required to assemble them into an image. Until recently, sensors were restricted to one or a few relatively broad wavelength bands by limitations of detector designs and the requirements of data storage, transmission, and processing. Recent advances in these areas have allowed the design of remote sensors that have spectral ranges and resolutions comparable to ground-based spectrometers.

IMAGE ACQUISITION

The radiant energy that is measured by an aerial or satellite sensor is influenced by the radiation source, interaction of the energy with surface materials, and the passage of the energy through the atmosphere. In addition, the illumination geometry (source position, surface slope, slope direction, and shadowing) can also affect the brightness of the upwelling energy. Together these effects produce a composite "signal" that varies spatially and with the time of day or season. To produce an image that we can interpret, the remote-sensing system must first detect and measure this energy.

Spectral Signatures

The spectral signatures produced by wavelength-dependent absorption and transmittance provide the key to discriminating different materials in images of reflected solar energy. The property used to quantify these spectral signatures is called *spectral reflectance:* the ratio of reflected energy to incident energy as a function of wavelength. The spectral reflectance of different materials can be measured in the laboratory or in the field, providing reference data that can be used to interpret images. As an example, Figure 7.5 shows contrasting spectral reflectance curves for three common natural materials: dry soil, green vegetation, and water. The reflectance of dry soil rises uniformly through the visible and near-infrared wavelength ranges, peaking in the middle-infrared range. It shows only minor dips in the middle-infrared range due to absorption by clay minerals. Green vegetation has a very different spectrum. Reflectance is relatively low in the visible range but is higher for green light than for red or blue, producing the green color we see. The reflectance pattern of green vegetation in the visible wavelengths is due to selective absorption by chlorophyll, the primary photosynthetic pigment in green plants. The most noticeable feature of the vegetation spectrum is the dramatic rise in reflectance across the visible near-infrared boundary, and the high near-infrared reflectance. Infrared radiation penetrates plant leaves and is intensely scattered by the leaves' complex internal structure, resulting in high reflectance. The dips in the middle-infrared portion of the plant spectrum are due to absorption by water. Deep, clear water bodies effectively absorb all wavelengths longer than the visible range, which results in very low reflectivity for infrared radiation.

Spatial Resolution

The spatial, spectral, and temporal components of an image or set of images all provide information that we can use to form interpretations about surface materials and conditions. For each of these properties we can define the *resolution* of the images produced by the sensor system. These image resolution factors place limits on what information we can derive from remotely sensed images.

Spatial resolution is a measure of the spatial detail in an image, which is a function of the design of the sensor and its operating altitude above the surface. Each of the detectors in a remote sensor measures energy received from a finite patch of the ground surface. The smaller these individual patches are, the more detailed will be the spatial information that we can interpret from the image. For digital images, spatial resolution is most commonly expressed as the ground dimensions of a picture element (pixel).

Spectral Resolution

The *spectral resolution* of a remote sensing system can be described as its ability to distinguish different parts of the range of measured wavelengths. In essence, this amounts to the number of wavelength intervals ("bands") that are measured and how narrow each interval is. An "image" produced by a sensor system can consist of one very broad wavelength band, a few broad bands, or many narrow wavelength bands. The names normally used for these three image categories are *panchromatic, multispectral,* and *hyperspectral,* respectively. Panchromatic images record an average response over the entire visible wavelength range (blue, green, and red). Because this film is sensitive to all visible colors, it is called panchromatic film. A panchromatic image reveals spatial variations in the gross visual properties of surface materials but does not allow for spectral discrimination. Some satellite remote-sensing systems record a single very broad band to provide a synoptic overview of the scene, commonly at a higher spatial resolution than other sensors onboard. Despite varying wavelength ranges, such bands are also commonly referred to as panchromatic bands.

Multispectral Images

To provide increased spectral discrimination, remote-sensing systems designed to monitor the surface environment employ a multispectral design: parallel sensor arrays detecting radiation in a small number of broad wavelength bands. The commonly used satellite systems use from three to six spectral bands in the visible to middle-infrared wavelength region. Some systems also employ one or more thermal infrared bands. Bands in the infrared range are limited in width to avoid atmospheric water vapor absorption effects that significantly degrade the signal in certain wavelength intervals. These broadband multispectral systems allow discrimination of different types of vegetation, rocks and soils, clear and turbid water, and some man-made materials. A three-band sensor with green, red, and near-infrared bands is effective at discriminating vegetated and nonvegetated areas. The high resolution visible (HRV) sensor aboard the French SPOT (Système Probatoire d'Observation de la Terre) 1, 2, and 3 satellites (20 meter spatial resolution) has this design. Color-infrared film used in some aerial photography provides similar spectral coverage, with the red emulsion recording near infrared, the green emulsion recording red light, and the blue emulsion recording green light. The IKONOS satellite from Space Imaging (4-meter resolution) and the LISS-II sensor on the Indian Remote Sensing satellites IRS-1A and 1B (36-meter resolution) add a blue band to provide complete coverage of the visible light range and allow natural-color band composite images to be created. The Landsat Thematic Mapper (TM) (Landsat 4 and 5) and Enhanced Thematic Mapper Plus (ETM+) (Landsat 7) sensors add two bands in the middle infrared (MIR). Landsat TM band 5 (1.55 to 1.75 μm) and band 7 (2.08 to 2.35 μm) are sensitive to variations in the moisture content of vegetation and soils. Band 7 also covers a range that includes spectral absorption features found in several important types of minerals. An additional TM band (band 6) records part of the thermal infrared wavelength range (10.4 to 12.5 μm). Current multispectral satellite sensor systems with spatial resolution better than 200 meters are compared in Tables 7.1 and 7.2. Note that DigitalGlobe successfully launched and deployed the QuickBird high-resolution satellite that began commercial operations in early 2002 with the offer of imagery at 0.61 m panchromatic and 2.44 m multispectral resolutions. See <http://www.digitalglobe.com>.

Hyperspectral Images

Multispectral remote sensors such as the Landsat Enhanced Thematic Mapper and SPOT XS produce images with a few relatively broad wave-

TABLE 7.1. Remote-sensing satellites in space

Platform/ sensor/ launch year	Picture element size	Image size (cross x along-track)	Spectral bands	Panchromatic cell size	Nominal revisit interval*
Ikonos-2 VNIR 1999	4 m	11 x 11 km	4	1 m	11 days (2.9 days†)
Terra (EOS-AM-1) ASTER 1999	15 m (Vis, NIR) 30 m (MIR) 90 m (TIR)	60 x 60 km	14	X	16 days
SPOT 4 HRVIR (XS) 1999	20 m	60 x 60 km	4	10 m	26 days (5 days†)
SPOT 1,2,3 HRV (XS) 1986	20 m	60 x 60 km	3	10 m	26 days (5 days†)
IRS-1C, 1D LISS 111 1995	23.6 m 70.8 m (MIR)	142 x 142 km 70 x 70 km Pan	3	5.8 m	24 days (5 days Pan†)
Landsat 7 ETM+ 1999	30 m	185 x 170 km	7	15 m	16 days
Landsat 4,5 TM 1982	30 m	185 x 170 km	7	X	16 days
IRS-1A, 1B LISSI,II 1988	36.25 m (LISSII) 72.5m (LISS 1)	148 x 148 km	4	X	22 days
Landsat 4, 5 MSS 1982	79 m	185 x 185 km	4	X	16 days
IRS-1C, 1D WiFS 1995	189 m	810 x 810 km	2	X	5 days

Source: Adapted from Smith, 2002.
Note: Ikonos-2: Space Imaging Inc., USA; Terra Landsat: National Aeronautics and Space Administration (NASA), USA; SPOT: Centre National d'Etudes Spatiales (CNES), France; IRS: National Remote Sensing Agency, India. For further details: <http://www.gsfc.nasa.gov/mission.html>; <http://Landsat7.usgs.gov/>; <http://eos.gsfc.nasa.gov>; <http://www.isro.org>; <http://www.spaceimaging.com>; <http://www.spotimage.fr>; <http://edcwww.cr.usgs.gov>.
*Single satellite, nadir view at equator
†With pointing capability

TABLE 7.2. Sensor bands of remote-sensing satellites

Platform	Visible (Vis) bands (μm)	Near IR (NIR) bands (μm)	Mid. IR (MIR) bands (μm)	Thermal IR (TIR) bands (μm)	Panchromatic band (μm)
Ikonos-2	B 0.45-0.52 G 0.52-0.60 R 0.63-0.69	0.76-0.90	None	None	0.45-0.90 B,G,R,NIR
Terra (EOS-AM-1)	G 0.52-0.60 R 0.63-0.69	0.76-0.86	1.60-1.70 2.145-2.185 2.185-2.225 2.235-2.285 2.295-2.365 2.36-2.43	8.125-8.475 8.475-8.825 8.925-9.275 10.25-10.95 10.95-11.65	None
SPOT 4	G 0.50-0.59 R 0.61-0.68	0.79-0.89	1.58-1.75	None	0.61-0.68 R
SPOT 1,2,3	G 0.50-0.59 R 0.61-0.68	0.79-0.89	None	None	0.51-0.73 G,R
IRS-1C, 1D	G 0.52-0.59 R 0.62-0.68	0.77-0.86	1.55-1.70	None	0.50-0.75 G,R
Landsat 7	B 0.45-0.515 G 0.525-0.605 R 0.63-0.69	0.75-0.90	1.55-1.75 2.09-2.35	10.40-12.50	0.52-0.90 G,R,NIR
Landsat 4,5	B 0.45-0.52 G 0.52-0.60 R 0.63-0.69	0.76-0.90	1.55-1.75 2.08-2.35	10.40-12.50	None
IRS-1A, 1B	B 0.45-0.52 G 0.52-0.60 R 0.63-0.69	0.77-0.86	None	None	None
Landsat 4, 5	G 0.5-0.6 R 0.6-0.7	0.7-0.8 0.8-0.9	None	None	None
IRS-1C, 1D	R 0.62-0.68	0.77-0.86	None	None	None

Source: Adapted from Smith, 2002.
Note: The rows in Table 7.2 are to be read in conjunction with the rows in Table 7.1.

length bands. By comparison, hyperspectral remote sensors collect image data simultaneously in tens or hundreds of narrow, adjacent spectral bands (as little as 0.01 μm in width). These measurements make it possible to derive a near-continuous spectrum for each image cell (Figure 7.6). After adjustments for sensor, atmospheric, and terrain effects are applied, these image spectra can be compared to field or laboratory reflectance spectra to identify and map surface materials such as particular types of vegetation or diagnostic minerals associated with ore deposits. Hyperspectral images contain a wealth of data and are difficult to interpret. Interpretation requires an understanding of the exact properties of the ground materials measured and how these relate to the data produced by the hyperspectral sensor. Further information on hyperspectral remote sensing is given in a tutorial at <http://www.microimages.com/>.

Plant Spectra

The spectral reflectance curves of healthy green plants also have a characteristic shape that is dictated by various plant attributes (Figure 7.7). In the visible portion of the spectrum, absorption effects from chlorophyll and other leaf pigments govern the curve shape. Chlorophyll absorbs visible light very effectively but absorbs blue and red wavelengths more strongly than green, producing a characteristic small reflectance peak within the green wavelength range. As a consequence, healthy plants appear green to the eye. Reflectance rises sharply across the boundary between red and near-infrared wavelengths (sometimes referred to as the *red edge*) to values of around 40 to 50 percent for most plants. This high near-infrared reflectance is primarily due to interactions with the internal cellular structure of leaves. Most of the remaining energy is transmitted and can interact with other leaves lower in the canopy. Leaf structure varies significantly among plant species and can also change as a result of plant stress.

Thus, species type, plant stress, and canopy state can all affect near-infrared reflectance measurements. Beyond 1.3 μm, reflectance decreases with increasing wavelength, except in two pronounced water absorption bands near 1.4 and 1.9 μm. At the end of the growing season leaves lose water and chlorophyll. Near-infrared reflectance decreases and red reflectance increases, creating the familiar yellow, brown, and red leaf colors of autumn.

Spectral Libraries

Several libraries of reflectance spectra of natural and man-made materials are available for public use. These libraries provide a source of reference spectra that can aid the interpretation of hyperspectral and multispectral images.

ASTER Spectral Library

This library has been made available by the National Aeronautics and Space Administration (NASA) as part of the Advanced Spaceborne Thermal Emission and Reflection Radiometer (ASTER) imaging instrument program (Figure 7.8). ASTER is one of the instruments on the EOS (Earth Observing System) AM-1 satellite and records image data in 14 channels from the visible through thermal infrared wavelength regions as part of NASA's Earth Science Enterprise program. The ASTER spectral library includes spectral compilations from NASA's Jet Propulsion Laboratory, The Johns Hopkins University, and the United States Geological Survey (Reston, Virginia). The ASTER spectral library currently contains nearly 2,000 spectra, including minerals, rocks, soils, man-made materials, water, and snow. Many of the spectra cover the entire wavelength region from 0.4 to 14 μm. The library is accessible interactively via the Internet at <http://speclib.jpl.nasa.gov>. It is possible to search for spectra by category, view a spectral plot for any of the retrieved spectra, and download the data for individual spectra as a text file. These spectra can be imported into an image processing spectral library. The ASTER spectral library can also be ordered on CD-ROM at no charge from the Web site.

USGS Spectral Library

The United States Geological Survey Spectroscopy Laboratory in Denver, Colorado, has compiled a library of about 500 reflectance spectra of minerals and a few plants over the wavelength range from 0.2 to 3.0 μm. This library is accessible online at <http://speclab.cr.usgs.gov/spectral. lib04/spectral-lib04.html>. Users browse individual spectra online or download the entire library. There is information at this Web site on the Airborne Visual and Infrared Imaging Spectrometer (AVIRIS) and a tutorial on imaging spectroscopy.

SATELLITE ORBITS FOR REMOTE SENSING

Consistent, long-term measurements are needed of the key physical variables that define earth-system processes. A full set of observations requires different orbits. For global coverage, polar orbits will view the entire earth over the course of many orbits over several days. Low-inclination orbits will permit observation of a portion of the earth over several days, with the observations on successive days being made at different times of day. Geostationary orbits permit continuous observations in time, but only for a limited view of the earth. Because a geostationary satellite progresses in its orbit at the same rate as the earth's rotational rate, it can provide a fixed view of the earth's sphere that is determined by its selected position above the equator.

As a satellite in a polar or near-polar (high-inclination) orbit passes over the earth, the earth's rotation shifts the satellite ground track westward, so that after a period, which varies for different types of satellites, the entire earth is covered and the cycle begins again. If orbits are sun synchronous, the satellite passes over each latitude at the same local time, providing consistent lighting and allowing easier comparison between data and images taken of the same area on different days.

GEOGRAPHIC INFORMATION SYSTEM (GIS)

A geographic information system is a computer-assisted system for the acquisition, storage, analysis, and display of geographic data. GIS technology integrates common database operations such as query and statistical analysis with the unique visualization and geographic benefits offered by maps (Burrough, 1990).

Maps have traditionally been used to explore the earth and to exploit its resources. GIS technology is an expansion of cartographic science that takes advantage of computer science technologies, enhancing the efficiency and analytical power of traditional methodologies (Coulson et al., 1991; Ballestra et al., 1996).

GIS has become an essential tool in the effort to understand complex processes at different scales: local, regional, and global. In GIS, the information coming from different disciplines and sources, such as traditional or digital maps, databases, and remote sensing, can be combined in models that simulate the behavior of complex systems. Remote sensing is a very important contributor of information to a GIS (Maracchi, Pérarnaud, and Kleschenko, 2000).

In agrometeorological applications, the preliminary basic information is often provided by historical archives of different disciplines such as geography, meteorology, climatology, and agronomy. Data collected directly in the field are very important, because they provide the ground truth. Meteorological stations, field data collection (ecophysical observations, agronomic practices, insect attacks, diseases, soil, etc.), and direct territorial observations are fundamental to all the possible agrometeorological applications (Maracchi, Pérarnaud, and Kleschenko, 2000). Where there is a lack of information, models can be used to complete the information to assist in the understanding of the real situation (Rijks, Terres, and Vossen, 1998). An important component is the incorporation of digital elevation models (DEMs) (Moore, Grayson, and Ladson, 1991), which are three-dimensional representations of the landscape. This allows consideration of many other parameters, including hydrology and sunshine duration and intensity.

In a GIS all this information can be linked and processed simultaneously, obtaining a syntactical expression of the changes induced in the system by the variation of a parameter. The technology allows for the contemporary updating of geographical data and their relative attributes, producing a fast adaptation to real conditions and obtaining answers in near-real time.

GIS Applications

Public agencies, research laboratories, academic institutions, and private and public services have established their own information systems incorporating GIS. Because of the increasing pressure on land and water resources and land-use management and forecasting (crop, weather, fire, etc.), GIS has become an irreplaceable and powerful tool at the disposal of decision makers.

In developed countries, agricultural and environmental GISs are used to plan the types and times of agricultural practices and regional management activities, and for monitoring devastating events and evaluating agricultural losses. Maracchi, Pérarnaud, and Kleschenko (2000) gave as an example the evaluation of fire risk in Tuscany, Italy. The final map produced was the result of the integration of satellite data with regional data, through the implementation of GIS technology (Romanelli, Bottai, and Maselli, 1998).

In developing countries, the data used for the production of information layers are often unreliable or even lacking. Implementation of GISs must take a different approach from those used in developed countries. A first phase should use sufficiently simple systems to answer specific problems. Completion of the different information layers should be gradual, eventually creating a fully operational GIS (Maracchi, Pérarnaud, and Kleschenko, 2000). An example of a preliminary information system is given by the SISP (Systéme Integré de Suivi et Prevision des rendements, an integrated information system for monitoring cropping season by meteorological and satellite data) for Niger. The SISP was developed to enable monitoring of the cropping season and to evaluate an early warning system with useful information about the evolution of crop conditions (Di Chiara and Maracchi, 1994). Longley and colleagues (2001) have recently written a useful textbook on the subject of GIS, and MicroImages tutorial on GIS is available at <http://www.microimages.com/>.

GLOBAL POSITIONING SYSTEM (GPS)

GPS is a satellite-based navigation system developed and maintained by the U.S. Department of Defense. A constellation of 24 satellites broadcasts continuous timing signals that GPS receivers are designed to monitor. When a GPS receiver detects at least three of these satellites above the horizon, the unit can derive its position on the earth's surface by triangulation and provide map coordinates for the user. Most GPS receivers can collect a stream of map coordinates collected at intervals and save them as a file for later use. Appropriate software can import such log files and use them as data sources for input into GIS mapping software. GPS receivers are used in a variety of scientific, commercial, and industrial applications and support varying degrees of accuracy. The U.S. military no longer purposefully degrades the accuracy of the GPS signal for civilian receivers, so accuracy depends primarily on the quality and configuration of the receiver. GPS-derived data can be used to establish geographic location for mapping features of interest and to track vehicles, field agents, or other moving entities.

Further information on GPS is available at <http://www.microimages.com/>, <http://trimble.com/gps/index.html>, <http://garmin.com/aboutGPS>, and <www.auslig.gov.au/geodesy/gps/>.

REMOTE-SENSING APPLICATIONS

Some examples of the use of remote sensing for agricultural applications are discussed in this section. Many of the examples have been taken from the Australian situation. Additional information is available in White, Tupper, and Mavi (1999).

Vegetation Cover and Drought Monitoring

Vegetation cover is important as an indicator of available fodder and to protect the soil resource from erosion. Cover can be estimated using remote sensing, field measurements, pasture and crop models, and farm survey data. Remote sensing has been used to assist in assessing the severity of drought across Australia, current satellite imagery being compared to that of previous years. It has also helped in determining the spatial extent of exceptional droughts (McVicar and Jupp, 1999). Another valuable use is in aiding the validation of temporal (including temporal/spatial) agronomic models, as has been incorporated into the National Drought Alert System known as Aussie GRASS (Australian Grassland and Rangeland Assessment by Spatial Simulation) (Brook and Carter, 1994; Wood et al., 1996).

The most commonly used measure of vegetative cover has been the Normalized Difference Vegetation Index (NDVI), which is based on differences in reflection of red and near-infrared (NIR) light. Chlorophyll pigments in leaves absorb red light, and changes in leaf structure can influence NIR reflectance. Foliage presence, as measured through the leaf area index, can be related to NDVI. LAI is m^2 leaf per m^2 ground. Thermal data, transformed to the Normalized Difference Temperature Index (NDTI), are also of value in assessing vegetative cover and drought monitoring (Bierwirth and McVicar, 1998). The Advanced Very High Resolution Radiometer (AVHRR) on the polar-orbiting NOAA satellites is a major tool used for vegetation monitoring. Other platforms include the polar-orbiting U.S. Landsat and French SPOT satellites. The NOAA, Landsat, and SPOT satellites orbit at about 700 km altitude. As a service to the agricultural and environmental sectors, the CSIRO Division of Marine Research produces a composite NDVI image of the whole of Australia every two weeks using data obtained from the Australian Centre for Remote Sensing (ACRES). A two-week compositing period is used to minimize cloud cover in the data. The composite images have a resolution of 1 km. They are made available to customers about ten days after the end of the two-week period. Historical data are also available going back to 1991.

AVHRR data are recorded and archived daily within the Bureau of Meteorology Research Centre (BMRC). A compositing pathway has been established using these data. To highlight changes in the monthly maximum value composite NDVI between sequential months, the Maximum Value Composite Differential (MVCD) has been developed (Tuddenham et al., 1994; Tuddenham and Le Marshall, 1996). The MVCD is based on the difference between two images recorded at approximately two-month intervals, with a log stretch to enhance the subtle difference in the NDVI signal that has occurred over that time. The changes may involve either a browning or greening of the vegetation cover during the two-month period. This can be useful to identify whether a season is atypical in terms of the timing of either seedling emergence or herbage drying off. More information can be found at the Web sites for the Bureau of Meteorology <http://www.bom.gov.au/nmoc/NDVI/> and Environment Australia <http://www.ea.gov.au/land/monitoring/>.

Use of Thermal Data in Drought Monitoring

In Australia, daytime thermal data are used to monitor regional environmental conditions. Jupp and colleagues (1998) jointly developed the Normalized Difference Temperature Index to remove seasonal trends from the analysis of daytime land surface temperatures derived from the AVHRR sensor. The NDTI has the form

$$\mathrm{NDTI} = T_{\infty} - T_s / T_{\infty} - T_0 \qquad (7.1)$$

where T_∞ is a modeled surface temperature if there is an infinite surface resistance, that is, ET is zero; T_s is the surface temperature observed from the AVHRR sensor; and T_0 is a modeled surface temperature if there is zero surface resistance; hence ET equals ET_p. T_∞ and T_0 can be thought of as the physically limited upper and lower temperatures, respectively, for given meteorological conditions and surface resistances. They define a range within which meaningful AVHRR surface temperatures must fall. If T_s is close to the T_0 value, it is an indication of conditions being "wet," whereas if T_s is close to the T_∞ value, dryness is indicated.

T_∞ and T_0 are calculated through the inversion of a resistance energy balance model. The parameters required at the time of satellite overpass are meteorological- and vegetation-related parameters. Required meteorological data include air temperature, solar radiation, relative humidity (or some other measure of vapor pressure), and wind speed. However, many meteorological stations record only daily air temperature extremes and rainfall. McVicar and Jupp (1999) have tested and extended strategies to determine air temperature, solar radiation, and relative humidity at the time of the satellite overpass. Wind speed can be obtained from daily wind run data, if available, or long-term climate surfaces.

Vegetation parameters, mainly LAI, are obtained from reflective data. For four dates in 1995, in cereal cropping and pasture environments in Victoria, relationships were developed between 1 m^2 in situ LAI measurements and the planetary-corrected albedo Landsat TM simple ratio (McVicar, Jupp, Reece, and Williams, 1996). These relationships were then used to scale the TM simple ratio to provide estimates of LAI at a 30 m^2 cell size for an entire TM scene. These data were then related to AVHRR simple ratio with a resampled cell size of 1 km^2 (McVicar, Jupp, and Williams, 1996). Hence 1 m^2 measurements of LAI were scaled to 1 km^2 estimates of LAI by using TM data as the intermediate scalar. For wooded areas 30 m^2 field sites were established and LAI measured, which was subsequently related to AVHRR vegetation indexes (McVicar, Walker, et al., 1996). This enables AVHRR reflective data to be scaled to estimates of LAI for cropping and pastures (McVicar, Jupp, and Williams, 1996) and wooded vegetation (McVicar, Walker, et al., 1996). Hence, the NDTI is calculated at the points, which are sometimes separated by distances of 500 km, where meteorological data are recorded to support the calculation. AVHRR-derived NDVI and T_s are used as covariates to interpolate the NDTI away from the ground meteorological stations using a spline interpolation algorithm called ANU_SPLIN (Hutchinson, 1995). This results in NDTI images. This has been done for ten years of AVHRR data focusing on the Murray-Darling Basin in southeast Australia.

The thermal data used in the NDTI calculations are affected by a few environmental parameters. The controlling parameter of the NDTI is the partitioning of the available energy into the latent and sensible heat fluxes. This partitioning is determined by the available moisture to be transferred to the atmosphere via ET. The amount of energy partitioned into the sensible heat flux is one determinant of the observed surface temperature. Consequently, the NDTI is more sensitive to changes in resource availability than the NDVI, which integrates the response of the environment to the resource. The NDTI has a greater ability to map the availability of water. This provides a measure of stress when plants are not yet responding to a reduction in chlorophyll content, thereby reducing the NDVI. More important is the ability of the NDTI to map moisture availability that will be influenced by rainfall that falls between meteorological stations. The NDVI will not be able to map these events with the same temporal resolution due to the time lag between rainfall and plant response.

The aim of producing the NDTI is to allow insight into the regional water balance. ET being common to both water balance and energy balance model formulations achieves this. In water balance models ET is defined in terms of volume of water, usually measured as milliliters per day. In energy balance models ET is defined in terms of energy, measured in watts per unit area. The water-balance-derived moisture availability can be used to determine the amount of net available energy (AE) at the earth's surface utilized by the latent heat flux. The remaining AE is partitioned toward the sensible heat fluxes. The sensible heat flux can then be physically inverted to provide a modeled surface temperature based on the water balance moisture availability, denoted, $T_{s\,WB}$. This can be compared to the AVHRR-derived surface temperature, denoted $T_{s\,AVHRR}$.

The residual between the two parameters $T_{s\,AVHRR}$ and $T_{s\,WB}$ is minimized using a global optimization technique called simulated annealing, which alters some water balance operating characteristics (McVicar, Jupp, Billings, et al., 1996). This allows daytime thermal observations to be linked to the water balance model by bringing the two temperatures into agreement over the ten years of data. The residual is minimized and expressed as

$$\sum\left(T_{S-AVHRR} - T_{S-WB}\right)^2 \tag{7.2}$$

Rangeland Monitoring

The Western Australian Department of Land Administration (DOLA) undertook a project over several years to monitor vegetation condition using time series analysis of the NDVI obtained from the AVHRR sensor (Smith,

1994). This has been applied to the extensive rangelands of Western Australia (Cridland, Burnside, and Smith, 1994). Stocking density and when to muster are important issues, exacerbated by the size of individual paddocks. Pastoralists want to muster livestock only once a year. Having an indication of the available feed can assist in the decision of when to muster.

Cridland, Burnside, and Smith (1994) analyzed the four years of NDVI data by plotting the NDVI signal as a time series. The height, in NDVI units, from a varying baseline to the maximum peak within the growing season is calculated. This green "flush" is the response of the landscape to rainfall. The baseline was varied to account for the influence of perennial cover on the NDVI signal. The baseline is defined as the minimum value from the previous year.

The vegetation response or "flush" recorded as the maximum for a particular year is then considered relative to the absolute maximum "flush" within the four (or more) years of data. As well as indicating where and when grazing conditions are poor, both images may be used to highlight opportunities to increase stocking densities due to an increase in available feed. This can help place individual years within a historical context.

Monitoring Bushfire Activity

DOLA is also engaged in monitoring bushfire activity in northern Australia, at a continental scale on behalf of the Environmental Resource Information Network (ERIN), and at a finer resolution for the Fire and Emergency Services Authority of Western Australia and the Bushfire's Council of the Northern Territory. The Queensland Department of Natural Resources and Mines also has a fire monitoring program. Fire without follow-up rain can be ecologically devastating, so fire control and hotspot monitoring are very relevant to managing for climate variability, this being particularly important in the savanna country of northern Australia. See <http://www. dola.wa.gov.au>, <http://www.eoc.csiro.au/>, and <http://www.LongPaddock.qld.gov.au/SatelliteFireMonitor/>.

Remote-Sensing Applications by New South Wales Agriculture

NSW Agriculture uses spatial information technology to assist in the management of agricultural emergencies including bushfire and flood (Tupper et al., 2000). GIS, GPS, and remote sensing combine to add value to the outcomes of its emergency management activities.

Mapping Land Inundated by Floodwater

Northwestern New South Wales experienced major flooding during the La Niña event of spring 1998, which caused millions of dollars of damage to agricultural production and resources. NSW Agriculture needed to quantify the areas inundated by water to assist with planning for future floods and to validate current assistance programs. Various satellite data sources were investigated for mapping the flood. Data were required every few days, making temporal resolution an important issue. A multisensor approach using SPOT, Landsat, and Radarsat data could have produced accurate results at a property scale. The high purchase cost and high demand placed on human and computer resources for processing precluded the use of these data. NOAA AVHRR was a suitable alternative with six overpasses per day and its thermal imaging capability. Its 1 km spatial resolution was a drawback but was sufficient for regional-scale mapping.

Data from all NOAA satellite numbers 12, 14, and 15 overpasses were obtained over the period, giving a total of 415 images. The thermal bands 3, 4, and 5 represented absolute temperatures in 1/100 of a degree, while reflected bands 1 and 2 represented calibrated reflectance values. Discrimination between land and water was best in the thermal bands in the predawn images, with water consistently 8 to 10°C warmer than land. This is because water temperature remains relatively stable throughout the diurnal cycle, whereas the soil surface is then at its coolest (Sabins, 1997).

Over 5,000 km of rivers were mapped. The area inundated was 4.5 million hectares. There was good agreement between the maps derived from remote sensing and data from field surveys, with errors restricted to narrow rivers. NOAA AVHRR data were of no use for mapping narrow rivers or waterlogged soil. Higher spatial and spectral resolution imagery is required in these cases.

Near-Real-Time Fire Monitoring

In February 1999, a bushfire occurred in inaccessible country in central New South Wales. Major stock and property losses were incurred by landholders. Over 8,000 sheep were killed on 34 properties. The inaccessible nature of the area made determining the fire extent difficult for field staff. NSW Agriculture provided near-real-time monitoring of the fire over a four-day period. NOAA AVHRR data were chosen because they have high temporal resolution (four hourly) and good spectral resolution for active fire-front identification. These data were available within 50 minutes of the satellite overpass. The NOAA AVHRR was not designed for active fire monitoring and suffers limitations in terms of spatial resolution and radiometric saturation. However, it proved to be suitable for this task (Setzer and Verstraete, 1994).

Images were assessed individually using techniques and band combinations appropriate to the time of capture and the direction and extent of the smoke plume. Daytime images captured on the first day, when the fire was travelling northeast and the smoke was blown clear of the southern and western edges of the fire scar, presented an opportunity to map these edges using the visible and reflected infrared bands. On the third day, wind strength abated, allowing smoke to rise and remain above the scar, completely covering the area. The wavelengths of band 3 (3.55-3.93 μm) enabled identification of the fire front, even when the area was obscured by smoke. Unlike a cloud, which is impenetrable by radiant energy, smoke consists of very fine particles through which radiant energy can pass relatively unaffected (Sabins, 1997). Analysis of images when smoke was hanging over the fire was restricted to density slicing of band 3. This produced a clear picture of the active fire front. Nighttime overpasses were displayed (RGB 345) to produce clear images of the fire front and hotspots. GIS polygons were created by on-screen digitizing around fire fronts and, smoke permitting, fire scars for daytime overpasses. The creation of individual polygons enabled mapping of the total fire scar and the fire progression in four hourly increments. This technique provided field staff with access to maps of the fire's location in near-real time.

Assessing Agricultural Losses Following Bushfire

In December 1997, a fire swept across open farming and grazing country in southern New South Wales. Major losses were incurred by landholders. Remote-sensing and GIS technology were used to quantify agricultural losses. Landsat TM data were chosen due to their low cost per unit area and good spatial and spectral coverage. The limitation of Landsat data in this case was their temporal resolution (16 days). Two consecutive overpasses were obtained. The overpass before the fire occurred on December 9, with the next on December 25. Given the rate at which cereal crop harvesting occurred, an image for the morning of the fire would have been desirable to assess unharvested crop losses. Furthermore, with 16 days between images, clouds can delay postfire image acquisition until it is too late. Satellite imagery needs to be obtained as soon as possible after a fire. Spectral separability of fire scars decreases over a relatively short period due to vegetation regrowth and removal of ash by wind and rain (Eva and Lambin, 1998).

A simple differencing algorithm was applied, producing a seven-band difference image. Band 1 of the prefire image was subtracted from band 1 of the postfire image, and so on, for seven bands. Change bands were assessed to determine those that best differentiated the fire scar. Bands 2, 3, and 4 proved most suitable. This image was used to produce a polygon of the fire scar. The prefire image was classified to determine prefire land- cover types. The classes differentiated were harvested crop, unharvested crop, actively growing pasture, senescent pasture, and timber. Discrimination between improved and native pastures was not possible given the prevailing dry conditions.

Using the classified image, fire scar polygon, and property boundaries, a GIS analysis was performed to extract which properties were affected by fire, the area of each property affected, and the area of each land-cover type lost on each property. The data produced through GIS analysis were reported in both map and tabular format (Figure 7.9).

Environmental Resources Information Network

The Environmental Resources Information Network has, over a number of years, used a number of techniques to analyze changes in AVHRR-derived NDVI images. Recently all AVHRR data held by ERIN have been recalibrated using the method proposed by Roderick, Smith, and Ludwick (1996). A number of analytical tools have been used to interpret the NDVI data.

Mapping the divergence of NDVI relative to the long-term mean has been done at two-month intervals. Having such a fine temporal resolution is important for data to be placed into a historical context, as it allows changes due to vegetation phenology, inherent seasonal changes in solar radiation, and air temperature to be normalized. This is important for determining the divergence from "normal" conditions for the particular month rather than using yearly extremes. The analytical approach of the "flush" which has been applied to Western Australia, and follows the idea of determining the flush of NDVI at a pixel level on an annual basis, will be applied to the entire Australian continent. See <http://www.ea.gov.au/land/monitoring/index.html>.

Remote-Sensing Applications by Queensland Department of Natural Resources and Mines

Satellite-based information on vegetative cover is an important layer within a GIS devoted to monitoring seasonal changes in vegetation, land clearing, and the extent and severity of drought (Brook and Carter, 1994; Carter et al., 1996). Considerable emphasis has been devoted to field validation of NDVI data and model output with respect to pasture biomass and tree cover (Wood et al., 1996).

Monthly NOAA NDVI satellite data are presented as decile (relative) greenness maps in the same manner as rainfall is often reported.

A NOAA receiver is used for fire mapping. Common AVHRR processing software (CAPS) is used for postacquisition processing. Maps of fire scars are used to "reset" grass biomass in spatial models and to investigate fire frequency in grazed lands. Combining data on the area burned, with model biomass and nitrogen content, allows calculation of greenhouse gas emissions.

Calibration and Validation of Spatial Models

NDVI and thermal data provide a high resolution (spatial and temporal) data set that can be matched to a synthetic NDVI produced by biological models. NDVI data were compared to a model's synthetic NDVI signal to independently validate the model, both spatially and temporally. In the Aussie GRASS project, the NDVI imagery has been used to spatially fine-tune some of the pasture growth parameters. NDVI data are also being used with a generic algorithm to investigate optimization of model parameters such as transpiration use efficiency (Carter et al., 2000).

Tree, Land Use, and Soil Attribute Mapping

Long-term mean NDVI data have been used to map tree density and cropping areas on a national basis (Carter et al., 1996; McKeon et al., 1998). In Queensland the Statewide Landcover and Trees Study (SLATS) is mapping tree density, tree clearing rates, and some land use with Landsat TM imagery for the entire state. Data from this project have been used to upgrade existing NOAA-based tree maps used in spatial models. The data are also being investigated for mapping land degradation. Research is in progress to translate mean NDVI and air temperature data into tree biomass data for Australia.

See <http://www.LongPaddock.qld.gov.au/RainfallAndPastureGrowth/Qld/> (The Long Paddock Satellite Imagery) and <http://www. LongPaddock. qld.gov.au/SatelliteFireMonitor/>.

Agrometeorological Decision Aids Driven by Real-Time Satellite Data

In a NASA-sponsored program titled "Use of Earth and Space Science Data over the Internet," Diak, Bland, and Mecikalski (1996) developed a suite of products for agriculture that are based on satellite and conventional observations, as well as state-of-the-art forecast models of the atmosphere and soil-canopy environments. Earlier attempts to apply satellite data to agriculture were plagued by data and information difficulties, which made the information systems problematic and unreliable. This situation has changed, however, and the timely retrieval of the multiple near-real-time satellite and supporting data sets required for routine use by agricultural applications is now feasible. This will become increasingly so with future availability of such data on the Internet. Similarly, dissemination of the resulting data and analyses to end users is now possible via the Internet, satellite-based commercial data transmission services, and telecommunication services.

In the TiSDat (Timely Satellite Data for Agricultural Management) project, Diak and colleagues (1998) selected agricultural sectors in which data availability was thought to be hampering the full adoption of currently available knowledge to management decision making. These applications were also selected based on the availability of some form of decision framework that could be improved through the application of satellite data and modern computer modeling techniques. The crops initially targeted for decision support systems were of comparatively high economic value, and the associated growers were motivated and well organized. In each area selected, the use of an improved information base had the potential to have a positive impact on environmental quality.

The products included an irrigation scheduling product based on satellite estimates of daily solar energy, a frost protection product that relied on prediction models and satellite estimates of clouds, and a product for the prediction of foliar disease based in satellite net radiation, rainfall from ground-based measurements, and a detailed model of the soil-canopy environment.

Irrigation Scheduling

The optimal management of available water resources is extremely important. Excessive irrigation leads to leaching of fertilizers and other agricultural chemicals from the soil into the water and leakage of water to the groundwater resulting in a rising water table in some environments (Postel, 1993). Knowledge of daily available solar energy for evapotranspiration is fundamental to irrigation management. Satellite data can provide the high-quality estimates of incident solar energy at the surface required for evapotranspiration estimates with much greater spatial detail and cost effectiveness than can be achieved through a network of ground-based pyranometers. Usually, eight to twelve individual U.S. series of Geostationary Operational Environmental Satellites (GOES) images are used during the course of a day, at hourly intervals, to make instantaneous estimates of the solar energy conditions at the satellite image times. These estimates are then integrated over time to provide daily estimates of solar energy at a site. The model used to estimate solar energy was similar to that described by Diak and Gautier (1983), but modified for the newer generation of GOES satellites (*GOES-8* and *GOES-9*) (Diak, Bland, and Mecikalski, 1996; Menzel and Purdum, 1994).

Geographical maps of estimated insolation and evapotranspiration are produced every day during the growing season (<http://www.soils.wisc.edu/wimnext/water.html> and <http://cimss.ssec.wisc.edu>).

Frost Protection of High-Value Crops

Ready access to real-time satellite and surface data and forecast model predictions of minimum temperature can lessen frost damage, improve harvests, and reduce the use of water applied to prevent such damage. In Wisconsin, cultivated cranberries are the major frost challenge. Any improved information on impending frost conditions can aid in minimizing water usage and the resulting environmental and energy impacts. The minimum temperature forecast system relies on a combination of satellite cloud information and synoptic upper-air and hourly surface measurements of temperature, humidity, and wind speed. Several computer forecast models, based on the physics of the atmosphere and land surface, as well as a statistical adjustment procedure, are used to interpret the data sources and predict if freezing temperatures will occur overnight. Satellite-derived cloud data are assimilated into the University of Wisconsin-Madison Cooperative Institute for Meteorological Satellite Studies (CIMSS) Regional Assimilation System (CRAS) run in near-real time, with a forecast duration of 48 hours. A time series of prognostic information on the air temperature, humidity, and wind speed of the lower atmosphere and also downwelling thermal radiation is passed to a one-dimensional soil/vegetation model, called the Atmosphere-Land Exchange (ALEX) system. The CRAS-ALEX prediction provides the first estimate of temperatures for the day and is generally available at about noon local time. In the evening, several updates are made using timely satellite-derived cloud information and also surface-based measurements.

Real-time 10 km satellite-derived cloud cover from the *GOES-8* atmospheric sounding instrument are provided to the TiSDat effort through the cooperation of the National Oceanic and Atmospheric Administration Advanced Satellite Products Project at the University of Wisconsin Space Science and Engineering Center. Daily real-time cloud products are viewable both on the CIMSS and NOAA Web sites (<http://cimss.ssec.wisc.edu/> and <http://www.noaa.gov/>, respectively).

Foliar Disease Management

The last product involved foliar disease in potato, and it depended on a decision support system named WISDOM developed by the University of Wisconsin-Extension, which resided locally on growers' home computers. The threat posed by this type of disease depends significantly on temperature and humidity within the crop canopy and the presence of free water on leaves (Stevenson, 1993). Growers interfaced WISDOM with a server to obtain rainfall, meteorological data, surface radiation inputs, and canopy model output required by WISDOM for the blight models. Use of the early blight model within WISDOM reduces the number of fungicide applications used compared to conventional practices.

The TisSDat product for this application uses satellite, surface, and radar data inputs, coupled to a version of the ALEX adapted to potato, to provide WISDOM with the data required by the blight models. The relevant data sets include satellite-based hourly estimates of solar radiation and net longwave radiation, as well as surface-based measurements.

Further details about these three applications and additional references can be found in Diak and colleagues (1998).

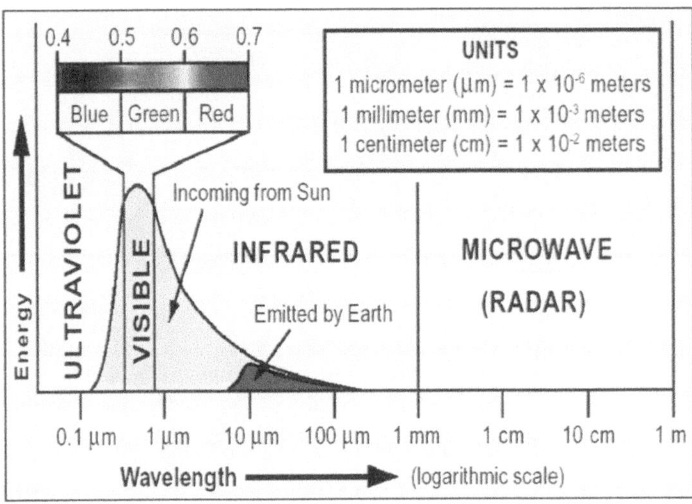

FIGURE 7.1. The portion of the electromagnetic spectrum that is useful in remote sensing of the earth's surface (*Source:* Reprinted from *Introduction to Remote Sensing of Environment (RSE) with TNTmips® TNTview®,* R.B. Smith, 2002, with permission from MicroImages, Inc.)

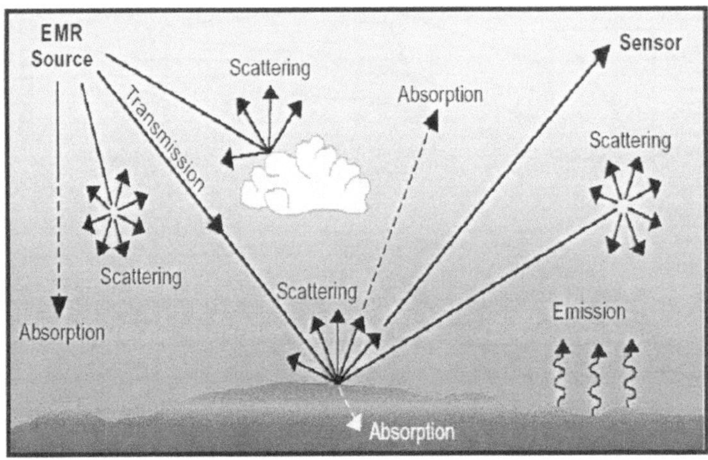

FIGURE 7.2. Typical EMR interactions in the atmosphere and at the earth's surface (*Source:* Reprinted from *Introduction to Remote Sensing of Environment (RSE) with TNTmips® TNTview®,* R.B. Smith, 2002, with permission from MicroImages, Inc.)

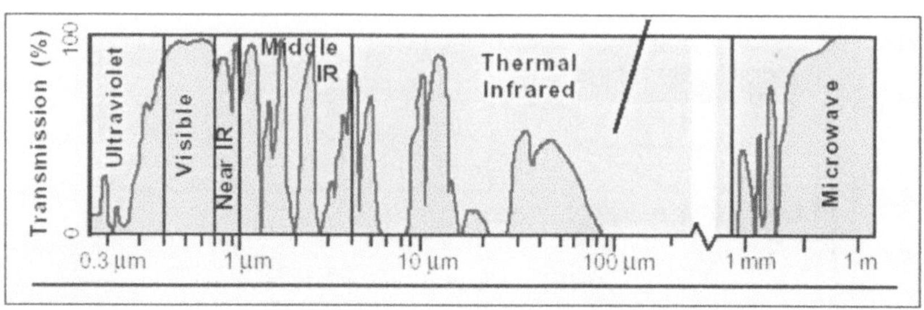

FIGURE 7.3. Variation in atmospheric transmission with wavelength of EMR (*Source:* Reprinted from *Introduction to Remote Sensing of Environment (RSE) with TNTmips® TNTview®,* R.B. Smith, 2002, with permission from MicroImages, Inc.)

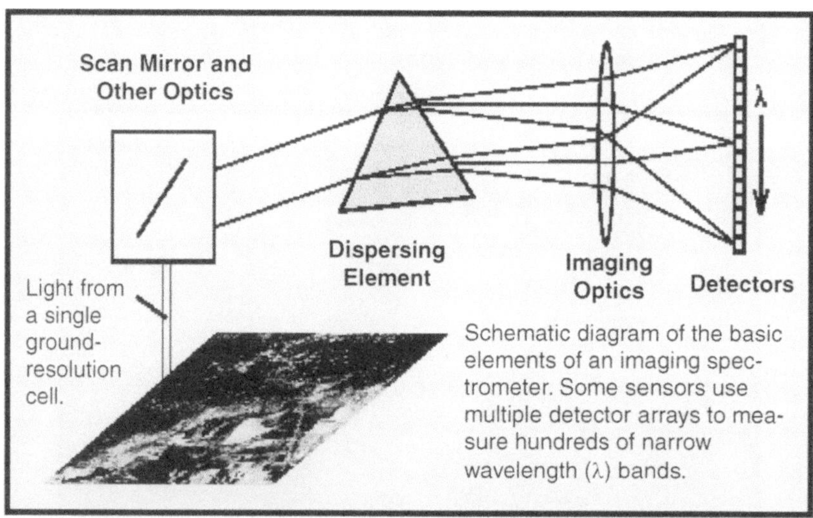

FIGURE 7.4. Schematic diagram of the basic elements of an imaging spectrometer (*Source:* Reprinted from *Introduction to Remote Sensing of Environment (RSE) with TNTmips® TNTview®,* R.B. Smith, 2002, with permission from MicroImages, Inc.)

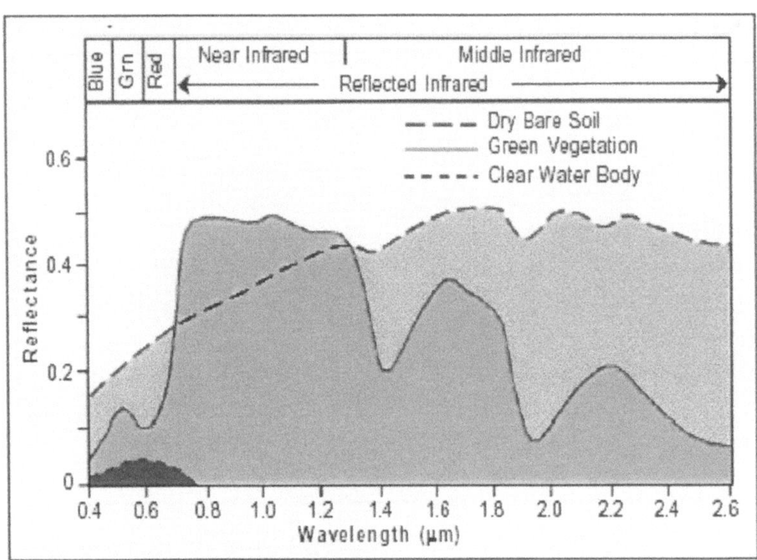

FIGURE 7.5. Spectral reflectance characteristics of common earth surface materials (*Source:* Reprinted from *Introduction to Remote Sensing of Environment (RSE) with TNTmips® TNTview®*, R.B. Smith, 2002, with permission from MicroImages, Inc.)

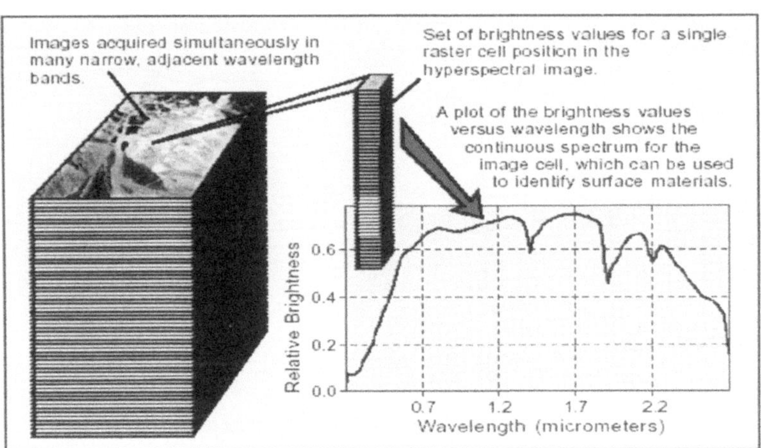

FIGURE 7.6. Hyperspectral remote sensors make it possible to derive a continuous spectrum for each image cell (*Source:* Reprinted from *Introduction to Remote Sensing of Environment (RSE) with TNTmips® TNTview®*, R.B. Smith, 2002, with permission from MicroImages, Inc.)

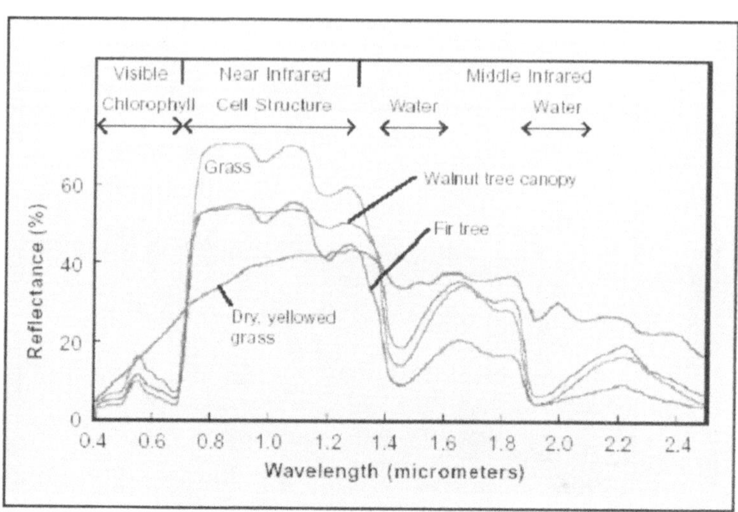

FIGURE 7.7. Reflectance spectra of different types of green vegetation compared to a spectral curve for senescent leaves (*Source:* Reprinted from *Introduction to Remote Sensing of Environment (RSE) with TNTmips® TNTview®,* R.B. Smith, 2002, with permission from MicroImages, Inc.)

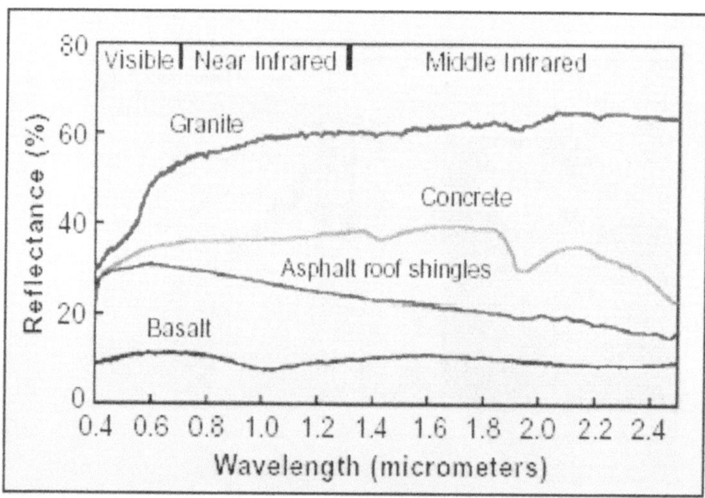

FIGURE 7.8. Sample spectra from the ASTER Spectral Library (*Source:* Reprinted from *Introduction to Remote Sensing of Environment (RSE) with TNTmips® TNTview®,* R.B. Smith, 2002, with permission from MicroImages, Inc.)

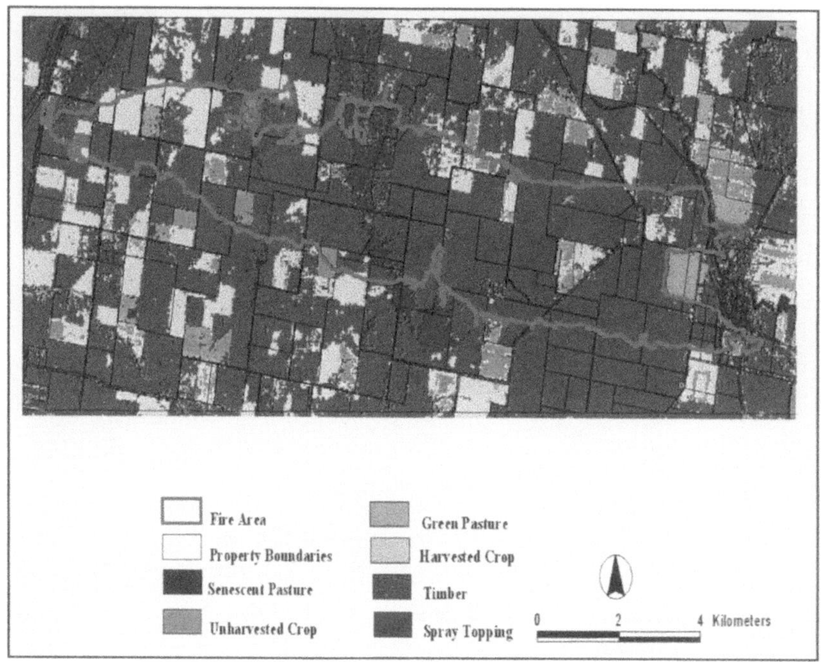

FIGURE 7.9. Classified, difference image showing fire scar, properties, and land-cover types

Chapter 8

Role of Computer Models in Managing Agricultural Systems

MODELING BIOLOGICAL RESPONSE TO WEATHER CONDITIONS

Attempts to relate agricultural production to weather go back to the start of domestication of plants and are still evolving. These were initially qualitative studies which were later followed by statistical analyses. The use of growth chambers and the development of precision instruments and their use in field observations provided quantitative estimates of how plant processes responded to variations in temperature, available water, and other environmental conditions. The quantitative measurements also provided an understanding of the microclimatic characteristics of biological systems. By the late 1960s and early 1970s extensive literature documented the response of plant growth and development to environmental conditions (Decker, 1994). These developments paved the way in the 1980s and 1990s for work on mathematical models of plant response and yields to varying environmental conditions.

The comprehensive development and use of plant and animal dynamic simulation models started with the availability of the computer in the early 1970s. By the end of the twentieth century several thousand computer-based plant and animal dynamic simulation models were developed to expand scientific insight into complex biological and environmental systems. Both simple and complex models are now available. In some cases, simple models are not appropriate because they are not programmed to address a particular phenomenon. In other cases, complex models are not appropriate because they may require inputs that are not practical to obtain in a field situation (Boote, Jones, and Pickering, 1996; Jorgensen, 1999). In a review of agrometeorology over the last two centuries, Decker (1994) described a progression from a descriptive science to a modeling approach based on analytical procedures using biological and physical processes.

Crop growth models have many current and potential uses for answering questions in research, crop management, and policy. Models can assist in synthesis of research understanding about the interactions of genetics, physiology, and the environment, integration across disciplines, and organization of data. They can assist in preseason and in-season management decisions on cultural practices, fertilization, irrigation, and pesticide use. Crop models can assist policymakers by predicting soil erosion, leaching of agrochemicals, effects of climatic change, and large-area yield forecasts. Their use has resulted in huge economic benefits.

MODELS

A model is a schematic representation of our conception of a system. A model brings into mind the functional form of a real object, such as children's toys, a tailor's dummy, and mock-ups of buildings and structures to be later constructed in the real forms. Models also construct objects or situations not yet in existence in real form. A model can also be referred to as a representation of a relationship under consideration and may be defined as an act of mimicry. Models can be broadly divided into statistical models and dynamic simulation models.

Statistical Models

Statistical models are those that do not require detailed information about the plant involved but rely mainly on statistical techniques, such as correlation or regression, relating to the appropriate plant and environmental variables (Norman, 1979). Most statistical models are crop-yield weather models, which are applied to estimate yield over large areas with variable success. The regression coefficients themselves are not necessarily related to the important processes and therefore are highly variable with crop type, region, etc. Many studies are required to produce the regression equations necessary for the widespread application of this kind of model. A great advantage of these simple crop-weather models is that they use readily available weather data. Although results are not very accurate, the statistical model is able to recognize the years that bumper crops and crop failures can be expected, several weeks prior to harvest.

Regression models are attractive because of their simple and straightforward relationship between yield and one or more environmental factors, but these are not accurate enough to be used for other areas and other crops. Despite this limitation, they are used extensively for the prediction of yield of a single crop over a large region, with a variety of soils, agronomic practices, and insect-disease problems. A combination of such factors is still beyond the dynamic simulation models. It is a technique well worth retaining in the arsenal of tools available to the agricultural climatologist (Whisler et al., 1986).

Dynamic Simulation Models

A dynamic model is the one whose output varies with time and in which processes are characterized. To characterize processes, the state variables must be known. State variables are those necessary to define the state of the system at a point in time. Dynamic simulation crop models predict changes in crop status with time as a function of biogenetic parameters (Hume and Callander, 1990).

Simulation means that the model acts like a real crop, gradually germinating and growing leaves, stems, and roots during the season. In other words, simulation is the process of using a model dynamically by following a system over a time period.

Dynamic models can be classified as preliminary models, comprehensive models, and summary models (Penning de Vries et al., 1989). Preliminary models have structure and data that reflect current scientific knowledge. These are simple because insight is at the exploratory level. A comprehensive model is a model of a system in which essential elements are thoroughly understood and much of this knowledge is incorporated. Summary models are abstracts of comprehensive models and are found at production levels. Comprehensive dynamic models predict yield much closer to reality than do the regression models. However, the more accurate the dynamic model is, the more information is required for initialization and about driving variables. In many cases these may not be available, hence the regression models may still be our best option.

Developing a comprehensive, dynamic crop simulation model requires a multidisciplinary team. Plant physiologists, agronomists, and soil scientists are needed to help define both the overall framework of the problem and the specificities of the environment and plant growth relationship. Entomologists and plant pathologists are required to define the insect and pathogen subsystems that are important parts in the crop ecosystems. An agrometeorologist selects and contributes data about weather and microclimate fluxes in and around the plant canopies. A computer programmer selects the computer language and develops the overall framework of the model (Ritche et al., 1986).

After designing the first version of the model and analyzing the results of its output, faults are invariably found, which require changes in the structure of the model. These changes require additional study and verification, and the loop continues. Furthermore, in the process of development, the initial representation of the modeling process is improved upon and the range of experimental data used is widened. In the final verification process, it is important to consider an independent data set that was not used in the development of the model (Guardrian, 1977).

Verification and Validation

Verification and validation are two commonly used terms in modeling. Verification is the test of truthfulness or correctness. By comparing historical data recorded for real world systems to what the computer gives as the output of the model, verification certifies that the functional relationships modeled are correct. If a model does not behave according to expectations, then some correction of the functional relationship may be necessary or coefficients may need to be adjusted. The latter is called calibration.

Validation is concerned with the comparison of model predictions with results from independent experiments. Models can be considered valid and useful even when there are some differences between experimental data and simulation output. A model is considered valid if the simulated values lie within the projected confidence band.

APPLICATIONS OF CROP MODELS

Crop simulation models play an important role at different levels of application, ranging from decision support for crop management at a farm level to advancing understanding of sciences at a research level. The main goal of most applications is to predict final yield in the form of grain yield, fruit yield, root or tuber yield, biomass yield for fodder, or any other harvestable product. Certain applications link the price of the harvestable product with the cost of inputs and production to determine economic returns. Another application of crop simulation models is in policy management. Whisler and colleagues (1986) and Hoogenboom (2000) described a wide range of major areas in which the application of models is well established.

Crop Breeding

Many simulation models use genetic characteristics in the form of rate coefficients or other system constants in crop growth. These coefficients or constants can be evaluated during the validation and sensitivity processes. Duncan and colleagues (1978) have shown, with a simulation model of peanut, dramatic yield increases solely from changes in flowering time and some aspect of photosynthate partitioning. Breeding work can be undertaken to benefit from these useful characteristics so as to produce high-yielding, insect-disease-resistant cultivars capable of competing with weeds.

Physiological Probes

With traditional scientific techniques, it is almost impossible to obtain data on the many physiological processes that are important from a crop physiology point of view. For example, Whisler and colleagues (1986) stated that turgor pressure in the cells is the driving force for leaf expansion, but unfortunately, it cannot be measured directly. These can only be inferred from measurements using a dynamic simulation model.

If the models are comprehensive in nature, they can be tested for various growth processes and can help in eliminating unrealistic hypotheses, saving time and energy. Furthermore, exercises using a model may give rise to many more experiments to test various hypotheses. This is the way by which models can be used to probe the physiology of plants, which is not experimentally accessible. Modeling and experimentation can be mutually supportive in developing our understanding of crop physiology.

Sequence Analysis

In the sequence or crop-rotation analysis, one or more crop rotations can be analyzed. In this mode, different cropping sequences are simulated across multiple years. It is critical that in a crop-rotation analysis water, nitrogen, and carbon are simulated as a continuum. The main goal of a crop-sequence application is to determine the long-term change of soil variables as a function of different crop-rotation strategies (Bowen, Thornton, and Hoogenboom, 1998). Several models have been specifically developed to study the long-term dynamics of nitrogen and organic matter in soil. APSIM (Agricultural Production Systems Simulation Model) is used to evaluate crop sequences in the northern grain belt of Australia (McCown et al., 1996). Others have been specially developed to study the long-term sustainability of cropping systems (Thornton et al., 1995).

Strategic and Tactical Applications

In strategic applications of crop simulation models and decision support systems, the models are mainly run to compare alternative crop management scenarios. This allows for the evaluation of various options that are available with respect to one or more management decisions (Tsuji, Hoogenboom, and Thornton, 1998). To account for the interaction of these management scenarios with weather conditions and the risk associated with unpredictable weather, simulations are conducted for at least 20 to 30 different weather seasons or weather years (James and Cutforth, 1996). In most cases, daily historical weather data are used as input, and the assumption is made that these historical weather data will represent the variability of weather conditions in the future. In addition, the biological outputs and management inputs can be combined with economic factors to determine the risk associated with the various management practices being evaluated (Thornton and Wilkens, 1998).

Hammer, Holzworth, and Stone (1996) calculated the benefits of seasonal forecasting for tactical management of nitrogen fertilizer and cultivar maturity of wheat at Goondiwindi, Australia. Using the SOI phase system, they found an increase in profit of 20 percent, or about $10.00/ha. They also showed the risk of making a loss could be reduced by as much as 35 percent. Marshall, Parton, and Hammer (1996) also used APSIM simulations for wheat to investigate how risk and planting conditions changed with an SOI-based forecast. Similar benefits of seasonal forecasts were observed on sorghum, sunflower, corn, and peanuts (Hammer, Carberry, and Stone, 2000; Meinke, 2000).

In tactical applications, crop models are actually run prior to or during the growing season to integrate the growth of a crop with the current observed weather conditions and to decide, on a daily basis, which management decisions should be made. In this regard, the uncertainty of weather conditions in modeling applications has to be managed. For any crop model run, only the weather data up to the previous day will be available. If the weather forecasts are provided in some type of quantitative format, they can also be included with the simulation. There are various methods for handling the uncertainty of future weather conditions. The first one is to use historical weather data and to run the system for multiple years. Instead of historical weather data, generated data can also be used. If multiple years of historical or generated weather data are used as input, a mean and associated error variable can be determined for predicted yield as well as for other predicted variables. Over time, the error will become smaller, as the uncertain weather forecast data are being replaced with observed weather variables. If two or more management alternatives are being compared, one can evaluate the risk associated with each management decision, using both mean and error values of each predicted variable.

Computer models and expert systems are extensively used in irrigation. Packages are available that deal with irrigation scheduling, irrigation system evaluation, crop planning and selection of crop varieties, and irrigation system operation.

In the area of pest and disease management, especially integrated pest management (IPM), the application of models has been shown to be very profitable (Pusey, 1997). As the application of pesticides is rather expensive, farmers are interested in minimizing their use, from both economic and environmental viewpoints.

Another area of application is in climate change impact assessment. As climate change deals with future issues, the use of general circulation models (GCMs) and crop simulation models provides a scientific approach to study the impact of climate change on agricultural production and world food security. Similarly, the issue of climate variability especially related to the variation in sea-surface temperature (SST) of the Pacific Ocean or El Niño/Southern Oscillation (ENSO) has opened an area in which crop simulation models also can play an important role. They can potentially be used to help determine the impact on agricultural production due to ENSO and recommend alternative management scenarios for farmers that might be affected, thereby mitigating the expected negative impacts of ENSO and capitalizing on the opportunities in better seasons.

Forecast Applications

The application of crop simulation models for forecasting and yield prediction is very similar to the tactical applications. However, in the tactical decision application, a farmer or consultant is mainly concerned with the management decisions made during the growing season. In the forecasting application of the crop models, the main interest is in the final yield and other variables predicted at the end of the season. Most of the national agricultural statistics services provide regular updates during the growing season of total area planted for each crop, as well as the expected yield levels. Based on the expected yield, the price of grain can vary significantly. It is important for companies to have a clear understanding of the market price so that they can minimize the cost of their inputs. Traditionally, many of the yield forecasts were based on a combination of scouting reports and statistical techniques. However, it seems that crop simulation models can play a critical role in crop-yield forecasting applications if accurate weather information is available, both with respect to observed conditions and to weather forecasts. The STIN (Stress Index) model (Stephens, Walker, and Lyons, 1994) has been officially used for forecasting Australian wheat production. Accurate applications of crop simulation models require, in many cases, some type of evaluation of the model with locally collected data. Especially for yield forecasting, it is critical that yields are predicted accurately, as policy decisions related to the purchase of food could be based on the outcome of these predictions. One option is to use remotely sensed data that are being used to estimate yield, based on a greenness index. A more advanced application would be to link physical remote sensing with crop simulation models. With this approach, the simulated biomass can be adjusted during the growing season, based on remotely sensed or satellite data, and yield predictions can be improved based on these adjusted biomass values (Maas, 1993).

Spatial Analysis

One of the limitations of current crop simulation models is that they can simulate crop yield only for a particular site for which weather and soil data as well as crop management information are available. One recent advancement is the linkage of crop models with a geographic information system (GIS). A GIS is a spatial database in which the value of each attribute and its associated *x*- and *y*-coordinates are stored. To describe a specific situation, all the information available on a territory, such as water availability, soil types, forests, grasslands, climatic data, and land use are used. Each informative layer provides to the operator the possibility to consider its influence on the final result. However, more than the overlap of various themes, the relationships of the various themes is reproduced with simple formulas or with complex models. The final information is extracted using graphical representation or precise descriptive indexes (Hartkamp, White, and Hoogenboom, 1999; Maracchi, Pérarnaud, and Kleschenko, 2000). This approach has opened a new field of crop modeling applications at a spatial scale, from the field level for site-specific management to the regional level for productivity analysis.

Seasonal Analysis

In seasonal analysis applications, a simulation model is used to evaluate a management decision for a single season. This can include crop and cultivar selection; plant density and spacing; planting date; timing and amount of irrigation applications; timing, amount, and type of fertilizer applications (Hodges, 1998); and other options a particular model might have. Model applications can also include investment decisions, such as those related to the purchase of irrigation systems.

SIMULATION MODELS RELEVANT TO AUSTRALIAN FARMING SYSTEMS

During the twentieth century, world agriculture passed through three revolutionary eras. The first was the mechanical (1930-1950), the second, seed-fertilizer (1960-1970), and the third, information technology (in the closing decades of the century). Within the span of the last two decades of the century, several thousand computer-based plant and animal dynamical models were developed worldwide which have expanded scientific insight into the complex interactions between environmental and biological systems. Australian science and scientists made a substantial contribution in this expansion. Currently, scores of modeling groups and hundreds of individuals are actively engaged in this pursuit. Several crop and pasture models developed in Australia are in use on an international level.

To describe and discuss even a fraction of the models developed in Australia is beyond the scope of this book. The information given on the topic is not conclusive, nor is the list of sources of this information exhaustive. A summary of a limited number of models that are thoroughly tested and are currently in use is given in Table 8.1. Tree and crop models are described first, followed by pasture and animal models.

DECISION SUPPORT SYSTEMS (DSS)

Decision support systems (DSS) are integrated software packages comprising tools for processing both numerical and qualitative information. A DSS points the way for better decision making in the cropping and pastoral industries. It offers the ability to deliver the best information available, quickly, reliably, and efficiently.

The choices of planting time, varietal selection, grazing strategies, and fertilizer, irrigation, and spray applications are complex decisions to be made at the farm level. These are important and decisive because they cannot be postponed, are irreversible, represent a substantial allocation of resources, and have a wide range of outcomes, with consequences that impact the farm business for years to come. They are also hard decisions because they are characterized by uncertainty, mainly due to the highly variable climate. They are complex both in terms of the number of interacting factors and the trade-offs between risk and reward. A successful decision support system focuses on such decisions. A key element in the success of a DSS is the development of trust in its reliability and the willingness and ability of the targeted users to utilize the system.

TABLE 8.1. Crop, pasture, and animal production simulation models

Model	Output	Reference
BIOMASS Tree model	Growth, dry matter, yield	McMurtrie and Landsberg, 1991
CenW Tree model	Photosynthetic carbon gain, water use, nitrogen cycling	Kirschbaum, 1999
APSIM Agricultural Production Systems SIMulator	Carbon, water, and nitrogen balances of agricultural systems, crop rotations, interspecies competition, etc.	McCown et al., 1996
SIMTAG Simulation model for *Triticum aestivum*	All development phases, plant dry matter, grain yield	Stapper, 1984
Wheat model	Soil moisture content, crop growth rate, canopy leaf-area index, crop biomass, grain number, grain yield	Wang and Gifford, 1995
QBAR Barley model	Soil water balance, phenology, leaf area, biomass production, grain yield	Hook, 1997
AUSCANE Sugarcane model	Yield of millable cane stalks, sugar content of cane	Jones et al., 1989

CERCOT Cotton model	Growth and production	Hook, 1997
OZCOT Cotton model	Soil water content, evapotranspiration, yield, yield components, fruiting dynamics, leaf-area index, nitrogen uptake	Hearn, 1994
GRASP Pasture model	Soil water status, pasture growth, death and detachment, animal intake, diet selection, utilization, live weight gain	McKeon et al., 1990
PGAP Pasture Growth and Animal Production Model	Green and dead dry matter on offer, growth index, animal intake, pasture growth, pasture senescence, cumulative pasture growth for the year	Curtis, Bowden, and Fels, 1987
DYNAMOF DYNAmic Management Of Feed	Soil moisture availability, digestibility of pasture, wool growth, lambs born and sold, sales of ewe and wether, fleece weights, fiber length, diameter and staple strength, net farm income	Bowman, Cottle, et al., 1993; Bowman, White, et al., 1993

Optimism for using computer-based DSSs for agriculture in Australia started in the late 1980s. This was a period of proliferation in the development of local DSSs and the development of agricultural software in Australia and overseas which preceded the widespread use of computers in agriculture. In the early 1990s less than 5 percent of farmers had computers, and a lack of farm computers was seen to be the major constraint on the greater adoption of DSSs. By the close of the twentieth century, the percentage of Australian farmers with computers has increased to almost 75 percent (Grains Research and Development Corporation [GRDC, 2001]). This has helped advisers and farm managers to use DSSs for their short- and long-range decision making. Like proliferation of the computer simulation models, the list of DSSs developed in Australia is very long. Some of the DSSs relevant to Australian farming systems and claimed to be in current use are briefly described in Table 8.2.

Examples of Potential Uses of Some Decision Support Systems

Australian RAINMAN

1. It is springtime, and so far, the season is going well. A farmer at Tamworth, New South Wales, has to plan for the summer crop season. He wants information on the chances of rainfall in the coming summer so that he can adjust crop management by modifying nitrogen fertilizer applications.

The information the Tamworth farmer needs is easily derived from Australian RAINMAN. Information about Australian RAINMAN can be obtained from Queensland Department of Primary Industries at <http://www.longpaddock.qld.gov.au>. RAINMAN (Version 3.3) has historical rainfall records for thousands of locations, including Tamworth.

Procedure. Open Tamworth location in RAINMAN, and choose "Seasonal Forecasts" from the "Selector" dropdown window; the "Seasonal Rainfall" dialogue window opens. Click on "Setting"; the "Seasonal Time Setting" window opens. Select:

Season—October to December
SOI phase—August-September

TABLE 8.2. Decision support computer packages relevant to Australian farming systems

DSS	Use	Reference
Australian RAINMAN	Australian RAINMAN is a computer package that helps examine historical records of rainfall and temperature, analyzes monthly and daily rainfall records, and forecasts seasonal rainfall based on SOI and Indian Ocean sea-surface temperature	Clewett et al., 1999
MetAccess	MetAccess allows users to summarize and analyze long-term daily recordings of weather events in a great variety of ways and to display the results in graphical or tabular formats. A particularly useful feature is its facility for calculating an estimate of the long-term probability of the occurrence of specified weather events or patterns at any locality for which weather records are available.	Donnelly, Moore, and Freer, 1997
HOWOFTEN	The DSS is useful for identifying when planting opportunities occur, when flooding rains are most likely, and if sufficient rainfall has fallen to fill a soil profile.	Paull and Peacock, 1999
PLANTGRO	PLANTGRO determines the suitability of sites/land units for growth of different plants and the length of growing season of each plant at a nominated site for different planting times. It predicts growth of the plant for the climatic and soil conditions of the site, taking into account the month of planting.	Hackett and Harris, 1996
HOWWET	HOWWET uses farm rainfall records to calculate how much fallow rainfall is actually stored in the soil. The accumulation of mineralized nitrogen is also estimated through the fallow. DSS is useful when deciding how long to fallow, selecting crop type and plant density, choosing precrop irrigation requirements, and estimating expected yields.	Dimes, Freebairn, and Glanville, 1993
MUDAS	MUDAS is a crop-livestock integration system useful for profit maximization on farms with more than one enterprise (such as wheat and sheep). It classifies seasons according to four general criteria: (1) the amount of rainfall received during the summer, (2) the timing of the opening rains, (3) the amount of rainfall received at the break, and (4) the amount of spring rainfall. It provides options to allow within-season, or tactical, adjustments on the basis of information to date. These include adjustment to crop area, machinery use, live weight of sheep, fertilizer use, grain storage, and pasture deferment.	Kingwell, Morrison, and Bathgate, 1991
TACT	TACT calculates probability distributions of wheat yields and gross margins at a given location. Using predominantly empirical relationships, the package predicts wheat performance on a daily time step. Predictions of the timing of major phenological events are also available.	Abrecht and Robinson, 1996

TABLE 8.2 *(continued)*

DSS	Use	Reference
STIN	STIN (STress INdex) is an empirical wheat yield model. It is based on the assumption that wheat yields are a function of soil moisture at sowing (sowing date) and the timing/amount of rainfall during the growing season. Outputs are soil moisture at sowing and final crop yield. It is used to forecast shire, state, and national wheat yields.	Stephens, Lyons, and Lamond, 1989
WHEATMAN plus BARLEYPLAN	WHEATMAN plus BARLEYPLAN estimates and compares outcomes resulting from different farming options involving wheat, barley, and chickpeas. It assists with the choice of planting time, variety, and fertilizer strategy. It can also be used to compare many production and economic factors (such as gross margins) between properties and/or paddocks.	Woodruff, 1992
PYCAL	PYCAL (Potential Yield Calculator) is a computer program to monitor current seasonal rainfall against the historic record and to estimate stored soil water and potential yield for a range of cereal, pulse, and oilseed crops.	Grains Development and Research Corporation, 1999
SOWHAT	SOWHAT is a decision support tool for wheat farmers in areas of Mediterranean climate. It builds on information from the years with a high probability of above-average wheat yields and those with a low probability of good yields. Subsequently, it uses this information to develop a sowing strategy. SOWHAT enables farmers to use their own farm records to develop their strategies for maximizing farm wheat yields.	Balston and Egan, 1998
Whopper Cropper	Whopper Cropper is designed to apply cropping systems modeling and seasonal climate forecasting to crop management. It meets the demand of extension professionals for access to the cropping systems modeling capability of APSIM (Agricultural Production Systems SIMulator). It provides information on the impact of climate risk on crop yields for crop management alternatives beyond the experience of individual farmers. Whopper Cropper's graphical user interface is designed to enable farmers to explore management strategies at the beginning of each cropping season.	McCown et al., 1996
SIRATAC	SIRATAC is a tactical pest management decision support system for on-farm use by cotton farmers. Output consists of pest populations expected to infest the crop each day over the next few days and whether the crop needs to be sprayed. Output also reports how many of the fruits currently on the crop will contribute to yield, how many more will be produced, what the yield and harvest date will be, and the yield loss that will be inflicted by pests currently infesting the crop if they are not controlled.	Brook and Hearn, 1990

DSS	Use	Reference
EntomoLOGIC	EntomoLOGIC is a pest management decision support system for cotton growers. It contains three modules: a Helicoverpa life cycle module for predicting pest pressure for the next three days, a Helicoverpa diapause module that predicts the number of pupae entering diapause and their expected reemergence dates, and a mite module that predicts mite pressure and yield loss. By using EntomoLOGIC's standard thresholds, sprays can be scheduled according to the insect pressure.	Larsen, 2001
CLIMEX	CLIMEX is an insect pest control decision support system. It compares locations of potential distribution, compares consecutive seasons at a location, matches climates of potential invasion, and measures greenhouse and irrigation impact on insect population. The system is also used for matching climates or locations for expanding the production of a particular crop or an animal.	Skarrat, Sutherst, and Maywald, 1995
Grazfeed	Grazfeed assists agricultural advisers and producers in stock grazing and supplementary feeding decisions. It estimates cattle and sheep production (meat, wool, and milk) obtainable from a particular pasture and indicates the extent to which a chosen supplement might improve production or the amount of supplement required to reach a given level of production.	Freer and Moore, 1990; Moore, Donnelly, and Freer, 1997
GrassGro	GrassGro predicts pasture growth and production from a mob of sheep or cattle grazing a paddock or group of paddocks in any enterprise in a Mediterranean climate. Applications of GrassGro in use are to (1) compare the current season's pasture production with the same seasonal periods in the past, (2) explore the likely effects of different plant characteristics on the productivity of grazing animals, (3) compare stocking rates, (4) estimate gross margins from an enterprise, and (5) use as a strategic planning tool for the management of breeding stock and supplementary feed needs.	Donnelly, Freer, and Moore, 1994
SheepO	SheepO is a package for estimating wool production, lambing performance, stock sales, and gross margins, and in developing medium- to long-term (strategic) management plans. The submodels within SheepO use empirical relationships for simulating processes. The pasture model calculates the green and total available pasture at ten-day intervals. Green digestibility model implements adaptations to relationships developed for several grasses.	Whelan et al., 1987; McLeod and Bowman, 1992; Bowman, Cottle, et al., 1993; Bowman, White, et al., 1993
SummerPak	SummerPak helps in the management of sheep by assisting with decisions on stocking rates and supplementary feeding. It operates on a daily time-step and simulates feed intake, animal and wool growth, and herbage on offer.	Wang and Orsini, 1992

TABLE 8.2 *(continued)*

DSS	Use	Reference
FEEDMAN	FEEDMAN helps beef producers to compare feeding options for growing cattle in terms of forage utilization, animal performance, market options, and economics. It calculates monthly forage growth and sustainable stocking rates in response to monthly rainfall, soil nitrogen status, and tree density.	Rickert et al., 1996
RANGEPACK	RANGEPACK is a strategic assessment tool that follows a herd or flock through successive years to evaluate the lagged effects of climatic fluctuations on herd numbers, allowing the user to follow the gradual implementation of a new strategy.	Stafford Smith and Foran, 1989
DroughtPlan	DroughtPlan is a series of procedures and decision support tools for producers in property management planning activities. DroughtPlan helps producers to make decisions about stocking rate strategies, breeding, buying, selling, and feeding options, and enterprise selection in the face of climatic variability.	Stafford Smith et al., 1996
Aussie GRASS	Australian Grassland and Rangeland Assessment by Spatial Simulation (Aussie GRASS) is a spatial modeling framework for assessing the condition of Australia's grazing lands. Aussie GRASS gives information on pasture growth that allows estimates of long-term safe stocking rates.	Carter et al., 2000

Close "Seasonal Time Setting" and return to the "Seasonal Rainfall" window. Select:

> Average SOI from the "Methods of Seasonal Analysis" (Any one of the four methods can be chosen.)

RAINMAN generates a table (can be a graph also) that gives the information about rainfall at Tamworth during the coming season under several scenarios (see Table 8.3).

Rainfall at Tamworth is highly variable from season to season. From historical records it is established that like any other place in eastern Australia, the amount of rainfall in a particular season is normally influenced by the SOI prevailing in the two months prior to the commencement of a season. The average rainfall at Tamworth for the October to December period is 203 mm. This year, SOI during the months of August and September remained

TABLE 8.3. Chance of rainfall in summer (October to December) at Tamworth, using average SOI in August-September

Rainfall period: October to December	**SOI below −5**	**SOI −5 to +5**	**SOI above +5**	**All years**
Percent years with at least				
340 mm	0	0	14	4
290 mm	0	9	31	13
270 mm	6	16	37	19
250 mm	9	23	37	23
200 mm	43	49	63	51
150 mm	63	74	89	75
83 mm	89	98	100	96
Percent years above median 201 mm	43	44	63	50
KS/KW probability tests	0.899	0.197	0.976	0.990
Significance level	NS	NS	*	**
Years in historical records	35	43	35	113
Highest recorded (mm)	283	334	425	425
Lowest recorded (mm)	64	53	87	53
Median rainfall (mm)	175	196	217	201
Average rainfall (mm)	173	198	237	203

Note: NS = nonsignificant result; in RAINMAN a result with probability less than 90 percent; * = 95 percent probability of effect being real; ** = 99.9 percent probability of effect being real.

above +5. Therefore, the most appropriate values for this year are in the scenario "SOI above +5." This indicates a 63 percent chance of rainfall exceeding 200 mm in the coming October to December period. In other words, the odds favor an above-average rainfall in the coming season.

2. It is autumn in Walgett, New South Wales. To obtain the maximum yield, a farmer wants to sow his wheat crop at the first available opportunity. He wants to know the likely date of the first rain in the season so that he can be ready to complete the sowing operation in one go.

The information the Walgett farmer needs can be easily obtained from Australian RAINMAN. RAINMAN has historical rainfall records for Walgett and thousands of other locations.

Procedure. Open Walgett in RAINMAN, and choose "Daily Forecasts" from the "Selector" dropdown window; the "Daily Forecasts" window opens. Click on "Setting"; the "Critical Rainfall Event Setting" window opens. Select:

Rainfall period start date as May 1 and end date as October 30
SOI start month March and end month April
Minimum rain in a rainfall event as 25 mm and maximum duration of rain event as three days

Close the "Critical Rainfall Event Setting" window and return to the "Daily Forecasts" window. Select:

SOI phases from the "Methods of Daily Analysis" (Any one of the four methods can be chosen.)

A table (Table 8.4) is generated (can be a graph also) that gives the information about the rainfall event at Walgett during the coming season under alternative scenarios.

Time of occurrence of the first effective rainfall event at Walgett is highly variable from season to season. From historical records it is established that, like any other place in eastern Australia, the first effective rainfall event and overall rainfall in a particular season is normally influenced by the SOI phase during the two months prior to the commencement of a season.

The average date of the first rainfall event (rain > 25 mm in three consecutive days) at Walgett is July 15. This year, SOI value in March was –7.8 and in April –8.6. These values represent a negative SOI phase in March and April. Therefore, the most appropriate forecast for the coming season is in

TABLE 8.4. Chance of first rainfall event of the May to October season at Walgett, using SOI phases in March-April (Event = 25 mm in three days; rainfall period: May 1 to October 31)

SOI phases (March to April)	**SOI falling**	**SOI negative**	**SOI neutral**	**SOI rising**	**SOI positive**	**All years**
Percent years event occurs by						
May 8	0	6	0	9	4	3
May 15	8	18	16	14	11	13
May 22	13	29	20	18	19	19
June 1	25	53	40	36	26	35
July 1	50	71	60	50	48	55
August 1	67	71	64	86	63	70
September 1	71	71	68	91	67	73
October 31	83	71	80	100	81	83
Percent years after median June 24	46	71	60	45	44	52
KS/KW probability tests	0.508	0.818	0.475	0.826	0.739	0.618
Significance level	NS	NS	NS	NS	NS	NS
Years in historical record	24	17	25	22	27	115
Longest time to event (days)	1 year +	1 year +	1 year +	1 year +	1 year +	1 year +
Shortest time to event (days)	7	3	7	1	2	1
Median date of first event	June 30	May 30	June 3	July 3	July 2	June 24
Average date of first event	July 22	July 10	July 16	June 28	July 27	July 15

Note: NS = nonsignificant result, in RAINMAN a result with probability less than 90 percent.

the "SOI negative" scenario. Values in this column indicate a higher than average chance (53 percent) of the first rain event in the coming season occurring by June 1.

MetAccess

1. A farmer at Cobar, New South Wales, wishes to reduce water losses from his farm's water reservoir through a new technique. He is looking for information on the likely period during which weekly evapora-

tion may exceed 70 mm in the coming summer. He also wants information on approximate dates by which the first event is likely to occur in the season.

The information required by the farmer at Cobar is easily generated with MetAccess. Information about MetAccess can be obtained at <http://www.hzn.com.au>. MetAccess (Serial# 2001, Version Sept99) has historical records for Cobar.

Procedure. Open Cobar from MetAccess Weather Files; choose "Outputs" from the menu bar and "Probability" from the dropdown window. The "Probability" dialogue box opens. Select:

Evaporation greater than 70 mm in a seven-day interval
Probability—simple
Period from September 1 to March 31
Graph

A graph (Figure 8.1) is generated (can be a table also) which shows during the period November 24 to February 16 more than 50 percent chance of weekly evaporation exceeding 70 mm.

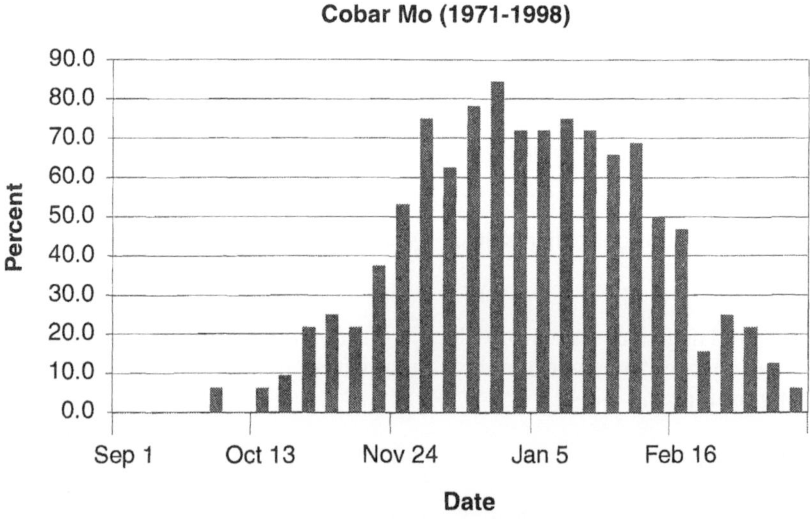

FIGURE 8.1. Probability of total evaporation greater than 70 mm over seven days at Cobar

Choosing cumulative probability instead of simple probability, a different graph (Figure 8.2) is generated. The graph shows that the probability of total evaporation exceeding 70 mm in any previous seven-day period begins to rise rapidly from the last week of October. By the first week of December, there is a 90 percent chance that the total evaporation has exceeded 70 mm for at least one seven-day period.

2. A horticulture manager at Hillston, New South Wales, observed that whenever daytime temperature suddenly rises above 36°C for two consecutive days during the period November to early December, his cherry crop is heavily damaged. To take measures to minimize this damage, he wants information on the most likely timings when such events may occur during the November to early December period of the coming season.

The information a horticultural manager needs is easily generated with MetAccess. MetAccess has historical records for Hillston.

Procedure. Open Hillston from MetAccess Weather Files, choose "Outputs" from the menu bar and then "Probability" from the dropdown window. The "Probability" dialogue box opens. Select:

Temperature greater than 36°C in a two-day interval
Probability—simple
Period from October 1 to December 31
Graph

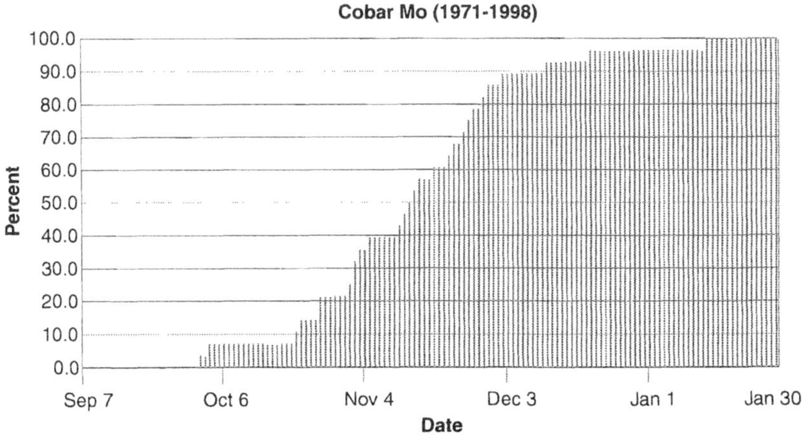

FIGURE 8.2. Probability of total evaporation greater than 70 mm in any seven-day period from September onward at Cobar

A graph (Figure 8.3) is generated (can be a table also) which shows that the chances of this event (maximum temperature exceeding 36°C for two consecutive days) cannot be ruled out altogether, even in the early days of November. The highest chance of first such event is, however, in the period November 27 to 30.

3. A farmer at Dubbo, New South Wales, suffers repeated losses in wheat crop production as his crop is subjected to severe frost at anthesis. He is interested to know the most likely date of the last frost in the coming season to readjust the sowing dates so that his crop is not caught up in a frost event when at anthesis.

The information needed by the farmer at Dubbo is easily generated with MetAccess. The DSS has historical records for Dubbo.

Procedure. Open Dubbo from MetAccess Weather Files, choose "Outputs" from the menu bar and then "Probability" from the dropdown window. The "Probability" dialogue box opens. Select:

Minimum temperature less than –2°C in a one-day interval
Probability—simple
Period from May 1 to October 31
Graph

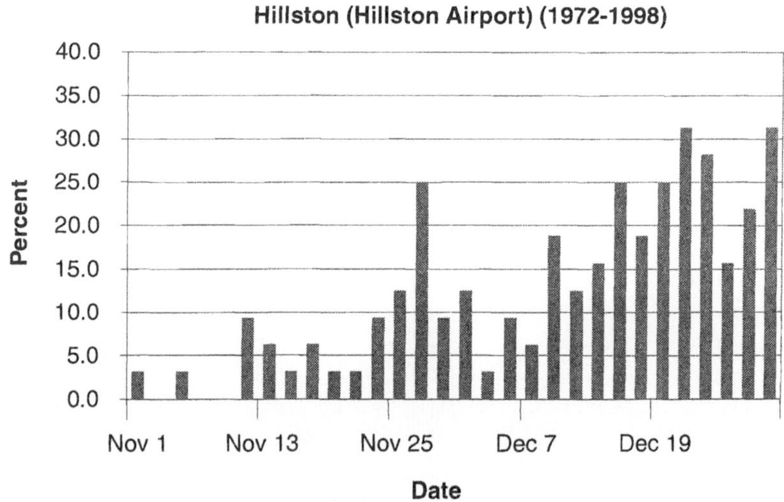

FIGURE 8.3. Probability of average maximum temperature greater than 36°C over 2 days at Hillston

A graph (Figure 8.4) is generated which reveals that the greatest frequency of frost incidence of this magnitude (temp < –2°C) is in the period from July 11 to August 25. However, the last frost can occur as late as September 14.

The anthesis time of a wheat crop that corresponds to the highest yield potential is also the period of maximum frost risk at Dubbo. As the season progresses, both frost risk and yield potential are strongly reduced. The farmer, while choosing the sowing date to avoid frost risk, should carefully consider the trade-offs between risk and production potential.

CLIMEX

1. Environment Protection Australia (EPA) has to develop long-term strategies at the national scale to control cane toad that is causing immense damage to the natural ecosystems. For this it needs the following basic information:
   a. Current known or potential spatial distribution of cane toad in Australia
   b. Potential spatial distribution under a global warming scenario

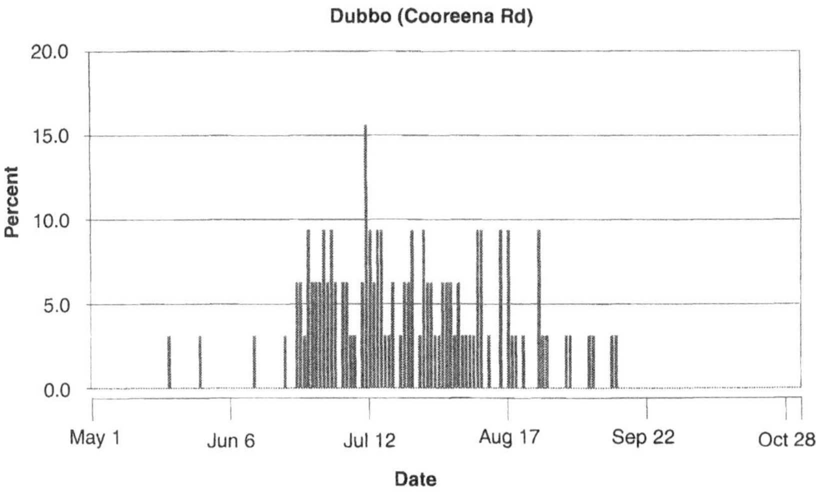

FIGURE 8.4. Probability of average minimum temperature less than –2°C over one day at Dubbo

The information required for (a) is easily available with the CLIMEX model. Information about CLIMEX can be obtained from CSIRO Entomology (<http://www.ento.csiro.au/climex/climex.htm>). The CLIMEX model (Version 1.1) has historical records for thousands of locations on all continents.

Procedure. Select "Compare Location" from the menu options of CLIMEX. The "Compare Location" dialogue box appears. Select:

Cane toad in "Species" and Australia in "Location Set"
Run model

A map (Figure 8.5a) or table can be generated. Figure 8.5a shows the locations where the cane toad has the required environment for survival and population buildup under the present climate conditions. Filled circles in the map indicate favorable locations—the larger the circle the more favorable the location. Crosses indicate locations not favorable to the long-term survival of cane toad.

The information required for (b) is also generated with the CLIMEX model.

Procedure. Select "Preference" from menu bar and then "Scenario" and "Geenhouse" from the dropdown windows. The "Select Greenhouse Scenario" dialogue box appears. Select "Edit," and the "Greenhouse Scenario" dialogue box opens. Edit greenhouse scenario maximum and minimum temperature and percent rainfall change (in this example, a 2°C rise in maximum and minimum temperature and 10 percent increase in rainfall is added, both for winter and summer). Select "Compare Location" from the menu options of CLIMEX. The "Compare Location" dialogue box appears. Select:

Cane toad in "Species" and Australia in "Location Set"
Click on "Greenhouse"
Run model

A table or a map (Figure 8.5b) is generated. Figure 8.5b shows the potential distribution of cane toad under a global warming scenario. Comparing the two maps shows that under the greenhouse effect, cane toad will spread in all coastal regions of South and Western Australia and in many inland areas of all states of mainland Australia.

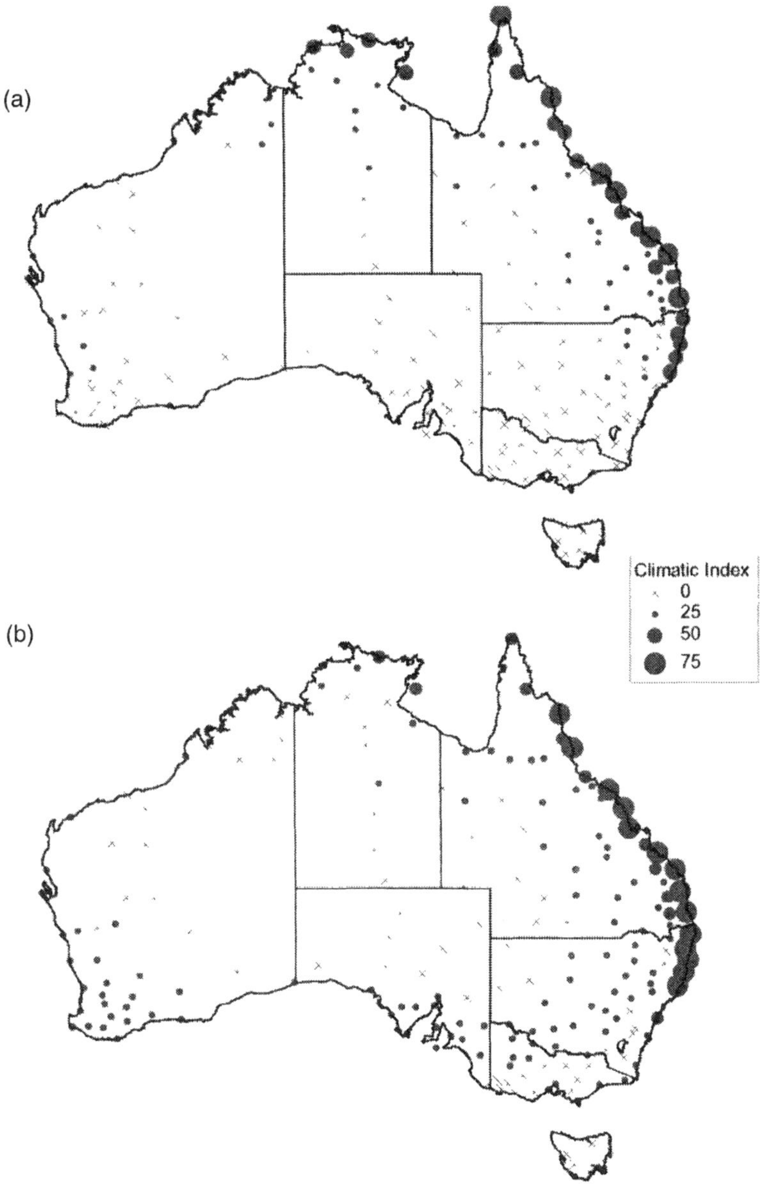

FIGURE 8.5. Cane toad in Australia: (a) present distribution and (b) distribution under greenhouse effect scenario

2. A Company based at Orange, New South Wales, is engaged in a highly profitable apple production and processing business. The company is eager to expand its business to other areas. To buy more farms on which to grow quality apples, it wants information on sites/places throughout Australia that match the climate of Orange.

The information that the company based at Orange needs can be easily generated with CLIMEX. CLIMEX has historical records of climates for thousands of locations on all continents. CLIMEX allows comparison of the average climates of different locations. This is done by measuring similarities in temperature, rainfall, and relative humidity. CLIMEX allows the operator to select a location (called "Target Location") and compare its climate with the climate of each location in a location set. The climate similarities in each pair of locations are measured by a match index between 0 and 100.

Procedure. Select "Match Climates" from the menu options of CLIMEX. The "Match Climates" dialogue box appears. Select:

Orange in "Target Locations" and Australia in "Matching Location Set"
Run model

CLIMEX generates a map or a table (Table 8.5) that shows 228 places having some sort of similarity with Orange in terms of individual weather elements and overall climatic conditions. Table 8.5 provides the information that the company based at Orange required. It shows the 15 top places (out of 228) in Australia that are more than 70 percent similar to Orange in terms of climate.

Use of Decision Support Systems (DSS)

Much effort and money have been invested into the development of decision support and expert systems throughout the world. However, the rate of adoption by farmers does not seem to keep pace with the development of the systems (Kuhlmann and Brodersen, 2001). The adoption of decision support systems in developing countries at the individual-farmer level is negligible and is likely to remain so at least in the near future. At the dawn of the twenty-first century, not even a small portion of the farming community of developing countries is aware of computers and their use in agriculture. Illiteracy and abject poverty are the most significant impediments. The small size of farms will remain a major obstacle to the adoption of computers and

Table 8.5. Locations in Australia with climatic conditions similar to those of Orange (match index range 0 to 100)

Continent	Country	Location	Total Oceania Australia New South Wales Orange	Maximum Temperature	Minimum Temperature	Rain Amount	Rain Pattern
Oceania	Australia	Taragala	86	89	97	90	87
		Tumba-rumba	80	72	91	92	85
		Launceston	76	81	81	88	81
		Glen Innes	75	66	79	94	82
		St. Helens	74	69	79	88	84
		Colac	73	73	71	92	81
		Armidale	72	63	80	89	82
		Goulburn	72	75	83	77	85
		Heywood	72	70	70	95	78
		Warragul	72	73	68	86	85
		Bathurst	72	71	90	72	89
		Cann River	71	60	76	90	84
		Canberra	71	72	90	72	87
		Tenterfield	70	56	72	95	81
		Bright	70	59	94	76	84

decision support systems. Furthermore, each developing country has uniquely local needs and uniquely local solutions to farm problems.

A limited number of systems appear to have been adopted for regular use in the agricultural industries of developed countries. Reasons for the poor rate of adoption have been highlighted in many studies (Parker, Campion, and Kure, 1997). In the United Kingdom, the reasons given for the lack of uptake of model-based systems are (1) lack of a computer base among the population; (2) system complexity; (3) use of inputs that the grower cannot easily provide; and (4) failure to show cost benefits. As a consequence, few DSSs launched in the United Kingdom during the 1990s made any impact on the industry (Parker, 1999).

In the Netherlands, many attempts have been made at the government level to introduce knowledge-based systems on farms, but dissemination speed is very low. Extension services tended to slow down the dissemination of DSSs, rather than promoting these products, for fear of competition from these systems. Furthermore, the use of knowledge-based systems by advisers is still quite low. Experiences (not quantified) also show that the use of these systems by farmers tends to result in extra work and a higher level of support required (Kamp, 1999).

Computer and DSS adoption studies in the United States (Ascough et al., 1999) commonly found that higher levels of farm size, farm income or sales, land ownership (tenancy), and education had positive effects on computer adoption. Increased age had a negative effect. The most frequently cited reasons for lack of adoption were high cost, lack of confidence or skill, not enough time, and small farm size. Other factors shown to have an impact on adoption were farm complexity, debt-asset ratio, exposure or perception that risk is important, and farm type (crop or livestock).

An Australian study (Lynch, Gregor, and Midmore, 2000) revealed that out of the 34 systems for which the information was maintained, only five have been in use. That is, 85 percent of the systems registered were not in use. Stubbs, Markham, and Straw (1998) examined attitudes and perceptions of farmers across five states of Australia as to how they view the computer as a tool in their decision making. Their main findings were

1. for many farmers the computers were seen as time wasters;
2. the majority of farmers are of the noncomputer generation and may see no reason to change their current habit of bookkeeping;
3. for many producers with small holdings, they could not justify the cost in terms of money and time;
4. many failed to see any benefit; and
5. determining which type of computer to buy and what software to use was a major obstacle for many farmers.

From these conclusions it appears that intelligent support systems are not particularly compatible with the current practices or attitudes of farmers.

P. T. Hayman and W. J. Easdown (personal communication) enumerated physical, economical, sociological, and farm management factors that have reinforced or hindered the adoption of WHEATMAN in the northern grain belt of Australia. These factors are also applicable to the other decision support systems. The reinforcing factors are the rapidly increasing access to powerful PCs on farms; the optimism of government agencies and willingness to substantially support DSS development and extension; development of DSSs with a team approach with active involvement of the end users; increasing ease of use of computers and development of user-friendly software; pressure on grain farmers to increase productivity as their profit margin is squeezed; and climate risk forcing careful decisions based on scientific information.

There are numerous factors on the limiting side. Of the farmers who own PCs, it is estimated that less than 22 percent are using them for farm management. The process for testing and releasing early versions of the programs leaves it open to criticism. Positive responses from naive end users create potentially unrealistic perceptions of a program's utility.

Results of surveys (Lewis, 1998; Ascough et al., 1999) suggest that individuals and organizations interested in the promotion of a DSS may enhance the success of its diffusion by

1. targeting farm businesses that already operate manual farm management information systems,
2. transferring appropriate information and knowledge to establish a farm record system that provides management information prior to DSS adoption,
3. targeting young primary industry decision makers who have a relatively high demand for management information to compensate for their relative lack of farming experience,
4. targeting those farm businesses in which spouses provide support in farm management, and
5. targeting farms with higher sales, larger acreages, and more enterprises (both cropping and livestock systems).

Such farms should experience more net benefit in DSS adoption than smaller farms. It is also suggested (Lynch, Gregor, and Midmore, 2000; Cain et al., 2003; Dorward, Galpin, and Shepherd, 2003) that in terms of software development, involvement of the end users in the decision-making process and close participation of the marketing organization are also crucial factors that can influence the acceptance and adoption of the software.

In developing countries, adoption of DSSs at the individual-farmer level is likely to remain at a slow pace until simple, rugged, dust-resistant, and low-cost devices are available. Governments can encourage the adoption of computers and decision support tools by purchasing these in bulk and then selling them to young, educated farmers at a discounted price. DSSs could also be given to village councils for use in community halls or village council offices. Agricultural extension staff could play a major role in disseminating the DSS-derived information to groups of farmers.

Chapter 9

Agroclimatological Services

WEATHER AND AGRICULTURE

Weather plays the dominant role in farm production. Weather is always variable, and farmers have no control over this natural phenomenon. Climate variability persisting for more than a season and becoming a drought puts great pressure on land and vegetation. Normal land-use and management systems become imcompatible with prevailing climate, and farm production is drastically reduced. Abnormalities such as drought and associated farm losses are not very frequent, but losses due to short-term climate variability and sudden weather hazards such as flash floods, untimely rains, hailstorms, and severe frost do occur year after year. Losses in transport, storage, and due to parasites, insects, and diseases are the indirect results of abnormalities in weather conditions and are a recurring feature. It has been estimated (Mavi, 1994) that, directly and indirectly, weather contributes to approximately three-quarters of annual losses in farm production.

Complete avoidance of all farm losses due to weather factors is not possible. However, losses can be minimized to a considerable extent by making adjustments through timely and accurate weather forecast information. When specifically tailored weather support is available to the needs of farmers and graziers, it contributes greatly toward making short-term adjustments in daily farm operations, which minimize input losses and improve the quality and quantity of farm produce. The seasonal weather outlook also provides guidelines for long-range or seasonal planning and selection of crops and varieties most suited to the anticipated weather conditions (Mjelde et al., 1997).

WEATHER AND CLIMATE FORECASTING

Three types of weather forecasts are prepared by the weather forecasting agencies in most of the countries of the world. These are the short-range forecast valid for 48 hours, the medium-range or extended forecast valid for five days, and the long-range or seasonal forecast valid from a month to a season. Each of these forecasts has a role to play in agriculture. Whereas short-range forecasts are most valuable in daily farm operations, medium-range and seasonal forecasts are important in longer-term farm operations and planning. Based on these forecasts, farmers can make the best use of favorable weather conditions and adjustments can be made for adverse weather.

Short-Range Weather Forecast

A short-range weather forecast is based on a detailed analysis of the physical processes occurring in the atmosphere. It incorporates information about current weather conditions and forecast information on high and low temperatures, wind velocity and direction, time and amount of precipitation, relative humidity, sunshine duration, and sudden weather hazards. This forecast information is available through television, radio, and newspapers and via the telephone from the forecasting agencies. The information is sufficiently accurate and can be effectively used for many field operations including spraying, hay making, sheep shearing, nitrogen top dressing, and preventing damage from frost.

Extended Weather Forecast (Up to Five Days)

The basis for preparing extended forecast information is similar to that of the short-range forecast, but the forecast is not very detailed. An extended forecast contains generalized information including change of weather type, sequence of rainy days, extended wet and dry spells, and general weather hazards such as cold and heat waves. The forecast information is sufficiently accurate and available from meteorological centers. In Australia, the National Climate Centre (NCC) and Special Services Units of the Bureau of Meteorology prepare extended forecasts. The extended weather forecast is most effective and useful in agriculture as it gives sufficient lead time for both planning and executing farm operations.

Seasonal Climate Outlook or Long-Range Weather Forecast

A seasonal climate outlook or long-range weather forecast is essentially a statistical product relating past climatic data with phenomena such as Southern Oscillation Index and sea-surface temperature. Of late, coupled ocean-atmosphere general circulation models (OAGCMs) are being increasingly used to make long-term forecasts by modeling the circulations and interactions of the ocean and the atmosphere. The seasonal forecast emphasis is on abnormalities in rainfall and temperature. Seasonal forecasts are prepared in every country by its national meteorological center. In Australia, the Bureau of Meteorology and Queensland's Department of Primary Industries prepare seasonal forecasts. The products are available at the Internet sites of the respective organizations and can be obtained by fax as well.

While using the forecast information, it is important to bear in mind that weather forecast accuracy is inversely related to the lead time of the forecast. The shorter the lead time, the greater the accuracy of the forecast. Weather forecasts for longer time spans become more and more generalized, and their accuracy decreases as the lead time increases. This happens because regional-scale changes in atmospheric patterns occur suddenly, which cannot be accounted for in the methodologies used for making long-range forecasts. A 24-hour forecast is more accurate and comprehensive than a 48-hour forecast. A five-day forecast is less accurate and less specific than a 48-hour forecast. Similarly, a long-range or a seasonal forecast is much more generalized and less accurate than a five-day forecast.

TAILORING CLIMATE INFORMATION FOR AGRICULTURE

There are excellent sources of information on general weather, and this information is readily available. Generalized forecasts have, however, limited use in farming. Weather information for agriculture needs to be tailored to meet the needs of farmers and graziers (see Figure 9.1). It should not be a repackaging of the general weather forecast of the national forecasting centers. It should be a tailored product that can be effectively used in growing crops, managing animals, and controlling pests and diseases. A comprehensive agroclimatological forecast or a farm advisory is an interpretation of how expected weather in the future and weather conditions accumulated to the present will affect crops, livestock, and farm operations.

An agroclimatological forecast usually has five components: weather synopsis, interpretation of weather for crops, interpretation of weather for farm operations, interpretation of weather for livestock, and interpretation of weather for crop pests and diseases.

- *Weather synopsis:* This is the description of locations and movements of low pressure systems, high pressure systems, upper air troughs, fronts, and associated weather with these systems. This information is derived from synoptic observations, prognostic charts, and visible and infrared imageries from meteorological satellites. For seasonal forecasts the inferences are drawn from historical data, sea-surface temperature, SOI values and phases, and other relevant tele-connections.

- *Interpretation of weather on crops:* Interpretation of weather conditions on crops takes into account the impact of weather on germination, growth rate, freeze protection, and irrigation demand. The cumulative effect of weather encountered and anticipated is used to determine dates of harvest, duration of harvest, and quality and storage capabilities of grains, fruits, and vegetables.
- *Interpretation of weather on farm operations:* Interpretation of weather on farm operations takes into account the drying rate of soil, evaporation losses, effect of heat, cold, and wind on applications of chemicals and fertilizers, and the drying rate of curing, wetting, and rewetting grains and hay.
- *Interpretation of weather on livestock:* Various combinations of heat and moisture in the atmosphere cause comfort or discomfort to animals. Indices are available that express the combined effects of temperature and humidity on animals. The indices provide indications of heat stress, cold stress, shelter requirements, and the effect of weather on meat, milk, and egg production. These indices are used to give timely warnings of anticipated weather dangerous to the health and safety of livestock.
- *Interpretation of weather for crop pests and diseases:* A close relationship exists between many animal and plant diseases, insect pests, and weather. The incidence of these diseases and pests is forecast in the light of accumulated and anticipated weather. Simulation, synoptic, and statistical techniques are used for forecasts which pertain to the probable development, intensity, spatial and temporal spread, or suppression of diseases.

IMPACTS OF WEATHER ON SPECIFIC INDUSTRIES AND THE ROLE OF FORECAST INFORMATION

An overview of the necessary decisions and the associated climate information required by agricultural industries is presented in Table 9.1 (p. 225). This table is based on information from a large number of personal communications, publications, and written comments from those engaged in specific industries (Mavi, 1994; O'Sullivan, D. B., personal communication).

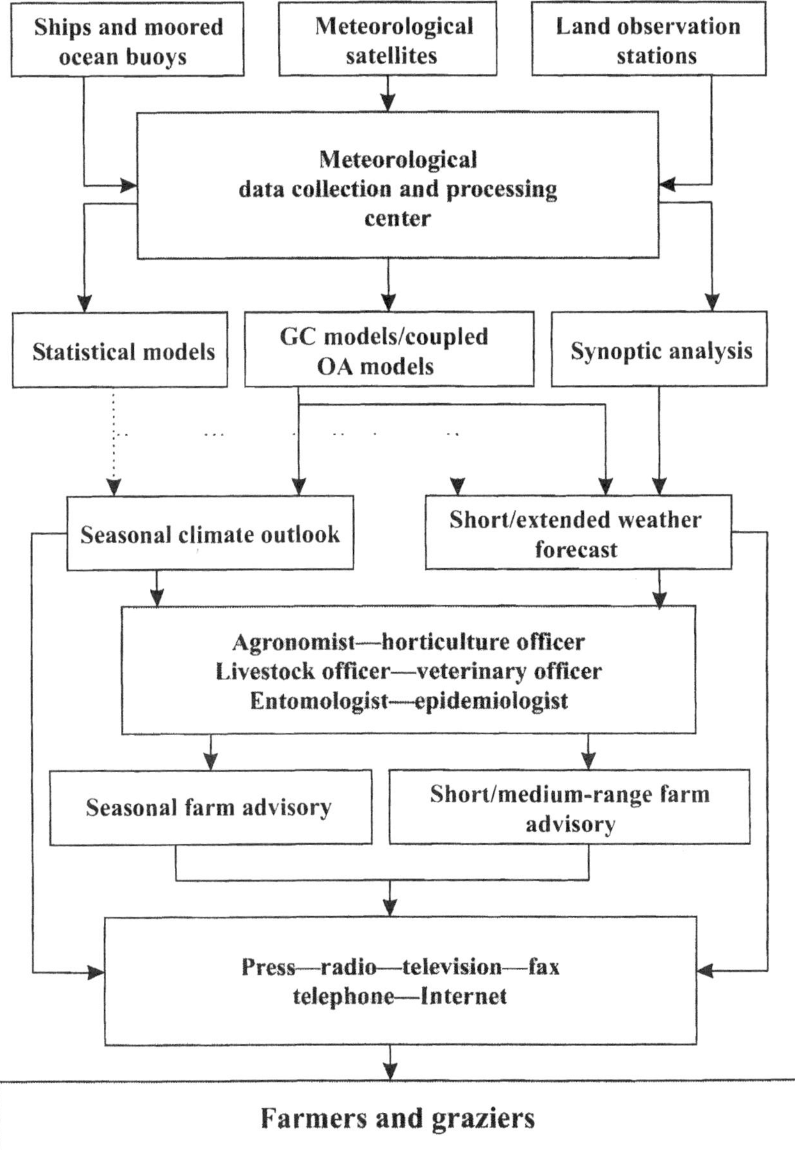

FIGURE 9.1. Weather information flow to farms

AGROCLIMATOLOGICAL INFORMATION SERVICES IN AUSTRALIA

Bureau of Meteorology

The principal information base for describing the climate of Australia, its variability, and its long-term trends is provided by the Bureau of Meteorology (1997a) which has national responsibility for meteorological (including climate) monitoring. It operates the official national climate observing infrastructure and the National Climate Centre, the latter being the custodian of Australia's historical climate records.

Short-term weather and seasonal climate forecasts are made available through the media, including newspapers, radio, television, and the World Wide Web (Bureau of Meteorology, 1997b). The former include the daily and three- to five-day rainfall and temperature forecasts released every few hours, along with specialized forecasts concerning inclement weather that threatens horticultural and agricultural crops and the survival of livestock (especially shorn sheep and newborn lambs and calves). Frost risk forecasts provide information on overnight minimum air temperatures of –2°C or lower when it is expected over significant areas. This information is available in all the states from June to mid-August.

The National Climate Centre currently provides a wide range of climate-related services and products, including data, maps, predictions, and consultative services (Beard, 2000). The most important agriculture-related products are three-month seasonal outlooks of both total rainfall and seasonal average maximum and minimum temperatures, together with enhanced information on the Southern Oscillation Index and the likelihood of El Niño or La Niña events.

In addition to the general weather forecast and the products from the National Climate Centre, the Bureau of Meteorology also issues specialized weather forecasts (FARMWEATHER) for agriculture through its Special Services Units located at each state capital. FARMWEATHER is a detailed rural forecast available on demand. It combines a weather graphics page, a recent satellite picture, together with an expert opinion composed by meteorologists in plain English. The information describes the weather outlook to four days ahead for particular regions, allowing effective short-term management decisions to be made by farmers. FARMWEATHER is available via fax for more than 20 regions throughout mainland Australia.

State-Sponsored Agroclimatological Services

The extension and advisory services of the various state departments of agriculture and primary industries provide advice, backed up by ongoing scientific research, to assist producers manage their properties and farm businesses while exposed to a variable climate and various forms of risk.

NSW Agriculture provides climate services for producers throughout the state. Seasonal rainfall and weather outlook for the next three months is an important component of the *Regional Review,* a monthly update of seasonal conditions and outlook for agriculture in New South Wales. Farmers and other users can look at climate information on the department's external Web site in the *Regional Review.* Climate workshops are one of the most important climate information delivery channels of the NSW Agriculture Department.

NSW Agriculture also runs an irrigation and disease forecast service for the fruit growers in Northern Rivers region of the state. It provides quantitative data on evaporation, rainfall, and soil and air temperature for six stations in the region pertaining to the past week. Other information is on water used by the various fruit trees, with an advisory for irrigation. A section of the advisory is on fruit tree disease status, warnings, and advice on spray applications (NSW Agriculture, 1997). In southern NSW, the Riverwatch service is designed for updated information on the state of the Murrumbidgee, Tumut, and Murray Rivers pertaining to height, trend, temperature, salinity, and turbidity for the next four days (Bureau of Meteorology, 1997c).

The Queensland Centre for Climate Applications (QCCA) is a joint venture of the Department of Primary Industries (QDPI) and Department of Natural Resources and Minerals (QDNRM). It provides a range of climate services and tools to farmers (Balston, 2000; Bureau of Meteorology, 2001).

The "Long Paddock" is the QDPI/QDNRM climate Web site and contains information on rainfall, rainfall probabilities, sea-surface temperatures, Southern Oscillation Index, the seasonal climate outlook, and drought status. The SOI phone hotline consists of a two-minute recorded SOI message-containing information on the SOI, recent rainfall, SSTs, seasonal rainfall outlook, and ENSO status, updated weekly.

The Aussie GRASS project outputs are produced regularly for Queensland, New South Wales, South Australia, Western Australia, and the Northern Territory. These include recent and current pasture production conditions relative to previous seasons. Seasonal climate outlook indicators include the average SOI, variations in sea-surface temperatures, forecast rainfall, pasture conditions, fire risk, and curing index, updated monthly. Most Aussie GRASS products are available on the Long Paddock Web site.

In addition to these services, conducting climate workshops for farmers is an important activity of the Queensland Department of Primary Industries.

The South Australian Research and Development Institute (SARDI) has developed practical climatological information resources, services, and tools to assist land managers to understand and manage the effects of climate variability on their farming enterprises (Truscott and Egan, 2000). The tools and services SARDI provides or supports include climate risk management workshops for delivery across southern Australia, and climate risk management decision support trials incorporating the Climate Risk and Yield Information Service.

In Western Australia, climate services provided by Agriculture Western Australia (AGWA) have focused on the development of decision support tools that enable farmers to prepare and respond to climate variability. Modeling of agricultural production has progressed to the point at which farmers can use management tools to assist decision making, or utilize the delivery of timely information on crop development and yield potential (Tennant and Stephens, 2000). Decision support tools to respond to climate variability that have been developed in AGWA include TACT, MUDAS, PYCAL, NAVAIL, STIN, SPLAT, and FLOWERCAL. Most of the modeling developments have been used largely within western Australia, but some have been extended to other states.

The information service is a weekly "fax back" delivery system that provides information on stored soil water, the progress of the season, expected yields, and other information to participating farmers in western Australia. The STIN model is used to produce soil moisture (at seeding) and wheat yield forecasts for every wheat-growing shire in Western Australia. This is mapped on a monthly basis and supplied to ProFarmer which distributes their magazine to major grain trading, marketing, and transport agencies. Output can also be accessed from the AGWA web site.

Private Agroclimatological Services

Several private weather forecasting services provide advisories to farmers and some other sectors of economic activity (Anonymous, 1996; Lyon, 1997; Jones, 1997). These forecasting services use similar techniques as those of the Bureau of Meteorology to prepare the forecast information. There are also private agricultural, environmental, and natural resource management consultants who are skilled in advising on how to cope with short-term and seasonal variability in climate, production, and product prices. Many are registered with the Australian Association of Agricultural

Consultants (AAAC), a section of the Australian Institute of Agricultural Science and Technology.

USE AND BENEFITS OF CLIMATE FORECAST INFORMATION

The National Farmer's Federation of Australia, in its publication *New Horizons* (White, Tupper, and Mavi, 1999), endorses the view of many farmers that improved seasonal forecasting is a high research priority to assist them in managing their properties. This has also been highlighted in several surveys (Stone and Marcussen, 1994; Elliott and Foster, 1994; Nicholls, 1985). Managers of water and other climate-sensitive sectors of the economy also claim that they would like to see significant advances in skill levels and lead times in seasonal forecasting (Albrecht and Gow, 1997). Another survey conducted by QCCA (Paul, Cliffe, and Hall, 2001) revealed that many graziers do use the forecasts to aid their stocking and stock-trading decisions, even though the reliability of forecasts remains an issue in many areas. Farmers in Queensland have certainly reacted to adverse SOI information by sending cattle to market, thereby reducing stocking rates on their properties.

Surveys of grain growers in New South Wales and Queensland have shown that farmers have used four-day weather forecasts to plan their sowing and spraying operations. They have also used frost risk information to switch crops and crop cultivars, and they have used the seasonal rainfall outlook to increase their nitrogen application rates and the area sown to crop. Meinke (2000) has cited specific examples in which farmers and several agencies in Queensland use the model-based information.

The Queensland University of Technology surveyed (Hastings and O'Sullivan, 1998) primary producers, cattle producers with some dairy farmers, croppers, and others in agricultural production in southeast Queensland. The aim was to gauge producer opinions of the impact of seasonal climate patterns and seasonal forecasting. Two other surveys to get feedback on the needs and use of climate information were conducted in New South Wales (Albrecht and Gow, 1997; Crichton et al., 1999). Combined and generalized, the surveys revealed that producers are vitally interested in climate information and predictions of important weather parameters such as rainfall and frost. There is a large need for relevant and user-friendly information about climate in rural Australia. The surveys suggest that there is room to improve official forecasts to build more confidence and also to establish a better understanding of official forecasts.

Many investigations are available in which economic benefits from agroclimatological services are quantified (Adams et al., 2003). Additional but unquantified economic benefits of agroclimatological advisories are through checking land degradation from wind and water erosion and decreasing environmental pollution from fertilizer leaching and chemical spray drifts.

The value of the weather forecast depends on the ability of the user to effectively translate such information into economic values and profit margins at the individual-farm level. Katz and Murphy (1997) have cited studies showing savings from frost forecasts for orchards in the range from $US 667 to $1,885 per hectare. In maize production the savings range from $17 to $58 per hectare; in wheat production, a perfect forecast resulted in a savings of $196 per hectare; and in grape production, an accurate three-week forecast resulted in a net profit of $225 per hectare.

The value of a seasonal outlook depends on the skill or accuracy of the forecast and its marginal value relative to other readily available sources of information to the manager of a particular production system. Effective application of seasonal climate forecasts of reasonable accuracy leads to decisions that generate improved outcomes. To be effective, however, the decision changes must produce positive changes in value by improving the relevant aspects of targeted performances. If the information is ignored or it does not lead to changed decisions, it has no economic impact or value (Freebairn, 1996). If the forecast is inaccurate, then the information is likely to have a negative value in the current season.

In a research study, the Kondinin Group has looked at the accuracy and reviewed the usefulness of seasonal climate forecasts for on-farm decision making in southern and western Australia (Buckley, 2002). The study revealed that long-term climate forecast models can predict rainfall for three-month periods with accuracy levels that are better than a guess. Most of the models were more accurate with a lead time of zero and one month. Longer lead-time forecasts were not accurate enough to use for on-farm decision making. Furthermore, the forecast accuracy is very low during the critical time of autumn, which means climate forecasts are best used as only a small component of the farm decision-making process.

The benefits of seasonal forecasts vary between industries and across regions (Hammer, Carberry, and Stone, 2000). Soils and vegetation exposed to high climate variability in pastoral areas can benefit through destocking in advance of drought so as to avoid overgrazing, stock losses, and accelerated erosion. Crop producers can assess whether to sow or fertilize a crop if the chance of a harvest is significantly diminished. Demands for irrigation water can be better estimated.

The value of seasonal forecasts to crop producers can be significant, but it varies with management and initial conditions, as well as with cropping systems and location. The forecasts can influence decisions on what crop, when and what area to sow, and whether to irrigate and/or fertilize a crop. Hammer and colleagues (2001) have cited case studies of how the applications of climate prediction at field/farm scale to dryland cropping systems in Australia, Zimbabwe, and Argentina have improved the profits of the farmers.

In the northern part of the Australian grain belt, significant increases in profit (up to 20 percent) and/or reduction in risk (up to 35 percent) can be achieved with wheat crops based on a seasonal forecast available at planting time (Hammer, Holzworth, and Stone, 1996). This can be achieved through tactical adjustment of nitrogen fertilizer application or cultivar maturity, with significant financial benefits (Marshall, Parton, and Hammer, 1996).

Petersen and Fraser (2001) suggest that a seasonal forecasting technology which provides a 30 percent decrease in seasonal uncertainty increases annual profits of the farmers in Western Australia by about 5 percent. In northwestern Victoria, if the seasonal forecast suggests adequate soil moisture in October, then a sunflower crop can be sown with a high probability of a good harvest (Jessop, 1977). In a similar way, seasonal forecasts can be used to determine whether a particular cereal, oilseed, or legume crop should be sown, based in particular on the probability of a favorable harvest.

The El Niño-Southern Oscillation has a dominant effect on climate in a number of the world's large-scale crop production areas. The SOI information contributes some skill to improving management decisions in Australia (Carberry et al., 2000). By changing between fallow-cotton, sorghum-cotton, or cotton-cotton rotation based on SOI phase in the August to September period preceding the next two summers, the average gross margins for the two-year period increased by 14 percent over a standard fallow-cotton rotation. At the same time, soil loss from erosion was reduced by 23 percent and cash flow was improved in many years. Clewett and colleagues (1991) used a crop model to show that growing crops in seasons with a strongly negative SOI before planting were unprofitable, compared to seasons with a strongly positive SOI before planting. SOI data can therefore be used to adjust the management strategy according to the level of climatic risk.

Dudley and Hearn (1993) used a SOI model to examine irrigation options for cotton growers in the highly variable, summer rainfall environment of northern NSW. The study demonstrated that if irrigators knew the current SOI before the commencement of each cotton season, more profitable timing of investment in plant and equipment might result. These benefits might be extended to suppliers of farm inputs and to processors.

Rangelands in the eastern half of Australia are particularly sensitive to the climatic events of ENSO, with consequences for stocking rate and land degradation. A policy of reducing stocking rate on the basis of El Niño forecasts can significantly reduce environmental degradation in adverse seasons (McKeon and White, 1992; Stafford Smith et al., 1996; Clewett and Drosdowsky, 1996).

Bowman, McKeon, and White (1995) examined the value of seasonal outlooks to wool producers in northern and western Victoria, assuming forecast accuracy for the next 12 months of 60, 80, and 100 percent. They concluded that the more accurate the seasonal forecast, the better was the long-term financial performance of the farms through reduction in livestock deaths and protection of the natural resource base.

TOWARD OPTIMUM UTILIZATION OF CLIMATE INFORMATION AND FORECAST PRODUCTS

Anthropogenic climate change, climate variability, and environmental degradation issues are among the big challenges of the twenty-first century. Greater responsibility has been imposed on farmers for climate-related risk management, and they must increasingly rely on climate forecast information for operational and strategic decision making. Advance warning of hazards and extreme climate anomalies at different time scales is therefore extremely important for them. Such early warning information can also form a crucial component of national/regional disaster preparedness systems, which will help to minimize loss of life and property, including damage to agricultural investments (Ogallo, Boulahya, and Keane, 2000). Apart from the traditional weather information, agricultural systems would benefit from the following, among many others.

Agrometeorological Database

Crop-weather as well as animal-weather relationships are derived from historical records of both climate and agriculture. Such records are also used in deriving the basic statistics and risks that may be associated with any climate-based planning and operational decisions (Doraiswamy et al., 2000). Availability of long-period, high-quality climate and agricultural records are therefore crucial for maximum application of climate information and prediction services in agricultural planning and operations. For some agroecological regions, such records are not available.

The length and quality of the climate and agricultural records are key issues that should be addressed, as they provide the information base in any efforts to optimize applications of climate prediction products in agricultural planning and management. User-specific computerized databases in acceptable formats need to be generated.

Real-Time Climate Information

Many agricultural operations, services, and research studies require real-time weather information on a daily, weekly, or ten-day basis. This information can be generated through an efficient network of agrometeorological stations, which at this time is very poor in most countries (Gommes, Snijders, and Rijks, 1996). Wherever the weather stations are available, most of them do not follow the pattern of agroecological zones. Weather stations installed and maintained by the meteorological departments in various countries are usually located near towns and at airports, where recorded observations are not representative of the agricultural landscape (Ogallo, Boulahyab, and Keane, 2000).

Research

Optimum utilization of any climate prediction product in agriculture requires applied agrometeorological research with two basic components: interdisciplinary research and multiscale research (Hatfield, 1994). The topics include understanding of the local climate/agricultural systems and the associated linkages, especially with respect to extreme events, climate, and pest/disease linkages, and adaptation of agricultural systems to local climate variability. Improved and integrated data sources and interpolation methods, locally validated crop models, and regional numerical forecast models are realistic and attainable goals for the near future.

Enhanced research efforts are required on the determination of the scale and time at which seasonal predictions are suitable for application to agriculture and the environment and on the connection between the past and present weather and the upcoming predicted season.

Downscaling Short- and Medium-Range Weather Forecasts

The science and technology of short- and medium-range weather forecasting with computer models are now quite advanced. Availability of operational short- to medium-range weather forecast products is increasing day by day. For such products to be more useful and effective in agricultural applications, they must be downscaled to the regional, local, and ultimately individual-farm levels. However, most regional/local downscaling techniques require a good knowledge of regional/local climate processes. This knowledge is highly inadequate due to serious limitations of basic local meteorological data and research.

Downscaling forecasts to a local level is one of the most difficult tasks ahead. Several downscaling techniques have been developed in recent years (Von Storch, Zorita, and Cubasch, 1993; Hughes and Guttorp, 1994; Zorita et al., 1995; Kidson and Thomson, 1998). However, much more effort is needed to achieve the desired goals.

Seasonal Climate Prediction

Improvement in seasonal climate prediction is one crucial factor that could reduce the vulnerability of agricultural systems to severe impacts of extreme interannual climate anomalies. The science and technology of climate prediction within monthly, seasonal, to interannual time scales is still young and is currently under intensive investigation worldwide. The last decade of the twentieth century, however, witnessed a major advance in understanding the predictability of the atmosphere at seasonal to interannual time scales (Palmer and Anderson, 1993; National Research Council, 1996; Carlson, 1998). El Niño and Southern Oscillation are some of the known key drivers to interannual variability and have been associated with worldwide extreme climate anomalies, including changes in the space-time patterns of floods, droughts, cyclone/severe storm activity, and cold and heat waves. For some of these, agricultural application models have been developed which transfer projected ENSO signals directly into agricultural stress indices (Nicholls, 1985; Cane, Eshel, and Buckland, 1994; Glantz, 1994; Keplinger and Mjelde, 1995; Hammer, Holzworth, and Stone, 1996; Mjelde and Keplinger, 1998).

Fast development in computer software, communication technology, and advances in climate science during the past few decades suggest that useful model-based seasonal forecasts are possible in the near future (Serafin, Macdonald, and Gall, 2002). Results from computer models have demonstrated that it is possible to predict sea-surface temperatures and El Niño over time scales extending from a few months to over one year.

At present, numerous impediments are obstructing the optimal use of seasonal forecasts. Nicholls (2000) has reviewed these impediments and has suggested strategies to overcome these problems so as to improve the use of seasonal forecasts. The challenge to improve climate predictions for seasonal to interannual scales has been taken in the WMO program known as the Study of Climate Variability and Predicability (WMO, 1997a,b). It needs to be addressed at national levels as well.

Skilled Multidisciplinary Human Resources

The interdisciplinary nature of agrometeorological services is a weakness that has to be addressed (Hollinger, 1994). At present, skilled multidisciplinary human resources for integrated agrometeorological applications are relatively limited. If an agricultural meteorology scientist alone has to deliver the most effective products to users, then he or she must be fluent in both biological and physical sciences, so as to look at the world from a different perspective than the physical or biological scientist. There is a great need to strengthen and equip national and regional climate and agrometeorological institutions/units with human resources with multidisciplinary training.

Tailored Products

The perspectives of many meteorologists are based on long-standing traditions about the type of information expected by their agricultural clients (Seeley, 1994). There is a need to address the climate information requirements of specific sectoral agricultural users so that climate prediction centers can produce custom-tailored products. Information has value when it is tailored and disseminated in such a way that end users get maximum benefit from applying its content (Weiss, Van Crowder, and Bernardi, 2000). Areas of agricultural expertise that have prospered throughout the years are those with a product that is wanted and used in agricultural production. The future will see increased availability of real-time, high-resolution weather data. Opportunities for agricultural meteorology services will grow dramatically if agricultural meteorologists meet the challenge of making custom-tailored products, defined and presented in their clients' language, to meet their precise needs, and educate agricultural producers in using weather data in a variety of management decisions (Perry, 1994). Rijks and Baradas (2000) suggested that the identification of clients' needs could be made through a process of listening to people in the industry and through dialogue about the issues that could make their work safer, easier, and more reliable.

Forecast Services and Users' Interface

There is an overexpectation of forecast accuracy among users. The common perception is that both long- and short-range forecasts are not reliable enough to use in decision making (Crichton et al., 1999). The difficulty is to convince the users what forecast accuracy is attainable with the current state of the art. It is crucial that farmers have good knowledge of the skill and limitations of any climate prediction products. To achieve this, agroclimatologists have to take a more proactive role than they have at present (Blad, 1994). Extension education programs are needed to educate agricultural producers about agrometeorological products and the skill and limitations of any climate prediction product (Stigter, Sivakumar, and Rijks, 2000).

To conclude, reducing the risk associated with increased climate variability has a high potential for increasing productivity and quality while protecting the environment. Agroclimatological services generate the possibility of tailoring crop and animal management to anticipated weather conditions either to take advantage of favorable conditions or to reduce the effects of adverse conditions.

TABLE 9.1. Role of weather/climate forecast information in key decisions in farm industries

Industry	Key decision	Why weather and climate information is important	Climate/weather information required	Strategies to reduce losses/enhance profits
Management	Buying new property	Debt taken in unfavorable weather conditions can make repayment difficult.	Historical records of rainfall, wind, temperature, and frosts	Buy only if climate is favorable for the enterprise. Avoid areas that have high recurrence of drought, floods, and frost.
	Investment in new machinery	Purchase/hire of high-cost machinery requires good weather for maximum income to ensure easy repayments.	Seasonal climate outlook	Make large purchases in seasons when the outlook is normal or better than normal.
	Seasonal planning	Warmer weather conditions may cause crops to mature early. Excessively wet season requires planning for control of weeds, insect pests, and diseases.	Seasonal climate outlook	Book labor and contractors earlier to harvest crops.
	Managing labor and equipment	Labor and machinery will not be efficiently deployed under an extreme combination of high temperature and high humidity or low temperature and strong winds. A combination of temperatures of 30°C and above, coupled with 70 percent or higher relative humidity, causes discomfort for humans. Wind chill: Wind chill stress on the human body occurs when temperatures are very low and strong winds are blowing. The convective heat loss from the body becomes painfully extreme. Exposure to wind chill in wet clothing is most dangerous.	Short-range weather forecast	A forecast of mild and fair weather indicates that the entire period offers excellent conditions and maximum hours for field operation. Utilize such days for operations in which long ninterrupted working hours are required. Avoid deploying labor for a field operation in a period of extremely hot humid weather in summer and wind chill periods in winter. Choose an alternative operation for which a minimum of human labor is required. Make the best use of machinery and labor in mild and dry weather.

TABLE 9.1 *(continued)*

Industry	Key decision	Why weather and climate information is important	Climate/weather information required	Strategies to reduce losses/enhance profits
	Marketing produce	Potential profit changes with production and quality estimates/information.	State, national, and worldwide weather forecasts	Monitoring weather conditions in countries that are major producers of the same crop/commodity can give an estimate of the right time to market the produce.
Cropping	What crop(s) to plant	Select a crop that makes the best use of the climate.	Probabilities of rainfall and abnormalities in temperature	Select late-maturing crops if planting dates are early, short-maturing crops if season is short. Select crops with higher drought tolerance in dry seasons.
	Variety of crop to plant	Most crop species have a number of varieties available that vary in their length of growing season or resistance to heat, cold, frost, waterlogging, or disease.	Seasonal climate outlook	Choose a crop variety that best suits the seasonal conditions. Plant varieties that mature before the possibility of late frost. Plant a long-season variety if rainfall is likely to be evenly spread and a short-duration variety if probability is of less rainfall.
	When to plant a crop	Most crop seeds cannot germinate below 4.5°C during the winter season and below 10°C during the summer season. Follow-up rainfall may make the paddock too wet to plant or more rainfall may be needed to allow the crop to establish. Even a light rainfall after the crop has been sown adversely affects the germination rate through crust formation.	Extended weather forecast; probability of follow-up rainfall in short term	Mild temperatures above 4.5°C in winter and above 10°C in summer are ideal for sowing seed crops provided soil moisture is adequate and sowing dates for the crop are optimal. Plant early if outlook is for continued rain or plant now if only one or two planting opportunities in the season.
	Optimum depth at which seed should be sown to achieve an optimal rate of seed emergence	Under extremely dry weather, soil moisture will deplete at a fast rate because of high evaporation. Hence, upper soil profiles will dry rapidly, resulting in inadequate	Extended weather forecast	A forecast of dry, hot, and windy weather will suggest sowing the crop at a slightly deeper depth than normal to achieve the desired germination of seed.

	moisture for the seed to germinate. The second limiting factor will be extremely high temperature in the shallower depths of the soil, and the seed planted at shallow depth will be roasted or the emerging seedlings will be burnt.		
Fertilizing	Fertilizing with nitrogen can increase crop yield potential but only if there is sufficient rainfall.	Seasonal climate outlook	Fertilize only at the optimum rate if outlook for the season is favorable.
Fertilizer application	Temperature, rainfall, and wind speeds determine the efficiency of fertilizer application. Wind speed greater than 15 km/hour does not allow the finely particled fertilizer to hit the ground at the right place. The spread is uneven and a substantial amount is blown away from the target and wasted as drift.	Short-range weather forecast of temperature, wind, and rainfall	A forecast of mild, dry, and light wind is ideal for fertilizer application. Apply fertilizer when the forecast is for less than 10°C, with no or insignificant rainfall, and wind speed less than 15 km/hour. Avoid finely particled fertilizer application on days for which the wind speed forecast is above 15 km/hour.
Disease control	Many crop diseases are affected by weather. As an example, yellow spot in wheat can become prevalent in wet years, causing reduced production.	Seasonal climate outlook	Be prepared for disease control if the outlook is for a wet season. Monitor the crop and undertake a spray application when the first symptoms of disease become apparent.
Insect control	Many insect pests become a problem in only particular seasonal conditions. *Heliothis* in the caterpillar stage is an example. *Heliothis* moth can move in on storm fronts.	Seasonal climate outlook; extended and short-range weather forecast of rain, wind, and temperature	*Heliothis* have a life cycle of four weeks in heat wave conditions versus ten weeks in cooler conditions.Therefore, scouting and spraying of the crop is necessary more often in heat wave conditions.

TABLE 9.1 *(continued)*

Industry	Key decision	Why weather and climate information is important	Climate/weather information required	Strategies to reduce losses/enhance profits
	Weed control	Wetter years or wetter than average seasons may cause an increase in the number of crop weeds.	Seasonal climate outlook and extended weather forecast of rain, wind, and temperature	Spray earlier to ensure weeds do not get too large, and if using ground spraying, spray when damage to soil structure by machinery is least.
	Harvesting	Rainy spell delays/prevents harvest and creates problems in transport and storage of harvested grain. Rainfall at crop maturity reduces grain quality and increases grain moisture.	Extended weather forecast of rain, wind, and temperature	Harvest early to avoid rains. If rainfall is anticipated, postpone the operation until the next clear day, when soil moisture does not interfere with the operation. Budget for the timely use of grain dryers to reduce moisture levels.
Sugarcane	Replant or retain old ratoon	New plantings culminate in poor stands and stunted growth in dry seasons.	Seasonal climate outlook	New planting should take place only in a favorable season. Maintain old ratoon if conditions are unfavorable.
	Determining harvesting and crushing schedules	Rainfall reduces the commercial cane sugar content (CSC) and hinders transport of cane from paddocks.	Extended and short-range weather forecasts	Harvest highest yielding blocks first or blocks more susceptible to waterlogging if rain is likely.
	Trash blanket	Trash on ground in dry weather will preserve moisture.	Seasonal climate outlook	Do not burn trash in dry years; harvest green.
	When to burn cane prior to harvest	Weather affects the effectiveness and safety of using fire as a tool for cleaning cane.	Extended and short-range weather forecasts of temperature, relative humidity, and rain	Fire cane only on days with low fire danger.
Horticulture	Site selection	Climate records can determine if the area is suitable for particular crops.	Historical records of rainfall, wind, humidity, temperature, and frosts	Select climatic site that suits the requirements of the crop to be grown.
	Crop selection	Most crops have specific climatic and water requirements.	Historical records of rainfall, wind strength, humidity, and frost; num-	Select crops that suit the local area and are not subject to ad-

	Crop selection *(continued)*	*Low temperatures:* Number, duration, and severity significantly influence plant growth and product quality. For example, lettuce heads are affected by several light frosts in a row. *High temperatures:* Heat wave conditions and high night temperatures markedly affect crop quality. *Rainfall:* In many horticultural areas, if irrigation water is not limiting, rain can cause damage and an increase in disease prevalence for most crops.	ber of frosts/year, likely dates of first and last frosts of the season	verse conditions that require extra management costs such as disease and insect control. Plant frost-susceptible crops at optimum times to avoid frost. Select crops that are comparatively resistant to thermal stress.
Viticulture	Site selection	High temperatures and sunlight can burn berries. Frost can kill the whole shoot of the plant or, in autumn, kill leaves before fruit is mature or damage berries. Strong winds can cause damage to vines and fruit production. Low rainfall causes drought stress and lack of production. High rainfall can cause berry splitting, dilution of sugar levels, loss of berries at flowering, and fungal diseases.	Historical records of rainfall, strong wind, and temperature	Select a climatic area that is suitable for viticulture. Local microclimates can have a major effect on success of viticulture enterprises. Select an area that is not subject to severe frost. Select an area that does not have extreme variations in wind speed. In drier areas, budget for enough irrigation to sustain crops through dry periods. In wetter areas prepare for the possibility of fungal diseases and lower fruit quality.
	Varietal selection	Chardonnay has early bud burst, which is more susceptible to late frost. Ruby Cabernet produces quality wines in hot areas.	Historical records of rainfall, wind strength and direction, humidity, and temperature	Generally wines in warmer areas have fewer varietal characteristics than those in cooler climates.
	Disease control	High temperature and rainfall cause the development of fungal diseases.	Short-range weather forecast	Prepare to spray crops with fungicides if conditions suit growth of fungi.

TABLE 9.1 *(continued)*

Industry	Key decision	Why weather and climate information is important	Climate/weather information required	Strategies to reduce losses/enhance profits
	Harvesting	Warm temperatures enhance growth and harvesting is easier.	Seasonal climate outlook; extended and short-range weather forecasts	Plan to harvest earlier if seasonal outlook is of warm weather.
Orchards	Site selection	Rainfall records can determine whether irrigation is required. High wind areas can damage trees.	Historical records of rainfall, wind strength and direction, humidity, and temperature	Select site with adequate rainfall or potential for irrigation.
	Fruit types and cultivars	Severe frost can affect flowering and accumulated chilling hours needed for plants to set fruit.	Historical records of rainfall, wind strength and direction, humidity, and temperature; frost occurrence and chilling hours	Site should be protected from wind or budget for windbreaks.
	Disease control	Combinations of certain atmospheric conditions can cause diseases. For example, high humidity and temperature is conducive for fungal diseases.	Short-range weather forecast, extended weather forecast, and seasonal climate outlook	Be prepared for disease control if forecast is of wet season. Spray fungicide before disease occurs.
	Planting new orchards	Nursery plants and seedlings are highly sensitive to extremes of weather.	Short-range weather forecasts	Plant only when weather is mild.
	Whether to insure for hail damage or erect hail netting	Hail destroys crops or reduces crop value.	History and probabilities of hail; short-range weather forecast of hail storms	Erect hail netting; insure for hail damage.
Water/ Irrigation	Location and size of water storage	Climatic expectations determine the size and location of surface water storage to satisfy water needs.	Seasonal climate outlook; historical records of rainfall, evaporation, and stream flow	Build storage that can provide irrigation in dry periods with adequate stream flows.
	Water allocations	Weather will determine if storage or water source is replenished.	Seasonal climate outlook	Crop smaller areas when outlook is for dry conditions and water allocation is low. Adopt water-saving practices.
	Stock water	Hot dry weather increases stock water intake and increases evaporation from the stored water.	Seasonal climate outlook; extended weather forecast	If the seasonal outlook is for lesser rains, use water sparingly and budget water allocation between animals and paddocks.

	Planning irrigation schedules—amount of water and time of application of water	Evaporation affects crop water requirements. Rain after irrigation causes lodging, crop damage, and erosion. At high wind speed, water flow and spread is adversely affected and its efficiency is reduced.	Extended weather forecast	Use moisture meters for irrigation scheduling. Irrigate when plant and soil moisture is low. If no rain is expected, the irrigation should be normal. If a light rainfall is expected, it should be budgeted in the irrigation schedule.
Crop Spraying	When to spray crops for weeds, pests, and diseases (ground applications)	Temperature above 30°C results in significant loss of highly volatile chemicals through evaporation. *Wind direction:* Wrong wind direction can result in chemicals on nontarget area/object. *Wind speed:* At higher wind speeds the chemical does not hit the right target, there is a big loss of chemicals through drift, and there is the danger of air, water, and soil pollution. *Humidity:* Under extremely dry conditions, the water carrier may evaporate completely or leave a very fine dust of solid chemical. Under humid conditions, evaporation may not take place but the droplets may drift for several hundred meters downward.	Short-range weather forecast of wind direction and speed, relative humidity, temperature, and rainfall for the localized area Anemometer Whirling psychrometer in the field	Spray only in conditions of minimum wind, mild temperatures, and low humidity. Ideal temperatures for spray efficiency are between 5 and 27°C. No rain should be expected for about six hours subsequent to spray application. Wind speed thresholds determining spray efficiency are: < 8 km/hour—ideal; 8-15 km/hour—good; 16-22 km/hour—fair; 23-28 km/hour—marginal; > 28 km/hour—unfavorable. Use anemometer to determine local wind direction; use smoke makers to assist in determining wind direction.
	Spray and dust (aerial applications)	Poor visibility, either due to low clouds or fog, and strong wind speed are the two most important weather factors that adversely affect aerial spray operations. Low cloud ceiling and visibility are two great risks for aircraft flying. At high wind speed, dust and spray will miss much of the target	Short-range weather forecasts of visibility, wind direction and speed, relative humidity, temperature, and rainfall for the localized area Anemometer Whirling psychrometer in the field	Spray when the wind is light, temperature is mild, and humidity is high, visibility is more than 1.5 km, and the sky is clear or only high clouds are present. Use anemometer to determine local wind direction; use smoke makers to assist in determining wind direction.

TABLE 9.1 *(continued)*

Industry	Key decision	Why weather and climate information is important	Climate/weather information required	Strategies to reduce losses/enhance profits
		area and result in heavy drifting and pollution. Dew will dilute the chemical and decrease its effectiveness. At extremely high temperatures, a considerable amount of liquid chemical evaporates either in the air above the plant canopy or just after falling on the target.		
Grazing/Pastures	Optimum stocking rates	Climate determines the type and amount of grass and herbage growth.	Historical climatic records; seasonal climate outlook	If seasonal outlook is favorable, stocking rates can remain at current levels.
	The number of stock to carry during the dry season	Weather determines how much stock feed will be available.	Seasonal climate outlook	Lower stock numbers before dry conditions set in to avoid cost of feeding or sale of stock at low market prices.
	Burning pasture for weed control	Weather affects the effectiveness and safety of using fire as a tool. In the longer term, burning before a dry period may mean a shortfall in feed supplies.	Short-range weather forecast of temperature, relative humidity, and rain; seasonal climate outlook	Burn grass only on days with low fire danger; burn only small areas if the outlook is poor, so that there will be extra feed for dry periods.
	Fire breaks	Weather can affect the severity of the fire season leading up to fire occurrences.	Short-range weather forecast; seasonal climate outlook	Maintain firebreaks early in the season and increase preparedness on potentially dangerous days.
	Feeding and supplements	Dry periods result in little or no plant growth.	Seasonal climate outlook	Budget to feed or supplement stock; buy and stockpile feed.
	Weed control	Rainfall and temperature determine the intensity of weed infestation.	Short-range weather forecast	Control weeds with chemicals only if they are not stressed for effective chemical use.

	Pasture improvement	Pasture improvement is a costly program, and the aim is to maximize establishment of pasture. Ideal climatic conditions are required for pasture improvement.	Historical climate records; seasonal climate outlook	Undertake pasture improvement if the seasonal outlook is favorable.
Haymaking	When to cut hay	If hay becomes wet or takes a long period of time to dry it loses nutrition.	Extended weather forecast for local areas	Cut hay only in periods of at least four consecutive days of fine sunny weather.
	Drying rate of soil/straw/hay	On average, under normal weather conditions, temperature contributes 80 percent toward evaporation, and wind and saturation deficit another 20 percent. When any or all three forces are working abnormally, evaporation or evapotranspiration increases proportionally.	Extended forecast of temperature, wind, and evaporative loss of water from soil, plants, and water bodies	Earlier than scheduled irrigation may be required. Hay will be ready for stacking earlier than when there are normal drying days.
	Silage or hay	If there is a likelihood of rain and hay needs to be cut, the hay can be made into silage 24 hours after it is cut. Silage has, however, less cash value, as it is difficult to handle.	Extended forecast for local areas	Cut hay only in periods of at least four consecutive days of fine sunny weather.
	When to bale	High-quality lucerne hay must be baled with some moisture content, usually at night after dew has fallen.	Short-range weather forecast; likely dew point temperature	Bale only when there is sufficient moisture level to stop leaf shatter but not so much moisture as to cause mold in hay. Normally bale at night after dew has fallen.
	Marketing	Hay prices are usually low in good seasons and high in poor seasons.	Seasonal climate outlook	Stockpile hay if the outlook is for a dry season and sell in dry seasons at better prices.
Sheep and Wool	When to shear	Choose a time of year to shear when newly shorn sheep are not subject to extreme weather changes.	Climate history; seasonal climate outlook	Shear when rainfall is less likely or when major temperature changes do not occur. Increase area under cover for sheep.

TABLE 9.1 *(continued)*

Industry	Key decision	Why weather and climate information is important	Climate/weather information required	Strategies to reduce losses/enhance profits
	When to muster for shearing	Rapid temperature changes can cause sheep losses after shearing. Wet sheep cannot be shorn; early warning of rain may allow more sheep to be put under cover.	Rainfall and temperature forecasts; sheep weather alerts	Graze sheep in areas protected from extremes of hot or cold. Muster early and shed as many sheep as possible.
	Lamb wind chill	Low temperatures below 15°C, coupled with rain and strong winds, cause hypothermia in lambs.	Short-range weather forecast of temperature, wind, and rainfall/snowfall	Retain lambs in-house for the critical period of high wind chill and/or assign the weakling stock to the more sheltered fields.
	Supplementary feeding	Lack of rain may necessitate early feeding of costly supplements to maintain growth and minimize production losses.	Seasonal climate outlook	Decrease stock numbers; buy feed supplements earlier at lower prices. Feed early to minimize losses.
	Treatment for fly control	Warm humid weather increases incidence of sheep becoming struck/infested with flies.	Seasonal climate outlook; extended weather forecast of prolonged periods of wet weather	Treat sheep with chemical before problems occur, or monitor sheep carefully in susceptible periods.
	Footrot	Wet conditions favor spread of footrot in sheep.	Seasonal climate outlook; short-range weather forecast	Plan to have sheep in paddocks less susceptible to prolonged wet conditions.
	Parasite control	Wet conditions allow an increase in the level of internal parasites.	Seasonal climate outlook	Pasture sheep in paddocks with less possibility of wet soil; drench sheep to decrease worm numbers coming into a wet season.
Cattle	Mustering	Wet conditions often make cattle handling difficult. In some cases it is not possible to truck stock after rain due to wet roads.	Short-range and extended weather forecasts	Arrange to muster when the outlook is for dry weather.
	Restocking	After drought, producers often buy stock to take advantage of extra paddock feed.	Seasonal climate outlook	Restock only if seasonal outlook is favorable. A break in the season may not last long, necessitating early sale or feeding of stock, causing losses.

	Weaning	Calves may need to be weaned from mothers earlier if there is a dry period and then sold or fed.	Seasonal climate outlook	Weaning calves early in dry weather stops stress on cows and allows them to go into calf for the following season.
	Parasite control	In wet conditions internal worms are more likely to increase in numbers.	Seasonal climate outlook	Treat stock early to avoid buildup of parasites, or pasture in areas where parasites are not such a problem.
	Animal feed requirements	Animals need greater than normal feed to maintain thermal balance in cold and chilly weather, whereas they eat less under hot and humid weather.	Extended weather forecast of temperature, rain, and wind	Estimate the amount of additional feed animals will need to perform normally under abnormalities in weather. Easy-to-use indexes or nomograms are available to quantify the feed requirements of the animals and their overall performance in extreme weather conditions.
Poultry	Heat stress mitigation	There is a rapid increase in death rate of the broilers of less than two months age when the temperature exceeds 30°C in a crowded poultry house. Mortality is high during heat wave conditions.	Short-range and extended weather forecasts of temperature, rain, and wind	With adequate ventilation, ample supply of drinking water, and well-spread birdhouses, temperatures up to 30°C can be tolerated without any stress. Any management measures that cause the birds to stand, move apart, and spread their wings to some extent will reduce heat stress.
Dairy	Heat stress mitigation	Hot weather causes heat stress in dairy cattle that results in a decrease in milk production.	Short-range and extended weather forecasts of temperature, rain, and wind	The resultant decrease in milk production and reproductive efficiency can be offset through a program consisting of cooling through shades, ventilation, spraying, and fans.
Pigs	Heat stress mitigation	Hot weather conditions cause heat stress to pigs that results in significant loss in body weight.	Short-range and extended weather forecasts of temperature, rain, and wind	Mist cooling the buildings during hot weather substantially reduces stressful conditions. It reduces the time of growth to market and produces a pig with less back fat.

Chapter 10

Using Climate Information to Improve Agricultural Systems

Natural variability is a characteristic of climate. It occurs on both long and short time scales. Seasonal fluctuations can be great—we experience droughts, floods, severe storms, and tropical cyclones. It has long been recognized that these variations in climate directly affect the growth, health, and survival of pastures, crops, and livestock. Farming operations testify to this with sowing, harvesting, lambing, calving, and shearing timed to maximize the best of seasonal conditions. What is now receiving greater attention is that climate sets the parameters for sustainable land use. A large part of reducing risk in agriculture and protecting natural resources from degradation is being aware of the climate record and seasonal fluctuations and forecasts so that production can be set at appropriate levels and land can be used to its capabilities.

With a high percentage of land being managed under some form of agriculture, the critical question is: How can climate information be used to improve decision making in agriculture? This information can be used in three main ways:

1. *Strategic purposes:* assessing production capability, farm layout, and choice of enterprises based on an interpretation of the local climate record
2. Tactical purposes: building planning and flexibility into the farming system for both levels of production and farming operations based on seasonal outlooks or forecasts
3. *Building resilience:* strengthening farming systems through diversification, risk management strategies, and off-farm income.

Advances in climate research and satellite and computer technology have improved the potential of farmers to prepare and adjust farming operations in a variable climate. For this potential to be realized, the complexities of the decision-making process in agricultural systems must be acknowledged and addressed. Adult learning processes must assume a higher profile to overcome the limitations of the dominant technology transfer model.

SETTING THE PLATFORM—PROPERTY PLANNING

Assessing Land Capability and Farm Layout

Property management planning (PMP) or whole-farm planning is an important tool for farmers because it attempts to integrate all the factors that drive rural life. It incorporates elements of physical planning, financial management, and the personal dimensions of life. Whole-farm planning encourages the farm management unit (family or company) to consider all these elements in formulating a vision for their life and their land. When the vision is articulated, it can be further refined into a series of achievable goals, because each element is analyzed for feasibility.

To take just one segment of the farming jigsaw, physical property planning encourages the stakeholders to use aerial photographs and overlays as an aid to identifying the natural resources (soils, water, vegetation, land classifications, and climate), infrastructure (fences, buildings, etc.), and enterprise characteristics of their farm. Problem areas or constraints are identified and a SWOT analysis is undertaken (identify strengths, weaknesses, opportunities, and threats) for each of these features.

The property planning process has been promoted by governments because it encourages farmers to take the long-term view and, by learning the capabilities of their land, to prevent environmental degradation of their natural resources and insulate themselves against the variability inherent in the climates of the earth.

In terms of climate, PMP examines factors such as rainfall, slope, and groundcover interactions, aspect, damaging winds and shelter, frost effects on pasture and crop growth, as well as the risks involved with extreme events of flood or prolonged drought.

Rainfall is the dominant factor in property planning because of its links with pasture and crop growth. A common planning theme is "using water where it falls." Simpson (1999) noted that runoff will vary from 2 percent to 12 percent of total rain, depending on soil type, topography, groundcover, and rainfall pattern. He made the point that runoff is that portion of annual rainfall that does not grow grass. The aspect of the land interacts with rainfall, wind, and solar influences. It has a major effect on the length of the growing season and pasture maturation. Aspect will also determine what species will survive and thrive (Simpson, 1999).

Drought has the potential to drastically reduce the productivity and longevity of introduced pasture. Well-managed native pastures show a natural resilience to climatic variations and are especially suited to particular land classes (Simpson, 1999). Farm advisers are now advocating that farm managers encourage a diversity of species, with special attention to native pastures, to take advantage of their ability to cope with climate features and, for some species, give high annual production.

A property plan allows the farm manager to site windbreaks in the most strategic places to achieve the following benefits:

- Wind speed reduction (The wind speed may be reduced for a distance of 15 to 30 times the height of the windbreak.)
- Wind erosion prevention during periods of susceptibility, such as crop fallows and drought conditions
- Reduced moisture evaporation directly from the soil and through plant transpiration
- Less temperature and wind chill stress on livestock and crops
- Reduced shattering and lodging in cereal crops
- Increased crop yield (Nicholson and Albert, 1988; Mavi, 1994)

Crop yields improve on the leeward side of shelterbelts due to improved microclimate and less physical wind damage to plants. The overall impact of a windbreak is shown in Figure 10.1.

These effects of windbreaks and shelterbelts have been demonstrated to increase wheat yields in Victoria by up to 30 percent and oat yields by up to 40 percent. Increased moisture availability, particularly during grain filling, was considered to be the dominant effect (Burke, 1991).

In the Esperance district of Western Australia, lupin crop yield increases of 27 percent between rows of trees made these windbreaks a profitable investment in terms of the crop yields alone. Other long-term benefits expected from the trees are timber products of posts, poles, and sawlogs (Burke, 1991).

The benefits of shelter for livestock have been demonstrated in numerous trials. For example, one Australian study from the New England Tablelands of NSW showed that cold stress can depress sheep liveweight gain by 6 kg and can depress wool growth by 25 percent (Lynch and Donnelly, 1980). In southern Victoria it has been calculated that the provision of shelter can increase milk production in dairy herds by 30 percent. Ten percent of this is due to greater efficiency of conversion of feed and 20 percent is due to the greater amount of feed available (Fitzgerald, 1994).

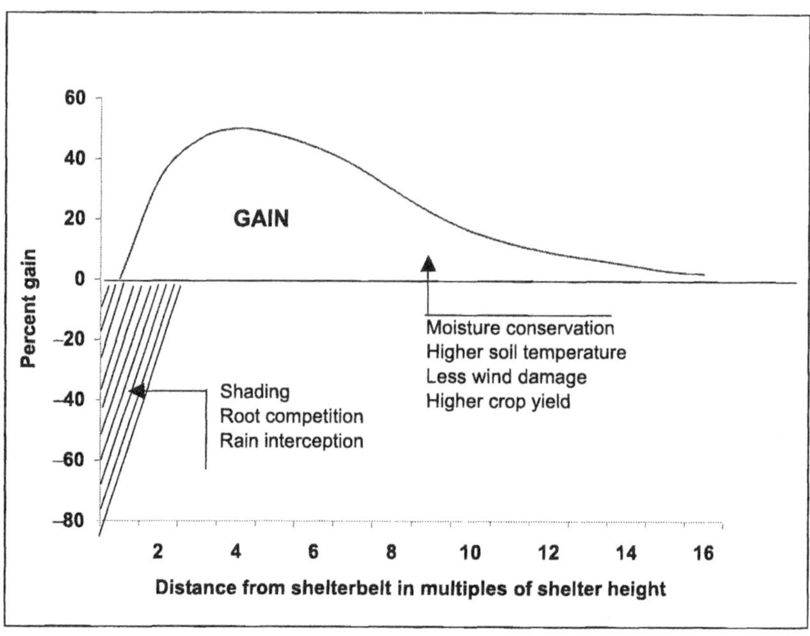

FIGURE 10.1. The effect of a shelterbelt on microclimate and crop yield (*Source:* Mavi, 1994.)

Shade and shelter provide protection for livestock from the effects of heat stress. A Queensland study (Davidson et al., 1988) showed that the provision of shade for dairy cows increased milk production for each cow by 2 kg per day, combined with an improvement in milk composition. A CSIRO (Jones, 2000) study in the Hunter Valley of NSW found that dairy cows without shelter produced 3 percent less milk than those with access to shelter. This loss represents 230 liters of milk per cow each year for a high-producing herd.

An understanding of the direction of the most damaging winds should also be incorporated into the design of farm infrastructure, such as the siting of sheds.

SUSTAINABLE PRODUCTION—SETTING THE ENTERPRISE MIX AND PRODUCTION LEVELS

Temperature and the quantity, variability, and seasonal distribution of rainfall will determine the type of crops and livestock enterprises suited to a given location. These parameters, combined with soil type and landform, also determine what levels of production (e.g., stocking rate or cropping area) are appropriate to avoid the threat of degrading the natural resource base on which the enterprise depends. In particular, these climate characteristics will determine the

- enterprise orientation (e.g., mainly cropping, integrated crop-livestock, or mainly livestock);
- ability to diversify into other enterprises (influenced by suitable plant cultivars and livestock breeds);
- season of cropping (summer, summer and winter, winter), which is influenced by rainfall distribution and the capacity of the soil to store water; and
- degree of flexibility in choosing rotations and enterprises (Tow and Schultz, 1991).

Grazing is the favored form of agricultural production for both low- and high-rainfall areas (although at vastly different intensities). This is because broadacre cereal cropping faces the risk of failure due to inadequate rainfall in semiarid environments and the risk of disease and operational interference because of too much rain in high-rainfall areas. Figures 10.2 and 10.3 provide examples of two contrasting regions in the cereal zone of Australia, their enterprise mix, and common rotations.

Assessing the characteristics of temperature and rainfall for a location is not only useful for choosing the most appropriate enterprise(s) but is also important when choosing the most productive plant cultivars. For example, a Western Australian viticulturist, Erland Happ, recognized that temperatures during the ripening period have a major influence on grape flavor (personal communication). He explored this further while investigating the purchase of another property. Happ believes that knowing the temperature range over the ripening period is critical in selecting grape varieties to plant and in choosing sites to grow grapes to maximize flavor. Before deciding on a property, he obtained hourly temperatures for an entire growing period. Happ calculated an index of "heat," in excess of 22 degrees, and compared this to other Australian and international sites known for their capacity to produce ultrapremium wines. This information enabled him to choose the most appropriate grape varieties for particular sites, based on temperature. This has led to the purchase of a second property which has particular advantages for early varieties.

Production levels should be based on an understanding of the local climate record, especially median rainfall and its distribution. Market prices and the calendar alone should not govern production. Available soil moisture and expected seasonal conditions and forecasts should also drive it.

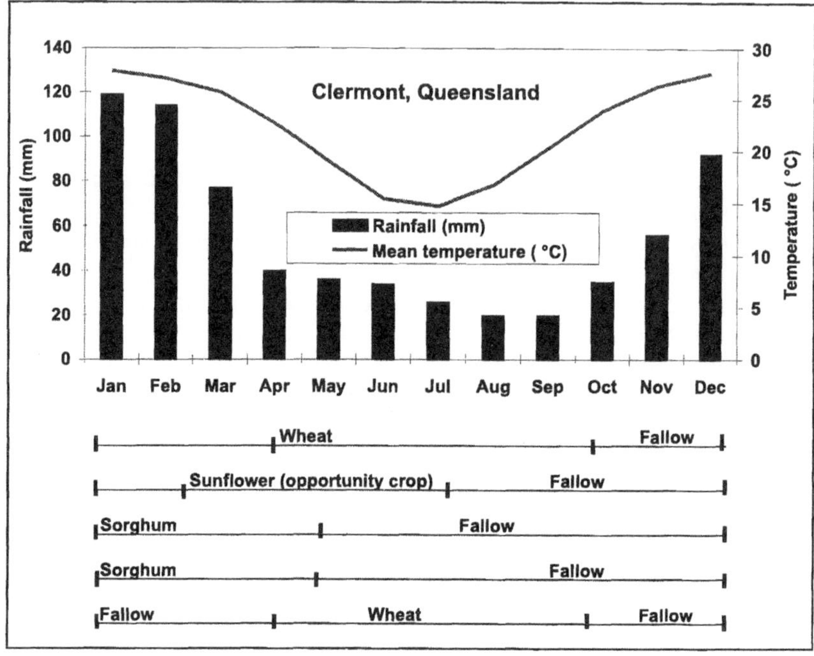

FIGURE 10.2. Climatic patterns and sequence of crop rotations in northern parts of the Australian grain belt (*Source:* Tow, 1991.)

MAKING EFFICIENT USE OF RAINFALL

Australia's rainfall and streamflow are among the most variable in the world (Standing Committee on Agriculture and Resource Management [SCARM], 1998). Lack of rain is the main factor limiting plant growth and agricultural production in general. The combined effects of rainfall, temperature, and evaporation determine the productive potential of crops and pastures. For those managing agricultural systems, the challenge is to respond to seasonal fluctuations in a timely and planned manner.

Rainfall is often described in terms of its annual average or seasonal quantity to characterize an area or to provide a regional indication of production potential. However, the amount of rain required for a productive pasture or crop varies from region to region. For example, in South Australia, 400 mm may be considered reasonable annual rainfall, while in Queensland it may be 700 mm. This is largely due to the decreasing "effectiveness" of rainfall for maintaining plant growth due to increasing evaporation (Tow, 1991).

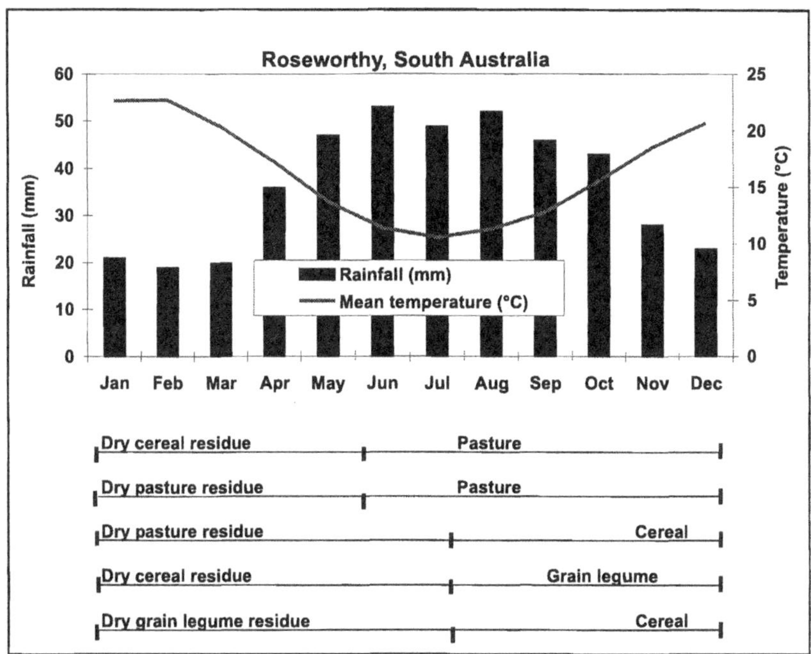

FIGURE 10.3. Climatic patterns and sequence of crop rotations in southern parts of the Australian grain belt (*Source:* Tow, 1991.)

Effective rainfall is related to the moisture available in the plant's root zone, allowing the plant to germinate, emerge, and maintain its growth. Soil moisture levels need to remain above the wilting point of plants, otherwise plants that cannot replace water lost by transpiration through their leaves will collapse or wilt. This results in death if replacement water is not added quickly. As a rule of thumb, the evaporation from an exposed soil surface is about one-third of that from the evaporimeter. By examining local figures for average monthly rainfall and evaporation, the number of months of effective rain can be assessed; this, when combined with temperature figures, shows the main growing season periods (Figure 10.4).

In northern cropping areas of Australia, effective rainfall may occur at different times throughout the year but not necessarily in a "growing season" block of five or more consecutive months. Storing soil moisture through fallows is critical in overcoming these "gaps" in effective rainfall.

Fallowing is the way most farmers make effective use of water. This involves conserving moisture in the soil between crops by killing (either by cultivation or spraying) any plants that would take moisture from the soil.

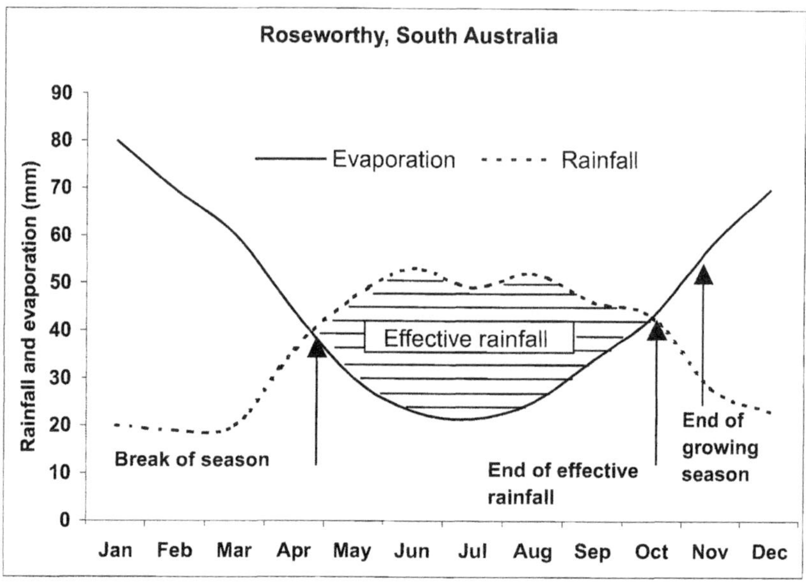

FIGURE 10.4. Rainfall, evaporation, and length of growing season at Roseworthy, South Australia (*Source:* Tow, 1991.)

The topsoil may dry out in this period, but a "bank" of water will remain in the subsoil for the following crop. The crop is sown when seasonal rain wets the topsoil, or in some cases, sowing machinery may be specially modified to sow into the existing moisture.

Farm Practices Affecting Water Use Efficiency

To make improvements in water use efficiency, the goal is to maximize and make productive use of soil moisture in the root zone while minimizing nonproductive losses from the root zone (Pratley, 1987). Nonproductive losses include evaporation, transpiration from weeds and volunteer plants, runoff, and deep drainage. Soil evaporation is a major component of nonproductive moisture loss (Perry, 1987). The relative amounts of these nonproductive losses will be determined by

- the quantity, variability, intensity, and seasonal distribution of rainfall;
- soil type, soil fertility, aspect, slope, and landform; and
- current land use.

Farm practices that improve water use efficiency include the following:

- Reduced cultivation or no-till practices—any soil disturbance by cultivation leads to increased evaporation.
- Crop residue or stubble retention reduces the impact energy of raindrops and prevents or slows the flow of water across the soil surface, improving infiltration. A mulch of dead material on the soil surface reduces temperature changes, lowering evaporation and preventing the germination of some weeds.
- Use of deep-rooted plant species—The better the plant can explore the soil profile by root growth, the more soil water will be used. Lucerne is the archetypical deep-rooted pasture plant.
- Use of perennial plants species, where possible, that respond to soil moisture year-round—Some native grasses are able to perform this task efficiently, e.g., *Danthonia* spp. and *Microlaena stipoides.*
- Maintaining groundcover at 70 percent or greater reduces runoff and erosion.
- An integrated pest management program and good crop nutrition help to maintain crops and pastures at maximum health and growth potential.
- Opportunity cropping rather than fixed rotations will take advantage of stored soil moisture and seasonal rainfall, as well as lowering groundwater levels, minimizing the risk of dryland salinity.

These factors can restrict efficient water use in a cropping system:

- Soil structural degradation or decline reduces water infiltration, soil water storage, and the optimum conditions required for the germination, emergence, and root growth of the cultivated plant.
- Poor crop nutrition slows the establishment and early productivity of a crop or pasture, thereby increasing the potential for evaporation and yield loss from weeds, pests, and diseases.
- Soil erosion removes the layer of the soil that contains the majority of the available essential nutrients—1 mm of topsoil lost through erosion is equivalent to 7.5 to 10 tonnes of soil per hectare.
- Soil acidification causes soil toxicity problems which stunt root growth and reduce the development and yield of the crop.
- The accumulation of free salts in the surface horizons can have a dramatic impact on vegetation, from stunting the growth of a pasture or crop to limiting what plant species can be grown.

- Poor weed control creates ongoing soil moisture and nutrient loss, reducing the productivity and quality of crops and pastures.
- Pests and diseases, depending on seasonal conditions and location, can reduce the vigor and productive performance of crops and pastures, increasing the nonproductive loss of water.
- Seeding significantly after the optimum sowing date can lower water use and the potential yield of a crop.

Assessing Water Use Efficiency

Assessing the water-limited potential yield of a crop is a useful performance benchmark to assess how well stored soil moisture and growing season rainfall are being used. The range of climatic factors in a given location (French, 1987) determines potential yield. Hayman and de Vries (1995) provide water-limited potential yield figures for a range of crop types (Table 10.1). These figures are approximately 75 percent of the "absolute potential" based on the French-Schultz model and are used because they represent a more realistic and obtainable "on-farm" target.

For example, the following calculations can be made for a wheat crop growing in the Narromine district of NSW receiving 266 mm of growing season rainfall (April-October) and assuming that one-third of the fallow period (December-March) rainfall of 176 mm is conserved soil moisture:

$$[(266 \text{ mm} + 59 \text{ mm}) - 110 \text{ mm}] \times 15 \text{ kg/ha/mm} = \quad (10.1)$$
potential yield of 3.2 tonnes per hectare
[(rainfall + fallow) – evaporation]

TABLE 10.1. Water-limited potential yield figures for various crop types

Crop type	Paddock evaporation (mm)	Potential "on-farm" yield (kg/mm/ha)
Wheat	110	15
Barley	90	18
Oats	90	22
Triticale	90	18
Canola	110	10
Grain legumes	130	12

Source: Adapted from Hayman and de Vries, 1995.

Comparing this with an actual yield of 2.8 tonnes/ha:

$$2800 \text{ kg/ha} \div [(266 \text{ mm} + 59 \text{ mm}) - 110 \text{ mm}] = 13 \text{ kg/ha/mm} \quad (10.2)$$

Calculate actual yield as a percentage of its potential:

$$(2.8 \text{ tonnes/ha} \div 3.2 \text{ tonnes/ha}) \times 100 = 88\% \quad (10.3)$$

Note: Calculations allow for a loss of 110 mm by direct evaporation and assume runoff and deep drainage over the growing period to be nil.

The previous comparison identifies that water use efficiency is 88 percent of the potential. This benchmarking exercise provides a general guide for farmers wanting to assess and improve their water use efficiency. Cornish and colleagues (1998) found that wheat farmers who had positive yield trends and showed higher productivity used rainfall more efficiently and more carefully managed soil fertility.

Making Tactical Adjustments—Farming to Season Type

Examining the climate record for many regions of Australia shows that there is no such thing as the typical or average season. The highly variable climate produces dramatic variations in crop yields from one season to the next. As a result, there can be large fluctuations in farm income between years. Egan and Hammer (1995) state that in some regions, the best three years in ten can generate up to 70 to 80 percent of income, while the poorest three years may result in a net loss of income.

The most critical decisions are made at sowing time, as growers commit farm resources (land, labor, machinery, finances, etc.) for the following season and beyond, with only limited opportunities for further modifications (Egan and Hammer, 1995). Farming to season type therefore requires a willingness to be flexible in production decisions, such as stocking rate, area to be cropped, and the level of inputs to be used (e.g., fertilizer), in order to reduce the risk to physical and financial resources and to maximize opportunities. These decisions should be based on climatic and soil conditions prior to sowing and on seasonal forecasts, which provide useful indicators of "season type" and yield prospects.

For example, Allen Lymn, a farmer in South Australia, has developed a risk management strategy to minimize the effects of climate on farming in his low-rainfall area. If his farm near Minnipa does not receive 40 mm of rain between April 1 and June 15, Lymn cuts back his cropping area and will even consider not sowing. In years when his farm receives between 40 and 100 mm, he sows an average area—about half the farm. In years with over 100 mm falling in this period, he increases the cropping area. By adjusting his cropping program depending on early-season rains, Lymn has increased the opportunity of gaining higher returns over a number of seasons.

Farm managers need to have strategies in place for both drier and wetter than "average" seasons. For a dryland cropping enterprise this may mean being prepared to alter the area sown, variety choice, time of sowing, and the amount of fertilizer used. The crop management options and strategies available to farmers depend on the region being farmed. Stored soil moisture is a critical factor in cropping decisions in northern grain-growing regions but is less significant for southern cropping areas, which depend more on growing-season rainfall and timing of the seasonal break (Egan and Hammer, 1995). For example, early research in northwest NSW into the relationship between wheat yields, time of seeding, and soil moisture demonstrated that as the depth of wet soil at planting increased, so did wheat yields, in an almost straight-line relationship (Figure 10.5). More recent research on the Liverpool Plains in NSW on the links between nitrogen fertilizer, climate forecasts, and stored soil water confirms this relationship (Hayman and Turpin, 1998). Although this relationship exists, there may be advantages in using, rather than storing, this soil moisture for long periods.

In the summer rainfall areas of Queensland and NSW, flexible cropping systems that adjust cropping in response to stored soil water give economic benefits. As rainfall increases, cropping frequency can also increase. Build-

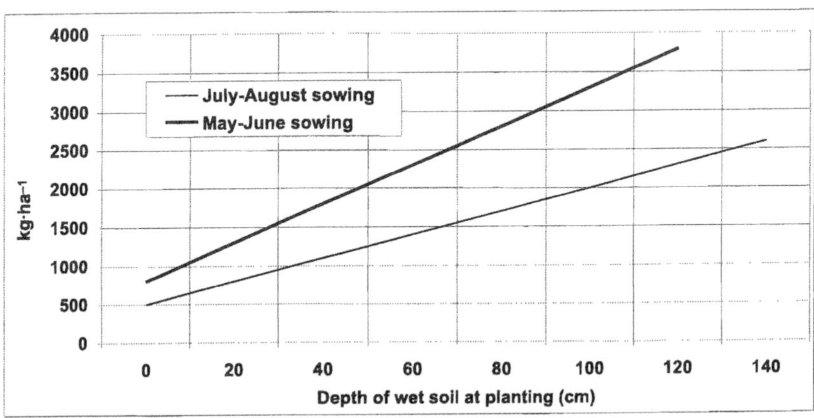

FIGURE 10.5. A generalized pattern of wheat yields in northern parts of Australian grain belt as determined by time of sowing and stored soil moisture (*Source:* Fawcett, 1968.)

ing flexibility into the system gives more than direct economic benefits. At the same time, soil erosion and soil salinity will be reduced. Storing summer rainfall in the soil profile by means of a long (up to six months) fallow will give high yields in the following wheat crop, but if the seasonal outlook is favorable, double or opportunity cropping may give higher returns. Recommendations for the northern cropping zones now consistently advocate the use of farming systems based on opportunity cropping rather than fixed rotations. In light of the threat in many of these areas from dryland salinity caused by rising water tables, this recommendation not only makes good economic sense but also has important environmental implications.

Using Forecasts and the Southern Oscillation Index in Decision Making

A way of managing rainfall variability is to examine the rainfall probabilities of a given location. These show the chances of receiving a particular amount of rain at a given time. Probabilities are like odds. If the chance of something happening is one in four, scientists will express it as a percentage—a 25 percent probability. Seasonal climate forecasts produced by the Bureau of Meteorology are usually given in terms of probabilities that can be linked to a property's rainfall history. Farmers will point out that within most decades about three years out of ten are poor years, four are average years, and three are good years. The probability of a poor rainfall season is 30 percent, an average season 40 percent, and a good season 30 percent. However, the Southern Oscillation Index shifts the probabilities. It works this way: Think of a wheel with equal segments (Figure 10.6, top left). The wheel is spun with an equal chance of landing on any segment. Now assume that the wheel has three segments, which represent a dry, average, or wetter than normal season. There is only one spin of the wheel per year. For eastern and northern Australia, in years when the SOI is very high, the probability shifts toward the wetter season category. This means there is a greater chance of landing on a good season when spinning the wheel. However, there is still a chance of landing on a poor season. For eastern Australia, when the SOI is very low, the probability shifts toward the dry category. When the wheel is spun this time there is a greater chance of landing on the "dry" segment. The odds never shift to give absolute certainty of a dry or wet year. We may not be able to obtain certainty, but we can obtain better chances, and "half a loaf is better than no bread." Managing climate risk is a process of assessing possibilities and turning them into probabilities.

For example, Stuart and Maxine Armitage (personal communication) farm an irrigated cropping property on the Darling Downs at Cecil Plains.

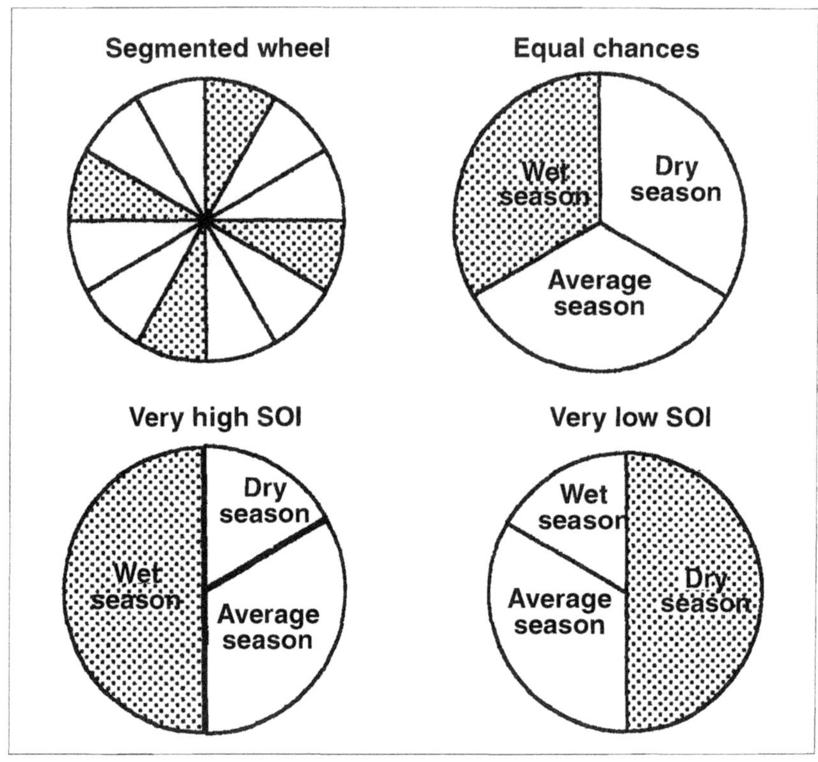

FIGURE 10.6. Probability of occurrence of average, good, and bad seasons (*Source:* Hayman and Pollock, 2000.)

They use climate probabilities and forecasts, taking into account the SOI, to determine for the season their enterprise mix, the area of crop to sow, and amounts of water needed for irrigation scheduling. Recently the Armitages had their irrigation dam half full and, faced with a potentially dry season, had some major decisions to make. The time was September and the Southern Oscillation Index was strongly negative in an El Niño year. Little rain had fallen during the winter, and there was no subsoil moisture. Using the computer software program Australian RAINMAN (Clewett et al., 1999), they found there was only a 20 percent chance of the 50 mm needed as planting moisture. The Armitages used the information to minimize their risk by making the following decisions. Rather than gamble on receiving planting rain, they used the water stored as prewatering to germinate and establish the cotton crop. They also reduced the crop area, because of the expected dry season and the reduced amount of water expected to be available for irrigation. By using the seasonal forecast and an assessment of water supply, the Armitages made a decision, which proved to be the right one for the season.

Seasonal forecasts can also help growers decide when to double crop. If, for example, the SOI is higher than 10 in May to June and some subsoil moisture is present in the profile, there is a good chance of a reasonable crop. If, on the other hand, the SOI is negative and subsoil moisture is low, the best decision is likely to be to fallow and not plant a crop (Hammer, Holzworth, and Stone, 1996; Greggery, 2000).

The following points should be considered when incorporating El Niño-Southern Oscillation events into seasonal decision making:

- Autumn is when to expect major changes in the movement and value of the SOI. The SOI usually becomes set as consistently negative, positive, or neutral by the end of May, and this phase can be used to indicate rainfall patterns over the next nine months, that is, until the start of the next autumn.
- A strong rise (to greater than +5) in the SOI in autumn indicates the probability of at least average rainfall for the following winter, spring, and summer.
- Positive (greater than +5) SOI readings in autumn months usually indicate the probability of above-average rainfall.
- Negative (lower than –5) SOI readings in autumn months usually indicate the probability of below-average rainfall.
- Look at the trend or phase of the SOI as well as individual numbers.
- Before using ENSO information in farm decision making, producers should be aware that (1) ENSO effects vary greatly between regions and with the time of year, (2) decisions are based on probabilities and not certainty, and (3) ENSO is only one of the factors controlling the climate.

The SOI is not a perfect forecasting tool. One of the limitations of forecasts is that they do not always provide enough lead time for farmers to prepare and adjust their winter cropping operations (Nicholls, 2000). However, this does not apply for summer cropping.

Climate forecasts and information only acquire value when decisions are modified in response to them (Hammer, 2000). However, overreaction to forecasts (e.g., selling large portions of a herd or not planting a crop) can be as detrimental to a farming system as ignoring or dismissing a forecast completely (Nicholls, 2000; Stafford Smith et al., 2000). For example, the *Australian Farm Journal* (February 1998, p. 10) reported about a farmer selling young sheep for $14 a head because of a decline in the SOI, only to find a short time later that they were worth $40 a head.

Planned and measured responses to seasonal forecasts should be applied with the understanding that forecasts are simply one more element to include in decision making. Examining the local climate record and using seasonal forecasts helps producers to bring realistic expectations into their decision making and increase their confidence when selecting management options (Clewett et al., 2000). The continued use of seasonal forecasts and the development of farming systems that respond to erratic rainfall and stored soil moisture have been shown to be superior to farming according to the calendar in Australia (Hammer, Nicholls, and Mitchell, 2000; Hayman, Cox, and Huda, 1996).

Sources of Weather and Climate Information

Climate information is delivered through a range of media, which are mainly sourced from national and regional meteorological centers. In Australia, the official source of climate information is the Bureau of Meteorology. The media range includes the Internet, facsimile services, television, radio, telephone services, newspapers, journals, and computer software packages. The media provide only generalized information, for a mainly "suburban" population. In farming, weather and climate information need to be more specific and related to current enterprises. Bayley (2000) provides a comprehensive review of the national weather and climate information and services available.

The most useful information for decision making will depend on location and the enterprises being operated. Table 9.1 (Chapter 9) summarizes some of the major agricultural enterprises, the key decisions to be made, and the climatic information required in decision making.

DEVELOPING RESILIENCE

Resilience refers to the *ability* a farming system has in withstanding unexpected and sometimes severe disturbances in the form of climatic extremes (e.g., prolonged drought), pests and diseases, changes in markets, and input costs. The management team determines this "ability" and the plans and strategies they have in place to reduce their exposure to these disturbances and, if need be, to "bounce back."

The resilience of different agricultural systems will depend on the intensity of production. For example, grazing systems that rely on high stocking rates are more vulnerable in dry seasons. The resilience of agricultural systems can be improved by the following approaches:

- Diversify crop and livestock systems. For example, diversification of crops and planting times reduce the impact of a poor season, especially for areas such as the northern wheat belt of eastern Australia (Ridge and Wylie, 1996a).
- Grow productive, deep-rooted perennial pasture species that can respond to and make efficient use of rainfall throughout the year. This includes maintaining native perennial pasture species, such as *Danthonia* (wallaby grass), which have grazing value and a tolerance of dry weather.
- Match livestock demand and pasture supply closely to reduce the prospect of feed deficits. Aligning peak animal demand (i.e., late pregnancy and lactation) with peak pasture supply will maintain pasture quality while reducing the need for supplementary feeding.
- Conserve and store fodder to overcome periods of feed deficiency. Fodder reserves should be increased as the stocking rate on a property increases; this will reduce the exposure to price changes for supplementary feeding in dry times.
- Use flexible rotational grazing rather than set stocking to make it easier to ration the feed supply and assess the amount of standing feed reserves available in dry times.
- Use different crop types and varieties and spread planting over a number of weeks to reduce the exposure of crops to weather damage at critical stages such as flowering and harvest time (Ridge and Wylie, 1996b).
- Use minimum- or zero-tillage systems that make efficient use of rainfall and that can outyield conventionally sown crops in dry years. Minimum-tillage cropping systems also provide greater flexibility in the timing of sowing operations.
- Use fallows to reduce a crop's reliance on growing-season rainfall.
- Provide shelter and windbreaks for livestock to reduce feed demand, improve animal production, and minimize livestock losses.
- Correct nutrient deficiencies and implement an integrated pest management program to improve the vigor of crops and pastures so they can better cope with climate variability.
- Use irrigation, where the opportunity exists, to overcome water deficits at critical growth stages of crops and pastures.
- Use netting to protect horticultural crops from hail damage.
- Carry out farm operations in a timely manner.

- Store or carry over grain or other products from productive years when prices are low to either use or sell at a later time.
- Reduce fuel loads and put in firebreaks before the bushfire season.
- Broaden the income base through off-farm investments and manage price risk by using forward contracts, futures, and hedging in response to commodity market fluctuations or seasonal forecasts.

MANAGING THE EXTREMES—DROUGHTS AND FLOODS

The experience of the extreme effects of climate, droughts, and floods generally provide the catalyst for change in management practices. However, many farmers tend to downplay the likelihood of these extreme events when not operating in them (Nicholls, 2000). This is made evident in the following quote: "We farmers should be planning for drought every day. What causes change for us is mostly drought. When we aren't in a drought we don't think about it" (Tim Wright, grazier, Uralla, NSW, personal communication).

The harsh experience of drought and floods shows that it is essential to have a strategy in place. These strategies should be based on local experience, an understanding of the local climate record, and the frequency of these events. Once developed, the strategies can be further guided and adapted according to seasonal forecasts and by calculating rainfall deciles to assess how dry or wet a month is compared to local historical rainfall figures. Deciles are calculated by ranking the rainfall record from lowest to highest recordings into 10 percent bands. Probability charts using rainfall deciles can be used to determine the probability of a certain amount of rainfall for making tactical decisions.

Droughts

Droughts are frequent but irregular events. Farmers and graziers generally refer to drought when the available rainfall over a period of time is not enough to give adequate plant growth and there is a major loss of agricultural production (see Chapter 5). Within some agricultural industries there can be differences of emphasis; references to a "feed" drought, "protein" drought, and "water" drought are common in the livestock sector.

During a drought, farm managers face a multitude of issues. By developing a personal plan, based on an assessment of resources and climatic risks, farm managers will be in a better position to manage a property through drought. A drought plan has four major components:

1. *Identification of the risks to enterprises.* This requires an estimate of the critical time(s) for rainfall and the absolute minimum needed to operate. This will depend on the enterprises. For example, breeding programs are at greater risk than stock trading.
2. *Analysis of the local climate to find the chances of the risk occurring.* This analysis is used to answer the following questions:

   - How often has the essential rain failed to come?
   - How long do low rainfall periods last on average?
   - Are they more likely at one time of the year?
   - What seasonal forecasting tools can be used?

3. *Identification of factors to monitor.* As well as measuring rainfall directly, the physical factors critical to an enterprise also need to be monitored. For example, limits need to be set on how much pasture loss can be accepted before some action must be taken to protect them. Factors to consider include

   - percentage of groundcover,
   - species change and weed percentage per paddock, and
   - dry matter produced in kg/ha.

4. *Formulation of an action plan.* The most difficult decisions involve livestock. Cropping tends to have more yes/no decisions, but in stock management there is a wide range of options. The most sensitive and the most forgiving areas under drought management practices need to be identified. These will include

   - areas to be protected from stock,
   - areas where stock can be fed,
   - areas that need additional fences or watering points,
   - priority areas for pest and weed control, including unwanted grazing by native animals,
   - which classes of stock to sell, and
   - whether to sell or agist stock.

Floods

Floods cause losses to stock and damage crops, pastures, and road and rail links. They also often put at risk the lives and economic well-being of rural communities. A flood management strategy should include the following elements:

- An identification of flood-prone land and the flooding history of the area
- An understanding of flood warnings and from where the flood patterns come
- Paddock and fence design to allow easy movement of stock, so that stock will not be trapped in floods
- "All weather" access tracks where possible
- An arrangement for fodder storage that is always accessible
- A flood evacuation plan that includes the relocation of stock and equipment
- Farm infrastructure checks, for example, the switching off of electrical supplies to buildings that may be flooded

THE DECISION-MAKING PROCESS— DEALING WITH RISK AND COMPLEXITY

Advances in weather forecasting and computer technology have improved the *potential* of farmers to prepare and adjust farming operations in response to a variable climate. For this potential to be realized, the complexity of decision making in agricultural systems needs to be acknowledged and the challenges facing farmers in accessing and learning to apply climate information need to be addressed.

According to Cousens and Mortimer (1995), many farmers regard weather and climate as unpredictable and beyond their control. It has therefore received little attention in its own right. Year-to-year variations are usually treated as background noise and are ignored.

PROVIDING CLIMATE TECHNOLOGY TO FARMERS

Although advances have been made in applying climate information to agricultural production and risk management, the communication of this information and technology to farmers holds particular challenges. Three of these challenges can be posed as questions:

1. Are the farmers, for whom the technology and tools are designed, able to learn in the context of how the information is presented?
2. Is the current extension model capable of changing practices?
3. Are we overstating the benefits of technology in providing farmers better control over their environment?

The answers to these questions cannot be explored in detail in this chapter, but some of the difficulties can be highlighted.

Property planning workshops and courses are conducted in each state, usually bringing farmers together to work in groups. This is an efficient way of dealing with the number of farmers involved but carries inbuilt difficulties and deficiencies. Shrapnel and Davie (2000) have identified that many farmers have personality profiles that, while enabling them to cope with the isolation that characterizes rural life, inhibit their ability to work with others in groups. They contend that the following elements are essential for the development of sustainable agriculture practices:

- Farmers must want to change.
- Farmers must know how to change.
- Farmers must have the necessary material resources to bring about change.
- Farmers must have the necessary psychological resources to bring about change.

These factors apply particularly to their ability to access and integrate information and technology in the area of climate. Climate workshops and technology transfer in group settings may not be relevant for farmers because groups are not the preferred method of learning for many rural people. So the questions remain: Are groups the best vehicle for communicating climate (or other) information to farmers? If not groups, how do researchers impart climate information?

The conventional method of promoting advances in technology to the rural community has been based on the linear extension model:

Research → knowledge → transfer → adoption → diffusion

This extension process has been widely criticized for its numerous false assumptions and limited applicability (Roling, 1988; Russell et al., 1989; Vanclay, 1992; Vanclay and Lawrence, 1995). Studies on the effectiveness of this model showed that research results were adopted by only a specific minority of farmers and that for the majority it was not a viable strategy for agricultural improvement. Nor, they contend, do different conceptual models such as those based on the idealized "farmer-led" model fare any better when it comes to adoption of technology.

The hypothesis of these social researchers is that farmers will do what they want to do anyway, and a more "helpful" role is to "work with" farmers in a supporting role, encouraging their enthusiasm and seeking to understand rather than to influence. This is obviously not the role of research or conventional extension workers. The authors also warn that this is not the only recipe for working with the rural community, but it is a system that does engender mutual understanding and trust between the various stakeholders.

The third question raised in this section is that of overstating the role of technology to farmers. The technocratic nature of traditional research and extension has been criticized for often reducing and masking the complexities of rural situations (Cornwall, Guijt, and Welbourn, 1994) and for its uncritical acceptance of technological innovation as a liberating agent (Buttel, Larson, and Gillespie, 1990; Vanclay, 1992; Vanclay and Lawrence, 1995). This criticism has been acknowledged and, in some cases, produced a reaction as described by Ridge and Wylie (1996b):

> In many cases, the researchers are not familiar with farm decision making and are not confident to express opinions on benefits to producers, while in other cases, the claimed benefits of a change in practice cannot be realized because researchers have failed to appreciate the practicalities of the situation that the farmer faces. (p. 11)

The desire of all those working in the area of climate studies is to make forecasting tools more reliable and more available to landholders. It is a noble aim, but a gap remains between availability and adoption. The existence of a form of technology and its use implies certain choices that are found in economic and social processes. Consequently, the form of technology may or may not act as a liberating agent, and this is dependent on the social and economic positions of those using or adopting it. The statement is supported by the findings of Hayman, Cox, and Huda (1996).

We are surrounded by tools for forecasting and reducing risks from climatic extremes and by decision support systems that are increasingly sophisticated and useful. How these tools and technologies are best made available and used by farmers is a vexed question that this section has attempted to highlight. Some fundamental philosophical questions are also raised that are being addressed and examined by many researchers in various fields of sociological studies.

COMMUNICATING NEW IDEAS AND PRACTICES— CREATING CHANGE THROUGH ADULT LEARNING

Improved strategies for communicating climate information, technology, and forecasts will take the following adult learning principles into account:

- Adults have a bank of past experience that should be respected and valued. This experience is both a helpful resource and a potential hindrance to new learning.
- Adults should be encouraged to identify their own needs and problems and to participate in the solution(s) that are designed to improve their circumstances.
- Adults learn best in environments that reduce any possible threat to their self-concept and self-esteem and that provide support for change and development.
- Adults are highly motivated to learn in areas relevant to their current developmental tasks and work roles, i.e., perception of immediate relevance.
- Feedback is essential for development and should be given promptly to motivate further learning.
- Adults learn best when they can set their own pace and when learning has immediate application in their lives.
- Learning programs should provide the opportunity for both autonomous "semi-independent" learning and for belonging to and participating in groups.

Rather than relying on the traditional linear extension model, a more appropriate approach may be to work with farmers through the decision-making/problem-solving process (Figure 10.7). Clewett and colleagues (2000) suggest that the information delivery and learning process needs to be flexible to allow for differences among individuals, their backgrounds, interests, and aspirations, as well as family and business circumstances. No one method or approach is adequate. For example, the promotion and extension of decision support software is important. However, this must be done on the understanding that although a very high percentage of farms have a computer, only 34 percent of Australian farmers are connected to the Internet at this stage (Australian Bureau of Statistics, 2000). In a recent survey of irrigators, 67 percent rated their personal computer skill as nil or basic (Keogh, 2000). This emphasizes the need for a range of strategies.

Education and training will need to play a large role if climate information and forecasts are to be accessed and used by those in agriculture. A research paper by Kilpatrick (1996) identified that farmers who undertake one or more education and training activities are three times as likely to be using a farm plan to make management decisions compared to farmers who undertake no training. The same study found that farmers who had taken further education and training (postschool) were more likely to make changes to land management practices to improve profitability and were generally more profitable.

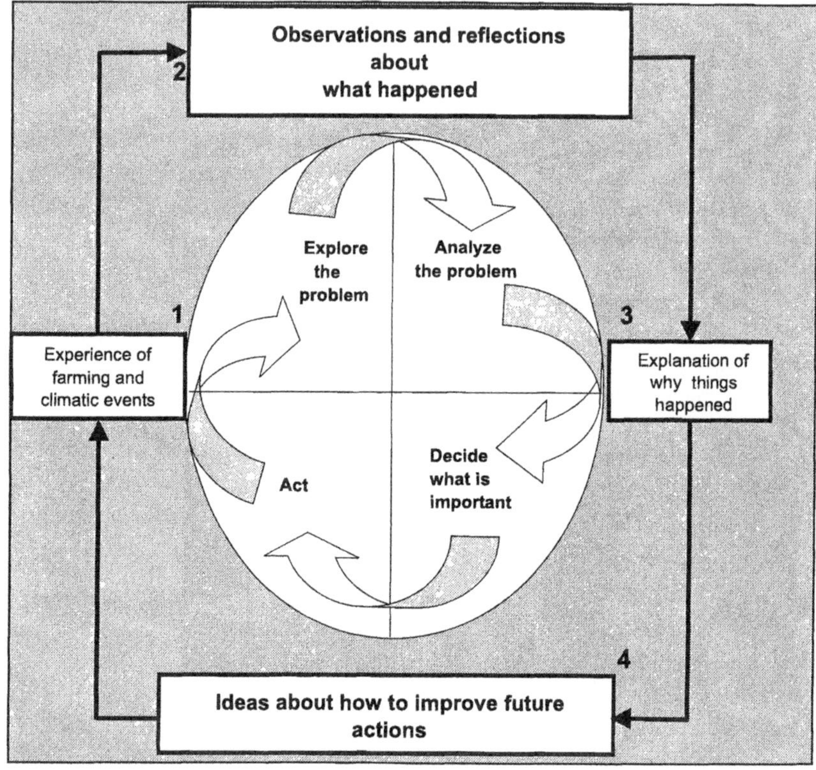

FIGURE 10.7. The cycle of problem solving (*Source:* Woodhill and Robins, 1998.)

In conclusion, sustainable agricultural production in a variable climate requires

- a knowledge and interpretation of the local climate and the weather and climate information available;
- plans in place to manage climatic extremes, such as drought and flooding;
- a knowledge that dry rather than wet years are the norm;
- the use of improved climatic information and tools to inform "on-farm" judgments and decision making; and
- a property management plan to set out actions and priorities to minimize risk from farming.

With the fine satellite and computer technology we now have at our disposal, it is tempting to assume climatology is a "perfect science." This technology, although significantly improving our forecasting ability, has provided more momentum for asking questions than providing conclusive answers. The vagaries of weather and unforseen events will remain.

Chapter 11

Climate Change and Its Impact on Agriculture

CLIMATE VARIABILITY AND CLIMATE CHANGE

Climate variability refers to variability observed in the climate record in periods when the state of the climate system is not showing changes. If the climate state changes, usually characterized by a shift in means, then the frequency of formerly rare events on the side to which the mean has shifted might occur more frequently with increasing climate variability (Salinger, Stigter, and Dasc, 2000).

Climate change is a movement in the climate system because of internal changes within the climate system or in the interaction of its components, or because of changes in external forcing either by natural factors or anthropogenic activities (International Panel on Climate Change [IPCC], 1996).

Natural variability is a characteristic of the global climate and occurs on both long and short time scales. Many climatologists believe that both long- and short-term climatic fluctuations are not random phenomena but organized events which are controlled by forces or energy resources either associated with the earth itself or with the planetary bodies of our solar system. Superimposed on these natural variations are changes induced by human activities. The release of greenhouse gases in the atmosphere in recent years is thought to be the cause of changing climatic patterns. Increases in global surface temperatures and significant interannual climate variability were observed in many regions of the globe during the later half of the twentieth century. The WMO (1998) has reported on warming trends, with proof of climate change and its continuation observed from many parts of the world.

Probable Causes of Climate Variability and Climate Change

External Causes

Astronomical periodicities. Many investigators of climate variability and change are of the opinion that astronomical periodicities influence the atmosphere directly or indirectly and bring periodic variations in climate. The most important periodicity is associated with the tidal forces of the sun, moon, and other planets (Munk, Dzieciuch, and Jayne, 2002). The time scale of many of the astronomical periodicities influencing the atmosphere cannot be demonstrated with certainty. However, it is widely accepted that they have a relationship with the sequence of ice ages and the intervening warmer and colder periods.

Sunspot cycle. The main known external causes of interannual climate variability and climate change are changes in solar output. The variation in solar output coincides with the most prominent and best-known solar disturbance of 11.08-year intervals, known as the sun spot cycle. Annual numbers of sunspots have been recorded for many years by astronomers, and these show the relative absence of sunspots from 1650 to 1700, the "Maunder minimum," with a slight decrease in solar output (Eddy, 1976). This minimum has been used by some to explain the Little Ice Age (1430-1850) in Europe. Estimates place the increase in solar irradiance between the Maunder minimum and now between 0.5 and 1.4 W m^{-2}, or an increase of 0.3 percent of the solar irradiance (IPCC, 1996).

Internal Causes

Greenhouse gases. Within the atmosphere there are naturally occurring greenhouse gases which trap some of the outgoing infrared radiation emitted by the earth and the atmosphere. The principal greenhouse gas is water vapor. The others are carbon dioxide (CO_2), ozone (O_3), methane (CH_4), and nitrous oxides (N_2O). These gases, together with clouds, keep the earth's surface and troposphere warmer than it would be otherwise. This is the natural greenhouse effect. Changes in the concentrations of these greenhouse gases will change the efficiency with which the earth cools to space. The atmosphere absorbs more of the outgoing terrestrial radiation from the surface when concentrations of greenhouse gas increase. This is the enhanced greenhouse effect—an enhancement of an effect that has operated in the earth's atmosphere for billions of years due to naturally occurring greenhouse gases (IPCC, 1996).

Human activities are changing the concentrations and distributions of greenhouse gases and aerosols in the atmosphere. The main human activities causing an increase in greenhouse gases are the combustion of fossil fuels and deforestation by forest burning. According to the IPCC (1996), global mean surface temperatures increased by 0.6°C since the late nineteenth century due to anthropogenic causes.

Volcanic aerosols. Volcanic activity can inject large amounts of sulfur-containing gases (primarily sulfur dioxide) into the stratosphere. Once reaching the stratosphere, some gases rapidly oxidize to sulfuric acid and condense with water to form an aerosol haze. Volcanic aerosols increase the planetary albedo, and the dominant radiative effect is an increase in scattering of solar radiation, which reduces the net radiation available to the surface/troposphere, thereby leading to a cooling. This can produce a large but transitory negative radiative forcing, tending to cool the earth's surface and troposphere for periods of up to two to three years. To have global effects, the latitude of eruption must lie between 30°N and 30°S. Eruptions poleward of these latitudes will affect the hemisphere only where the eruption occurs (IPCC, 1996). Because the impacts of volcanic aerosols last only a few seasons they increase the variability due to other effects.

The cryosphere. The changes in global snow and ice cover, other than in clouds, operate on long time scales except for seasonal snow cover. Monitoring of seasonal snow cover since 1972 shows that the extent of Northern Hemisphere snow cover has been less since 1987, particularly in spring (WMO, 1998). This might have decreased the regional surface albedo with a consequent temperature increase in the winter period for high latitude areas of the Northern Hemisphere (Sirotenko, 1999).

Land surface changes. Land surface changes, particularly large-scale afforestation or deforestation of areas, will affect the regional albedo and aerodynamic roughness. These will affect the transfer of energy, water, and other materials within the climate system. These effects are often more regional in their impacts on climate in the planetary boundary layer.

Internal Dynamics of the Climate System

Climate can vary because of the internal dynamics of the climate system. The most important source of this shorter time scale variability is from the El Niño/Southern Oscillation, North Atlantic Oscillation, and changes in sea-surface temperature.

El Niño/Southern Oscillation. Most of the internal variability of climate in the tropics and a substantial part of midlatitudes is related to El Niño/Southern Oscillation. ENSO is a natural phenomenon, and atmospheric and oceanic conditions in the tropical Pacific vary considerably, fluctuating somewhat irregularly between the El Niño phase and the opposite La Niña phase. In the former, warm waters from the western tropical Pacific migrate eastward, and in the latter, cooling of the tropical Pacific occurs. The whole cycle can last from three to five years.

As the El Niño develops, the trade winds weaken as the warmer waters in the central and eastern Pacific occur, shifting the pattern of tropical rainstorms east. Higher than normal air pressures develop over northern Australia and Indonesia with drier conditions or drought. At the same time, lower than normal air pressures develop in the central and eastern Pacific, with excessive rains in these areas, and along the western coast of South America. Approximately reverse patterns occur during the La Niña phase of the phenomenon.

The main global impacts that El Niño events cause are above-average global temperature anomalies. Since the mid-1970s El Niño events have been more frequent, and in each subsequent event, global temperature anomalies have been higher.

The North Atlantic Oscillation. Large-scale alternation of atmospheric pressure between the North Atlantic regions of the subtropical high (near the Azores) and subpolar low pressure (extending south and east of Greenland) determines the strength and orientation of the poleward pressure gradient over the North Atlantic and the midlatitude westerlies in this area. European precipitation is related to the NAO (Hurrell, 1995). When this is positive, as it was for winters in the 1980s, drier than normal conditions occur over southern Europe and the Mediterranean and above normal precipitation from Iceland to Scandinavia.

Sea-surface temperature. The atmosphere has very little memory, so after every short period everything that happened to it is forgotten by its internal mechanisms. Some investigators think it is the ocean that brings the atmosphere back onto the previous track. The ocean is sluggish and its time constant is entirely different from that of the atmosphere. So, once it adopts an abnormal thermal property, it transports the same from the surface to as deep as three or four hundred meters. Thus there is tremendous heat storage over the vast area of the ocean. Perhaps the long-lasting heat storage could in some complex way force the atmosphere to come back, from time to time, to a certain pattern. During the turbulent seasons when the wind is strong, the sea is stirred up and the anomalies of temperature work their way downward to deeper layers. With the onset of the season when the winds die down, the water generated in the turbulent season, whether cool or warm, stays at lower depths and remains hidden until the start of the next stormy season, when it is resurrected and brought up to the surface. Frequently there is a correlation between the sea-surface-temperature pattern of one winter and that of the next. From this, it could be concluded that if the sea temperature really affects the atmosphere, it might be one of the reasons why on occasions there are two years in a row with the same weather pattern.

OBSERVED CHANGE IN ATMOSPHERIC COMPOSITION AND CLIMATE

Carbon Dioxide

The concentration of carbon dioxide in the surface layer of the atmosphere was about 280 ppmv just before the industrial era started. This stood at 365 ppmv at the end of twentieth century. Thus the CO_2 concentration in the atmosphere has increased by about 30 percent in a span of 200 years. Burning of oil, coal, and natural gas and the clearing and burning of vegetation are the main causes of the rise. This gas makes the biggest contribution (about 70 percent) to the enhanced greenhouse effect (World Meterological Organization/Global Atmosphere Watch [WMO/GAW] 116, 1998).

Acidifying Compounds

Sulfur dioxide (SO_2) and nitrogen compounds are some of the major air pollutants emitted by industrial and domestic sources. Sulfur dioxide is further oxidized to sulfate, which exists in the atmosphere mainly as aerosols. Sulfate aerosols are found more in the Northern Hemisphere than in the Southern Hemisphere. Annual mean sulfur dioxide levels over land areas are estimated to be approximately 0.1 to 10 $ug{\cdot}m^{-3}$ (Ryaboshapko et al., 1998). Sulfate aerosols scatter (or reflect) sunlight, resulting in slight cooling at the earth's surface.

The main anthropogenic components of emissions of nitrogen compounds to the atmosphere are nitrogen oxides (NO_2), nitrous oxide (N_2O), and ammonia (NH_3). Nitrous oxide (N_2O), which is present in the atmosphere at a very low concentration (310 ppbv), is increasing slowly at a rate of about 0.25 percent per year. Despite its low concentration, it is an important greenhouse gas because of its longer lifetime (150 years) and much greater warming potential (about 30 times more than that of carbon dioxide). Burning of vegetation, industrial emissions, and effects of agriculture on soil processes have contributed to an increase of about 15 percent in the N_2O concentration in the atmosphere over the past 200 years (WMO/GAW 116, 1998).

Methane makes the next largest contribution to global warming—some 20 percent of the total. Although the annual increase in the methane load in the atmosphere is 1/100 that of carbon dioxide, its contribution to global warming is quite high (WMO/GAW 116, 1998). Its concentration has risen by about 145 percent over the past 200 years. The concentration of methane in the atmosphere (which is currently 1.74 ppm) is increasing at a rate of about 1 percent per year (WMO/GAW 116, 1998).

Tropospheric Ozone

Ozone is toxic for a wide range of living organisms. In the troposphere it is produced by a chain of chemical and photochemical reactions involving, in particular, nitrogen oxides, nitrous oxides, and volatile organic compounds (VOCs). Near the earth's surface, ozone concentrations are highly variable in space and time, with the highest values over industrial regions under suitable weather conditions. Global concentrations of ground-level ozone (yearly means) are about 45 $ug \cdot m^{-3}$ (Semenov, Kounina, and Koukhta, 1999). Measurements in Europe have shown that concentrations of ozone have increased from 20 to 30 $ug \cdot m^{-3}$ to 60 $ug \cdot m^{-3}$ during the twentieth century.

Anthropogenic emissions of chlorofluorocarbons (CFCs) and some other substances into the atmosphere are known to deplete the stratospheric ozone layer. This layer absorbs ultraviolet solar radiation within a wavelength range of 280 to 320 nm (UV-B), and its depletion leads to an increase in ground-level flux of UV-B. Enhanced UV-B negatively affects organic life in a number of ways. The current rate of increase of CFCs in the atmosphere is about 4 percent per year.

Ozone Hole

In 1985, large ozone losses were observed over the Antarctic region. NASA satellite observations showed that this ozone loss covered an extensive region, coining its name, the Antarctic ozone hole (Newman, 2000). The Antarctic ozone hole was subsequently shown to result from chlorine and bromine destruction of stratospheric ozone. The stratospheric chlorine and bromine levels primarily come from human-produced chemicals such as chlorofluorocarbons and halons, whose concentrations had been increasing throughout the 1970s and 1980s. Naturally occurring, extremely cold temperatures over Antarctica cause the formation of very tenuous clouds (polar stratospheric clouds, or PSCs). Certain chlorine and bromine compounds are then converted from benign forms into ozone-destructive forms when they come into contact with the surfaces of cloud particles. Hence, the massive ozone loss over Antarctica results from the unique meteorological conditions and the high levels of human-produced chlorine and bromine.

The Arctic stratosphere is considerably different from the Antarctic stratosphere. First, natural ozone levels in the Arctic spring are much higher than in the Antarctic spring. Second, Arctic spring stratospheric temperatures are much warmer than those in the Antarctic stratosphere. Because of the warmer Arctic stratospheric temperatures, polar stratospheric clouds are much less common over the Arctic than over Antarctica (Albritton and Kuijpers, 1999).

Changes in Temperature

Temperature anomalies that have been observed on a global and continental scale from the middle of the nineteenth century to the end of the twentieth century are shown in Figure 11.1. On a regional basis there are variations from these averages. The continent of Africa is warmer than it was 100 years ago (IPCC, 1996). Warming through the twentieth century has been approximately 0.7°C.

An average annual mean increase in surface air temperature of about 2.9°C in the past 100 years has been observed in boreal regions of Asia. During the cold winter season, mean surface air temperature increase is most pronounced at a rate of about 4.4°C/100 years (Gruza et al., 1997). In most of the Middle East region, the long time series of surface air temperature shows a warming trend. In Kazakhstan, the mean annual surface temperature has risen by about 1.3°C during 1894 to 1997 (IPCC, 1998). In temperate regions of Asia, covering Mongolia and northeastern China, temperature has increased at the global rate over the past 100 years. In Japan, the surface air temperature has shown a warming trend during the past century.

In tropical regions of Asia, several countries have reported increasing surface temperature trends in recent decades. The annual mean surface air temperature anomalies over India suggest a conspicuous and gradually increasing trend of about 0.36°C/100 years. The warming over India has been mainly due to increasing maximum temperatures rather than minimum temperatures, and the rise in surface temperature is most pronounced during winter and autumn (Rupakumar, Krishna Kumar, and Pant, 1994).

Warming trends in Australia are consistent with those elsewhere in the world. Australia warmed by 0.7°C from 1910 to 1990, with most of the increase occurring after 1950. Nighttime temperatures have risen faster than daytime temperatures (Whetton, 2001).

Most of Europe has experienced increases in surface air temperature during the twentieth century which, averaged across the continent, amounts to about 0.8°C in annual temperature (Beniston, 1997). The 1990-1999 decade has been the warmest in the instrumental record, both annually and for winter. Warming has been comparatively greater over northwestern Russia and

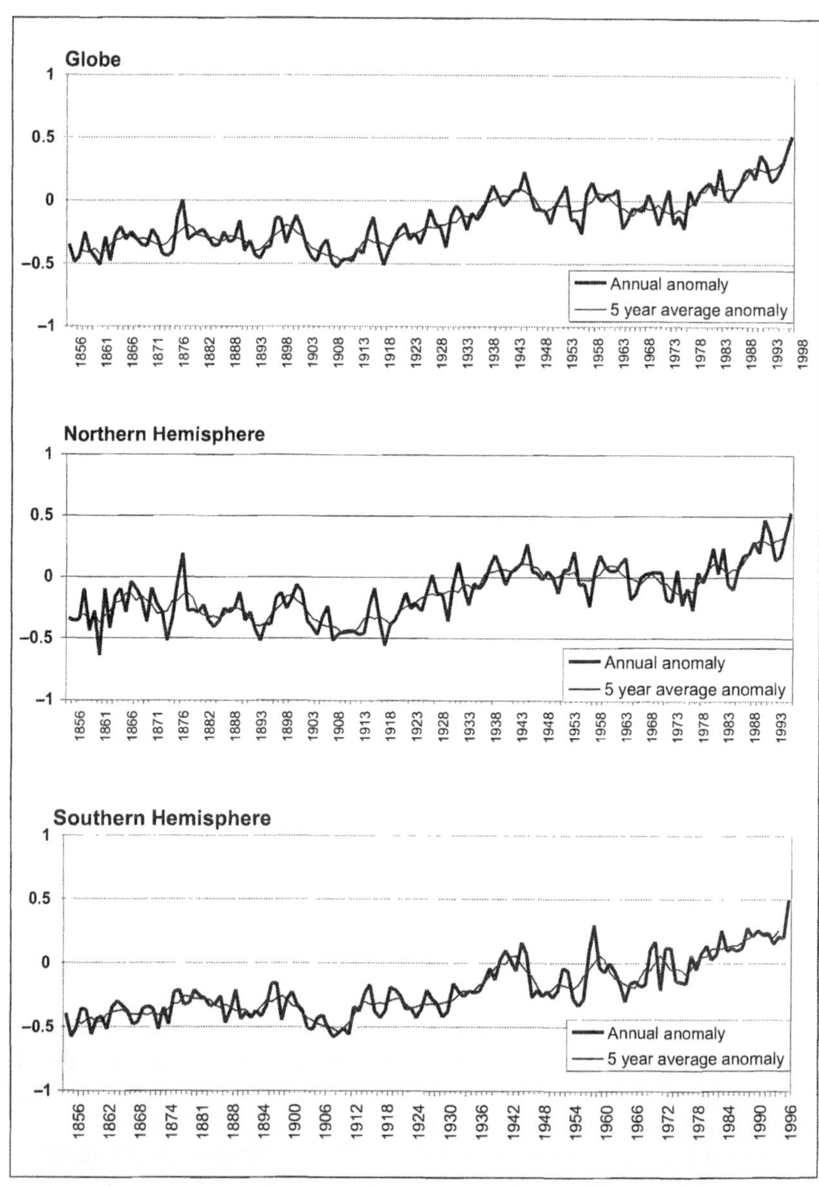

FIGURE 11.1. Trends in global and hemispherical temperature

the Iberian Peninsula, and stronger in winter than in summer. The warming in annual mean temperature has occurred preferentially as a result of nighttime rather than daytime temperature increases (Brazdil, 1996).

South American temperature records for many countries show temperatures have been warmer in the 1980s and 1990s, compared to the reference period from 1900 to 1940. Increasing trends have been found in the time series of daily mean and minimum air temperatures throughout Colombia, the Amazon region, and subtropical and temperate Argentina (Quintana-Gomez, 1999). North America, as a whole, has warmed by about 0.7°C/100 years, although this has been quite heterogeneous (Cubasch et al., 1995; Robinson, 2000). For example, the southeastern United States cooled slightly over that same period.

Significant warming in the Arctic since the beginning of the twentieth century has been confirmed by many different proxy measurements. Glaciers and ice caps in the Arctic have shown a retreat in low-lying areas since about 1920. Numerous small, low-altitude glaciers and perennial snow patches have disappeared. Greenland's ice sheet has thinned dramatically around its southern and eastern margins, many parts of which have lost 1.0 to 1.5 m per year in thickness since 1993 (Krabill et al., 1999). Snow cover extent in the Northern Hemisphere has reduced since 1972 by about 10 percent.

Summer sea ice extent has shrunk by 20 percent over the last 30 years in the Atlantic part of the Arctic Ocean (Walsh et al., 1998). Analysis of instrumental records has shown overall warming at permanently occupied stations on the Antarctic continent and Southern Ocean island stations. Sixteen Antarctic stations have warmed at a rate of 0.9 to 1.2 °C per century, and the 22 Southern Ocean stations have warmed at 0.7 to 1.0 °C per century.

Studies conducted by New Zealand Meteorological Service show that temperatures have been increasing by 0.1°C per decade in most of the small islands in the Pacific, Indian, and Atlantic Oceans and in the Caribbean Sea. Based on data from 34 stations in the Pacific from mostly south of the equator, surface air temperatures increased by 0.3°C to 0.8°C in the twentieth century.

Changes in Precipitation

Precipitation over North America increased by 70 mm per year during the later half of the twentieth century. These trends, like those of temperature, have been fairly heterogeneous. The largest increases have been in the northeastern and western coastal regions, with some regions of decreasing precipitation in the midcontinent (U.S. National Assessment, 2000).

Some parts of southern Mexico and Central America exhibit a trend toward less precipitation. In Colombia, long-term precipitation trends have been found with no preferred sign. For the Amazon region, recent studies based on the analysis of rainfall and river streamflow data show no significant trends toward drier or wetter conditions (Magana et al., 1997). In southern Chile and the Argentinean cordillera, a negative trend in precipitation and stream flow has been detected. Precipitation in subtropical Argentina, Paraguay, and Brazil exhibited an increasing tendency for the second half of the twentieth century (Magrin et al., 1999).

Trends in annual precipitation differ between northern and southern Europe. Precipitation over northern Europe has increased by between 10 and 40 percent in the twentieth century, whereas some parts of southern Europe have dried by up to 20 percent (Hulme and Carter, 2000). The time series of annual mean precipitation in Russia suggests a decreasing trend (Rankova, 1998).

For the long-term mean precipitation, a decreasing trend of about 4.1 mm/month during the last 100 years has been reported in boreal regions of Asia. In the arid and semiarid region of Asia, rainfall observations during the past 50-year period in some countries located in the northern parts of this region have shown an increasing trend on a mean annual basis. In Pakistan, the majority of stations have shown a tendency of increasing rainfall during the monsoon season. In the temperate region of Asia, covering the Gobi and northeastern China, the annual precipitation has been decreasing continuously since 1965. In the tropical region of Asia covering India and Sri Lanka, the long-term time series of summer monsoon rainfall has no discernible trends (Kothyari and Singh, 1996).

Most of the Arctic region has experienced increased rainfall since the 1950s. In the Antarctic region the rate of accumulation of ice shows increases in precipitation (Vaughan et al., 1999; Smith, Budd, and Reid, 1998).

In Australia, trends in rainfall are not very clear. The mean annual rainfall has increased by 6 percent (not statistically significant) since 1910. However, increases in the frequency of heavy rainfall and total rainfall are significant in many parts of southeastern Australia (Hennessy, Suppiah, and Page, 1999).

The average rise in sea level in the Australia-New Zealand region over the past 50 years is about 20 mm per decade (Salinger, Stigter, and Dasc, 2000). In the small island state regions of the Pacific and Indian Ocean, the rate of sea-level rise has also been approximately 2 mm per year.

OBSERVED IMPACT OF CLIMATE CHANGE

An accumulating body of evidence indicates that global warming (0.7°C over the twentieth century), especially during the last 50 years (0.43°C), has impacted a number of regions. The rates and patterns of climate change and impacts vary over the period as well. Anthropogenic and non-anthropogenic changes may have influenced greater changes in the climate system later in the century than earlier in the period. Impacts observed so far are primarily ecological in nature. These include changes in physiology, spatial distribution, species abundance and diversity, and timing of reproduction.

Much of the evidence of ecosystem changes to date has come from high-latitude (>40°N, 40°S) and high-altitude (>3,000 m) environments, and from species at their high-latitude range limits. Some evidence has been found in tropical and subtropical regions (both terrestrial and coastal ecosystems) and some in temperate oceans and coastal areas.

Hydrology

Observations in 1995-1996 show declines of glacier extent in western Antarctica and elsewhere, sea ice 40 percent thinner than 20 to 40 years ago in the Arctic, and shrinking of the area of perennial Arctic ice at a rate of 7 percent per decade (Vaughan et al., 1999). Satellite data over Northern Hemisphere extratropical lands show a retreat (about 10 percent reduction) of spring snow cover over the period 1973 to 1992 (Groisman, Karl, and Knight, 1994). Glaciers in Latin America have dramatically receded in the past decades. Many of them have disappeared completely. In 18 Peruvian glacial cordilleras, mass balances since 1968 and satellite images show a reduction of more than 20 percent of the glacial surface.

Changes in volume and areal extent of tropical mountain glaciers are among the best indicators of climate change. Himalayan glaciers that feed the Ganges River appear to be retreating at a fast rate. The estimated annual retreat of the Dokriani glacier (one of the several hundred glaciers that feed the Ganges) in 1998 was 20 meters compared to an annual average of 16.5 meters over 1993 to 1998. From observations dating back to 1842, the rate of recession of the snout (the point at which the glacier ice ends) has been found to have increased more than 2.5-fold per year. Between 1842 and 1935, the 26-kilometer long Gangotri glacier was receding at an average of 7.3 m every year, whereas between 1935 and 1990, the rate of recession had gone up to 18 m a year. Almost 67 percent of glaciers in the Himalayan and Tienshan mountain ranges have retreated since the 1970s (Fushimi, 1999).

Vegetation

Increased temperatures in mountainous regions appear to be causing plant species to move to higher altitudes. Approximate moving rates for common alpine plants are calculated to be between zero and four meters per decade (Grabberr, Gottfried, and Pauli, 1995).

In a short grass steppe in Colorado, with recorded temperature increases from 1964 to 1998, aboveground net primary productivity of the dominant grass is on the decrease (U.S. National Assessment, 2000).

Animals

A northward and upward shift has been detected in the range of checkerspot butterfly on the western coast of North America over the past century. In a sample of 35 nonmigratory European butterflies, 63 percent have ranges that have shifted north by 35 to 240 km during this century (Parmesan et al., 1999). The disappearance of 20 out of 50 species of frogs and toads in Costa Rica has been linked to recent warming (Pounds, Fogden, and Campbell, 1999). Earlier egg-laying dates have been found for 31 percent of 225 species of birds in the United Kingdom over the period 1971 to 1995 (Crick et al., 1997). In the Netherlands, the availability of caterpillar food has advanced by nine days over the period 1973 to 1995 (Visser et al., 1998).

In marine and littoral ecosystems, there is evidence of coral bleaching, declines of plankton, fish, and bird populations related to warming ocean temperature, mangrove retreat due to sea-level rise, and penguin species increases due to a decrease in sea ice.

Between 1974 and 1993, species richness of reef fishes has fallen and composition shifted from dominance by northern to southern species in the Southern California Bight (Holbrook, Schmitt, and Stephens, 1997).

Agriculture

Agriculture evidence of observed impacts is found in lengthening growing seasons at high latitudes, changing yield trends, and expansion of pest ranges. Carter (1998) observed that the growing season of the Nordic region (Iceland, Denmark, Norway, Sweden, and Finland) has lengthened over the period 1890 to 1995. Climate trends appear to be responsible for 30 to 50 percent of the observed increase in Australian wheat yields, with increases in minimum temperatures (decreases in frosts) being the dominant influence during 1952 to 1992 (Nicholls, 1997). Recent movement of agricultural pests and pathogens related to local climate trends is linked to global warming.

FUTURE SCENARIOS OF CLIMATE CHANGE

Uncertainties of Future Climate

Global climate toward the middle and later years of the twenty-first century is projected using general circulation and coupled atmosphere-ocean models (GCMs). However, many uncertainties currently limit the ability to project future climate change. Three main sources of uncertainty with regard to future climate are

1. future greenhouse gas and aerosol emissions;
2. global climate sensitivity due to differences in ways that physical processes and feedback are simulated (some models simulate greater global warming than others do); and
3. regional climate change that is apparent from differences in regional estimates in climate change from the same global warming.

The wide range of projected climate change suggests that caution is required when dealing with any impact assessment based on GCM results. O'Brien (1998) highlighted the fact that the earlier forecasts of greenhouse impact were exaggerated, and new studies are suggesting a postponement of the greenhouse effect. Consequently, there is more time than previously expected to adapt and to take technological action to alleviate global warming. Therefore, decision makers need to be aware of the uncertainties associated with climate projections while formulating strategies to cope with the risk of climate change.

Despite these uncertainties, GCMs provide a reasonable estimate of the important large-scale features of the climate system, including seasonal variations and ENSO-like features. Many climate changes are consistently projected by different models in response to greenhouse gases and aerosols and are explainable in terms of physical processes. The models also produce with reasonable accuracy other variations due to climate forcing, such as interannual variability due to ENSO and the cause of temperature change because of stratospheric aerosols.

Climate Scenarios

A scenario is a coherent, internally consistent, and plausible description of a possible future state of the world (IPCC, 1994). It is not a forecast; rather, each scenario is one alternative image of how the future can unfold. Scenarios are one of the main tools for the assessment of future developments in complex systems that are often inherently unpredictable, insufficiently understood, and possessing many scientific uncertainties. Scenarios are also vital aids in evaluating the options for mitigating future emissions of greenhouse gases and aerosols, which are known to affect global climate. There are three main approaches to climate scenario development:

1. *Incremental scenarios:* In this approach particular climatic or related elements are changed by realistic but arbitrary amounts. They are commonly applied to study the sensitivity of an exposure unit to a wide range of variations in climate.
2. *Analogue scenarios*: Analogue scenarios are both temporal and spatial. Temporal analogues use climatic information from the past as an analogue of the future climate. Analogue scenarios are based on past climate, as reconstructed from fossil records as well as observed records of the historical period. Spatial analogue scenarios are the climatic conditions in regions that are analogues to those anticipated in the study region in the future.
3. *Climate model output-based scenarios:* These are based on the results of general circulation model experiments. GCMs are three-dimensional mathematical models that represent the physical and dynamic processes responsible for climate. This is the most commonly used approach in climate change research.

A Generalized Global Climate Scenario of the Twenty-First Century

Based on multimodel output, the current range of twenty-first-century global surface temperature warming is 1.5 to 4.5°C, with a "best estimate" of 2.5°C (Kenitzer, 2000). The increases in surface temperature and other associated changes are expected to increase climate variability.

Climate models simulate a climate change-induced increase in precipitation in high and midlatitudes and most equatorial regions but a general decrease in the subtropics (IPCC, 1996). Across large parts of the world, changes in precipitation associated with global warming are small compared to those due to natural variability.

Global mean sea level is expected to rise as a result of thermal expansion of the oceans and melting of glaciers and ice sheets. The IPCC (1996) estimates sea-level rise from 1995 to the 2050s in the range of 13 and 68 cm and in the range of 15 to 95 cm to the year 2100, with a "best estimate" of 50 cm.

IMPACT OF CLIMATE CHANGE ON HYDROLOGY AND WATER RESOURCES

One of the major impacts of global warming is likely to be on hydrology and water resources, which in turn will have a significant impact across many sectors of the economy, society, and environment (Figures 11.2 and 11.3). Characteristics of many ecosystems are heavily influenced by water availability. Water is fundamental for human life and many activities, in-

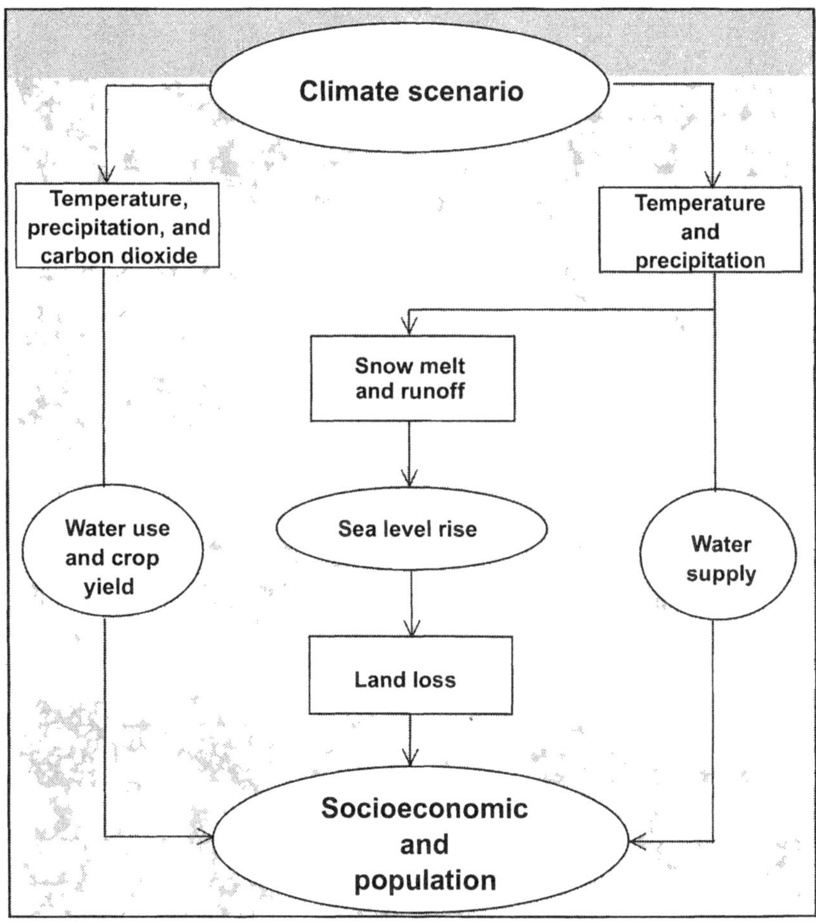

FIGURE 11.2. Impact of climate change on water resources and agriculture

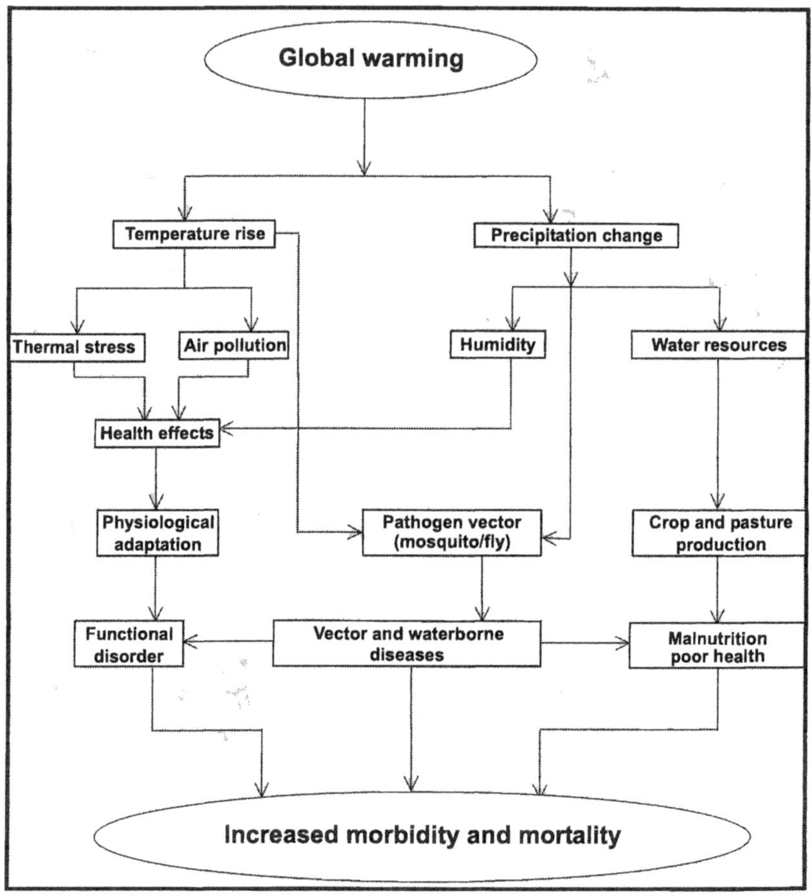

FIGURE 11.3. Impact of global warming on human and animal diseases

cluding agriculture, industry, and power generation. On the global scale, climate change is likely to worsen water resource stress in some regions but perhaps ameliorate stress in others. At the regional scale there are mixed signals.

Africa

A major impact of climate change over the African continent is a shift in the temporal and spatial distribution of precipitation. This will result in a shift of runoff or hydrological resources in both time and space.

Future climatic changes in the Nile River basin would be significant and possibly severe. For example, with 4°C warming and a 20 percent decrease in precipitation, Nile River flow decreases 98 percent. This represents a significant reduction in water supply (Gleick, 1993, 1998). Based on the results of the river flow responses, climate variables alone can cause a 50 percent change in runoff in the Gambia River catchment. In general, a 1 percent change in rainfall will result in 3 percent change in runoff (Jallow et al., 1999). For the Zambezi River basin, simulated runoff under climate change is projected to decrease by about 40 percent or more (Cambula, 1999).

Asia

Several simulation studies suggest that some areas of the Asian continent are expected to experience an increase in water availability, while other areas will have reduced water resources available.

The Himalayas have nearly 1,500 glaciers that provide snow and glacial-melt waters to keep the major rivers perennial throughout the year. Glacial melt is expected to increase under changed climate conditions, which would lead to increased summer flows in some river systems for a few decades, followed by a reduction in flow as the glaciers disappear (IPCC, 1998).

Large-scale shrinkage of the permafrost region in boreal Asia is also likely. Due to global warming, permafrost thawing will start over vast territories (IPCC, 1998). The perennially frozen rocks will completely degrade within the present southern regions. In the northern regions of boreal Asia, the mean annual temperature of permafrost and hence the depth of seasonal thawing (active layer thickness) will increase (Izrael, Anokhin, and Eliseev, 1997).

The average annual runoff in the river basins of the Tigris, Euphrates, Indus, and Brahmaputra Rivers would decline by 22, 25, 27, and 14 percent, respectively, by the year 2050 (Izrael, Anokhin, and Eliseev, 1997). Runoff in the Yangtze and Huang He Rivers has the potential to increase up to 37 and 26 percent. Increases in annual runoff are also projected in the Siberian large rivers: Yenise by 15 percent, Lena by 27 percent, Ob by 12 percent, and Amur by 14 percent.

Surface runoff is projected to decrease drastically in arid and semiarid central Asia under climate change scenarios and would significantly affect the volume of available water for withdrawal for irrigation and other purposes (Gruza et al., 1997).

In temperate Asia (Mongolia, northern China, and Japan), an increase in surface runoff seems likely, but a decline is possible in southern China. The hydrological characteristics of Japanese rivers and lakes are sensitive to climate change.

The perennial rivers originating in the high Himalayas receive water from snow and glaciers. Because the melting season of snow coincides with the summer monsoon season, any intensification of the monsoons is likely to contribute to flood disasters in the Himalayan catchments. Such impacts will be observed more in the western Himalayas compared to the eastern Himalayas, due to the higher contribution of snow-melt runoff in the west (Singh, 1998).

Australia

Global warming will adversely affect water resources in Australia. Although increases in stream flow are possible in northern Australia, decreases in stream flow seem likely in other parts of the country due to a decrease in rainfall. Estimated changes in stream flow in the Murray-Darling Basin range from 0 to –20 percent by 2030, and +5 to –45 percent by 2070 (Commonwealth Scientific and Industrial Research Organisation [CSIRO], 2001). Estimates also show large decreases in both the maximum and minimum monthly runoff. This implies large increases in drought frequency (Arnell, 2000). Another study (Kothavala, 1999) has also concluded that there will be longer and more severe droughts under doubled CO_2 conditions than in the control simulation.

Application of the CSIRO (1996) scenarios also suggests a possible combination of small or larger decreases in mean annual rainfall, higher temperatures and evaporation, and a higher frequency of floods and droughts in northern Victorian rivers (Schreider et al., 1996). A study of the Macquarie River basin in New South Wales indicated inflow reductions of 10 to 30 percent for doubled CO_2 and reduced stream flows if irrigation demand remains constant or increases (Hassall and Associates et al., 1998). There is also concern about the adverse effects of increased drought frequency on water quality through possible increases in toxic algal blooms (Murray-Darling Basin Commission [MDBC], 1999).

Europe

Calculations at the continental scale (Arnell, 2000) indicate that under most climate change scenarios northern Europe would see an increase in annual average stream flow, but southern Europe would experience a reduction in stream flow. In much of midlatitude Europe annual runoff would decrease or increase by around 10 percent by the 2050s, but the change may be significantly larger further north and south.

In Mediterranean regions, climate change is likely to considerably exaggerate the range in flows between winter and summer. In maritime western Europe the range is also likely to increase but to a lesser extent. In more continental and upland areas, where snowfall makes up a large portion of winter precipitation, a rise in temperature would mean that more precipitation falls as rain, and, therefore, winter runoff increases and spring snow melt decreases (Arnell, 2000).

South America

In some Latin American areas the availability of freshwater will be substantially changed by global warning, especially in areas where it is possible that the combined effect of less rainfall and more evaporation could take place and lead to less runoff (Marengo, 1995).

Hydrological scenarios for Central America show that a significant limitation of the potential water resources will occur due to an increase in evapotranspiration and changes in precipitation. Studies on vulnerability of hydrologic regions in Mexico and all Central American countries to future changes in climate suggest that potential changes in temperature and precipitation may have a dramatic impact on the pattern and magnitude of runoff, on soil moisture, and on evaporation (Arnell, 2000). In the Uruguay River basin a decrease in runoff during low-flow periods of the year is anticipated. Argentina could foresee a reduction in water availability from the snow melt in the high Andes and in central western regions (Marengo, 1995).

North America

In North America global warming may lead to substantial changes in mean annual stream flows, the seasonal distribution of flows, and the probability of extreme high or low flow conditions. Runoff changes will depend on changes in temperatures and other climatic variables, and warmer temperatures may cause runoff to decline even where precipitation increases (Nash and Gleick, 1993; Matalas, 1998).

Polar Regions

The Greenland ice sheet already suffers melting in summer over much of its margin. There is a trend toward an increase in the area and duration of this melt (Abdalati and Steffen, 1997). This is likely to continue. Airborne altimetric monitoring has shown that over the period 1993 to 1998, the Greenland ice sheet was slowly thickening at higher elevations, while at lower elevations, thinning of around 1 m/year was underway (Krabill et al., 1999). If warming continues, the Greenland ice sheet will eventually disappear, but this will take many centuries.

Over the Antarctic ice sheet, where only a few limited areas show summer melting, the likely response is toward a slight thickening as precipitation rates increase (Vaughan et al., 1999).

IMPACT OF CLIMATE CHANGE ON CROPS

Major impacts on crop plant growth and production will come from changes in temperature, moisture levels, ozone, ultraviolet radiation, carbon dioxide levels, pests, and diseases (Figures 11.2 and 11.3).

The effects of a temperature increase on photosynthetic productivity of crop plants will interact with the current rise in the atmospheric concentration of CO_2. Under elevated CO_2, the extra carbohydrates produced by increases in photosynthesis result in an increase in grain yield (Horie et al., 1996). Many researchers (Kimball et al., 1995; Samarakoon and Gifford, 1995; Horie et al., 1996; Pinter, 1996; Semenov, Kounina, and Koukhta, 1999) are of the opinion that the actual impact of elevated CO_2 on crop growth, and especially on yields, is likely to be significantly less than the estimates that are currently presented. It is suspected that a portion of the increase in grain yield driven by anthropogenic enrichment of the atmosphere may be suppressed by ozone.

Africa

On the African continent, global warming is likely to negatively alter the production of major food crops—rice, wheat, corn, beans, and potatoes. The high-altitude farming districts in Africa may have their altitudinal zonation wiped out and be forced to find new forms of agriculture. Wheat and corn associated with the subtropical latitudes may suffer a drop in yield due to increased temperature, and rice may disappear due to higher temperatures in the tropics (Odingo, 1990; Pimentel, 1993; Muchena and Iglesias, 1995). African agriculture is expected to survive and even become stronger where mixed cropping is currently practiced and where tree crops are predominant.

Asia

Global warming will affect the scheduling of the cropping season, as well as the duration of the growing period of the crop in all the major crop-producing areas of Asia. In general, areas in mid and high latitudes will experience increases in crop yields, while yields in areas in the lower latitudes will generally decrease (Lou and Lin, 1999).

In China, the yields of major crops are expected to decline due to climate change. The decline in rice yield is due to a shortening of the growth period, decrease in photosynthesis ability, and increase in respiration, demanding more water availability. The area under wheat is likely to expand in northern and western China. Climate change should be advantageous to wheat yield in northeastern China. However, in middle and northern China, high temperatures during later crop stages could result in yield reductions (Wang, 1996). A doubling of atmospheric carbon dioxide levels will substantially increase rice yields and yield stability in northern and north-central Japan (Horie et al., 1996; Rosenzweig and Hillel, 1998).

In India, while the wheat crop is found to be sensitive to an increase in maximum temperature, the rice crop is vulnerable to an increase in minimum temperature. The adverse impacts of likely water shortage on wheat productivity could be minimized to a certain extent under elevated CO_2 levels. They would largely be maintained for rice crops, resulting in a net decline in rice yields (Lal et al., 1998). Acute water shortage conditions combined with thermal stress should adversely affect both wheat and, more severely, rice productivity in northwest India, even under the positive effects of elevated CO_2 in the future.

The impact of rise in temperature and increases in atmospheric carbon dioxide on rice production in Bangladesh, Indonesia, Malaysia, Myanmar, the Philippines, South Korea, and Thailand suggest that the positive effects of enhanced photosynthesis due to doubling of CO_2 are canceled out for increases in temperature beyond 2°C (Matthews et al., 1995).

Australia

A study of global climate-change impacts on wheat crops across the Australian wheat belt shows that doubling CO_2 alone produced national yield increases of 24 percent in currently cropped areas but with a fall in grain protein content of 9 to 15 percent (Howden, Hall, and Bruget, 1999). However, if rainfall decreases by 20 percent, yields would increase for 1°C warming but decline for greater warming (CSIRO, 2001). With greater decreases in rainfall there would be much larger negative impacts, with cropping becoming nonviable over many regions, especially in Western Australia.

Banks (1996) estimated the broad impact of greenhouse effects on eastern Australian agriculture. According to these estimates, there will be an increase in summer growing grasses in the far west rangelands, and livestock stocking rates could increase. In the Tablelands, winter cereals will benefit from higher temperatures throughout the growing season and may provide more flexibility in sowing time. The productivity of perennial pastures will take advantage of increased rainfall. Lucernes will increase in importance as a component of pastures. In the coastal regions citrus crops will mature early. However, deciduous fruits that require vernalization with a significant frost period will be forced to higher latitudes.

More damage is expected to fruit crops from insect pests. Sutherst, Collyer, and Yonow (2000) examined the vulnerability of apples, oranges, and pears in Australia to the Queensland fruit fly under climate change. The results revealed that the range of the fruit fly would spread further south and the number of fruit fly generations in the sensitive area would progressively increase. One extra generation was experienced over the whole area with a 2°C rise in temperature. In this scenario, the damage costs due to Queensland fruit fly may further increase by $3.5 million to oranges, $5.6 million to apples, and $2.8 million to pears. Similar damage costs may increase due to light brown apple moth (CSIRO, 2001).

A projected decrease in frost will reduce frost damage to fruits. However, temperate fruits need winter chilling to ensure normal bud burst and fruit set. Warmer winters will reduce chilling duration, leading to lower yields and quality.

In New Zealand, generally drier conditions and reductions in ground water will have substantial impacts on cereal production in the Canterbury wheat and barley production area. Other grain-producing areas are less likely to be affected. Crop phenological responses to warming and increased carbon dioxide are mostly positive, making grain filling slightly earlier and decreasing drought risk. Rising temperatures would make the maize crop less risky in the south, but water availability may become a problem in Canterbury. Climate warming is decreasing frost risk for late-sown crops, extending the season and moving the southern production margin further south. Climate change may have mixed results on horticulture in New Zealand (Hall and McPherson, 1997).

Europe

Climatic warming will expand the area of cereals (wheat and maize) cultivation northward (Carter, Saarikko, and Niemi, 1996). For wheat, a temperature rise will lead to a small yield reduction, whereas an increase in CO_2 will cause a large yield increase. The net effect of both temperature and CO_2 for a moderate climate change is a large yield increase in southern Europe (Harrison and Butterfield, 1996). Maize yield will increase in northern areas and decrease in the southern areas of Europe (Wolf and van Diepen, 1995).

A temperature increase will shorten the length of the growing period and possibly reduce yields of seed crops (Peiris et al., 1996). At the same time, however, the cropping area of the cooler season seed crops will probably expand northward, leading to increased productivity of seed crops there. There will also be a northward expansion of warmer season seed crops. Analysis of the effect of climatic change on soybean yield suggests mainly increases in yield (Wolf, 1999).

The response of vegetable crops to changes in temperature varies among species. For crops such as onions, warming will reduce the duration of crop growth and hence yield, whereas warming stimulates growth and yield in crops such as carrots (Wheeler et al., 1996). For cool-season vegetable crops such as cauliflower, large temperature increases may decrease production during the summer period in southern Europe due to decreased quality (Olesen and Grevsen, 1993).

Potato and other root and tuber crops are expected to show increases in yield in northern Europe and decreases or no change in the rest of Europe (Wolf, 1999). Sugar beet may benefit from both the warming and the increase in CO_2 concentration (Davies et al., 1998).

For grapevines there is potential for an expansion of the wine-growing area in Europe and also for an increase in yield. The area suitable for olive cultivation could be enlarged in France, Italy, Croatia, and Greece due to changes in temperature and precipitation patterns (Bindi, Ferrini, and Miglietta, 1992).

South America

The impact of global warming and CO_2 increase on South American agriculture varies by region and by crop. Crop yield in the Pampas of Argentina and Uruguay is more sensitive to expected variations in temperature than precipitation. Under CO_2 doubling, maize, wheat, and sunflower yield variations were inversely related to temperature increments, while soybean would not be affected for temperature increments up to 3°C (Magrin et al., 1999).

Plantation forestry is a major land use in Brazil. Climate change can be expected to reduce silvicultural yields to the extent the climate becomes drier in major plantation states as a result of global warming (Gates et al., 1992; Fearnside, 1999).

North America

Estimates of the impacts of climate change on crops across North America vary widely (U.S. National Assessment, 2000; Brklacich et al., 1997; Rosenzweig, Parry, and Fischer, 1995). Most global climate-change scenarios indicate that higher latitudes in North America would undergo warming that would affect the growing season in this region. Estimates of increases in the frost-free season under climatic change range from a minimum of one week to a maximum of nine weeks (Brklacich et al., 1997).

For the North American prairies, Ontario, and Quebec, most estimates suggest an extension of three to five weeks. Although warmer spring and summer temperatures might be beneficial to crop production in northern latitudes, they may adversely affect crop maturity in regions where summer temperature and water stress limit production (Rosenzweig and Tubiello, 1997).

Drought may increase in the southern prairies, and production areas of corn and soybean may shift northward in Canada (Mills, 1994; Brklacich et al., 1997). Southern regions growing heat-tolerant crops such as citrus fruit and cotton would benefit from a reduced incidence of killing frosts resulting from a change in climate. Production of citrus fruit would shift northward in the southern United States, but yields may decline in southern Florida and Texas due to higher temperatures during the winter (Rosenzweig, Parry, and Fischer, 1995).

Mexican agriculture appears to be particularly vulnerable to climate-induced changes in precipitation, because most of its agricultural land is classified as arid or semiarid. On average, more than 90 percent of losses in Mexican agriculture are due to drought (Appendini and Liverman, 1993). Under the impact of global warming, the area presently suitable for rainfed maize production would shrink in northern and central regions of Mexico (Conde, 1997).

IMPACT OF CLIMATE CHANGE ON LIVESTOCK

Climate change may influence livestock systems directly by its effects on animal health, growth, and reproduction, and indirectly through its impacts on productivity of pastures and forage crops.

Africa

Domestic livestock in Africa (other than pigs) are concentrated in the arid and semiarid zones. The overwhelming majority of these animals feed predominantly on natural grasslands and savannas. In broad terms, changes in range-fed livestock numbers in any African region will be directly proportional to changes in annual precipitation. Given that several GCMs predict a decrease in mean annual precipitation of 10 to 20 percent in the main semiarid zones of Africa, there is a real possibility climate change will have a negative impact on pastoral livelihoods. Because the CO_2 concentration will rise in the future, its positive impact on water use efficiency will help to offset a reduction in rainfall of the same order. Simulations of grassland production in southern Africa indicate an almost exact balancing of these two effects for that region (Ellery, Scholes, and Scholes, 1996).

African cattle are mostly more heat tolerant than European cattle. In extremely hot areas, even the African breeds are beyond their thermal optimum. Under global warming, meat and milk production decline largely because the animals remain in the shade instead of grazing.

In the higher-altitude and higher-latitude regions of Africa, sheep are currently exposed to winter temperatures below their optimum. Mortality often results when cold periods coincide with rains. These episodes are likely to decrease in frequency and extent in the future.

Livestock distribution and productivity could be indirectly influenced by changes in the distribution of vector-borne livestock diseases, such as nagana, and the tick-borne East Coast Fever and Corridor disease (Hulme, 1996). Simulations of changes in the distribution of tsetse fly indicate that with warming it could potentially expand its destructive range.

Australia

Simulation studies conducted in Australia (McKeon et al., 1998; Hall et al., 1998) show that CO_2 increase is likely to improve pasture growth. There is also a strong sensitivity to rainfall, such that a 10 percent reduction in rainfall would balance out the effect of a doubling of CO_2 concentration. A 20 percent reduction in rainfall at doubled CO_2 concentration is likely to reduce pasture productivity by about 15 percent, liveweight in cattle by 12 percent, and substantially increase variability of stocking rates, reducing farm income. A substantial reduction in rainfall in many parts of Australia would tend to reduce productivity. However, in the far west rangelands of eastern Australia, summer growing grasses will increase as a result of increased summer rainfall, and rapid pasture growth will lead to higher stocking rates (Banks, 1996).

An assessment of the response of dairy cattle to heat stress in New South Wales and Queensland indicated significant increases in heat stress over the past 40 years. Physiological effects of heat stress include reduced food intake, weight loss, decreased reproduction rates, reduction in milk yields, increased susceptibility to parasites, and, in extreme cases, collapse and death (Davison et al., 1996; Howden, Hall, and Bruget, 1999). Jones and Hennessy (2000) modeled the impact of heat stress on dairy cows in the Hunter Valley in NSW under the climate change scenarios. They estimated the probabilities of milk production losses as a function of time. According to their estimates, under uncontrolled conditions, average milk loss from the cows without shade by the year 2030 will increase by 4 percent. By 2070, the milk loss will increase by about 6 percent of the annual production.

In New Zealand, productivity of dairy farms might be adversely affected by a southward shift of undesirable subtropical grass species, such as *Paspalum dilatatum* (Campbell et al., 1996).

Europe

Global warming may negatively affect livestock production in summer in currently warm regions of Europe (Furquay, 1989). Warming during the cold period for cooler regions is likely to be beneficial due to reduced feed requirements, increased survival, and lower energy costs. Impacts will probably be minor for intensive livestock systems where climate is controlled (confined dairy, poultry, and pigs). Climate change may, however, affect requirements for insulation and air-conditioning and thus change housing expenses (Cooper, Parsons, and Dernmers, 1998).

In Scotland, studies of the effect on grass-based milk production indicate that for herds grazed on grass-clover swards milk output may increase regardless of site, due to the effect on nitrogen fixation (Topp and Doyle, 1996).

South America

Ranching is a major land use in many parts of Latin America. In Brazil, Argentina, and Mexico, pastures occupy much more area than crops and livestock is almost exclusively raised on rangelands, with no storage of hay or other alternative feeds (Baethgen, 1997). Grass production in rangelands depends on rainfall, and reduced grass availability in dry periods limits cattle stocking rates over most of the region. In areas subject to prolonged droughts, such as northeastern Brazil and many rangeland areas in Mexico, production would be negatively affected by increased variability of precipitation due to climate change. In the Amazonian floodplains, higher peak flood stages would cause losses to cattle kept on platforms during the high-water period.

In Argentina, cattle are mainly fed on alfalfa and some other forage crops. A 1°C rise in temperature would increase alfalfa yields by 4 to 8 percent on average for most varieties, but there will be regional differences. Pasture yields would be reduced in areas north of 36°S and would be increased south of this latitude (Magrin et al., 1999).

North America

Estimates in livestock production efficiency in North America suggest that the negative effects of hotter weather in summer outweighed the positive effects of warmer winters (Adams, 1999). The largest change occurred under a 5°C increase in temperature, when livestock yields fell by 10 percent in Appalachia, the Southeast, the Mississippi Delta, and the southern plains regions of the United States. The smallest change was 1 percent under 1.5°C warming in the same regions. Livestock production could also be adversely affected by an increase in the frequency of blizzards in eastern Canada and the northeastern United States.

References

Chapter 1

Anonymous (2000). Climate risk. *Farming Ahead with Kondinin Group* 105: 28-41.

Blad, B.L. (1994). Future directions and needs for academic education in agricultural meteorology. *Agricultural and Forest Meteorology* 69: 27-32.

Bourke, P.M.A. (1968). The aims of agrometeorology. In *Agroclimatological Methods—Proceedings of the Reading Symposium* (pp. 11-15). Paris: UNESCO.

Decker, W.L. (1994). Developments in agricultural meteorology as a guide to its potential for the twenty-first century. *Agricultural and Forest Meteorology* 69: 9-25.

De Pauw, E., Göbel, W., and Adam, H. (2000). Agrometeorological aspects of agriculture and forestry in the arid zones. *Agricultural and Forest Meteorology* 103: 43-58.

Fleming, J. (Ed.) (1996). *Historical Essays in Meteorology—1919-1995.* Boston, MA: American Meteorological Society.

Hatfield, J.L. (1994). Future needs in agricultural meteorology: Basic and applied research. *Agricultural and Forest Meteorology* 69: 39-45.

Hollinger, S.E. (1994). Future directions and needs in agricultural meteorology/climatology and modeling. *Agricultural and Forest Meteorology* 69: 1-7.

Hoogenboom, G. (2000). Contribution of agrometeorology to the simulation of crop production and its applications. *Agricultural and Forest Meteorology* 103: 137-157.

Hoppe, P. (2000). Challenges for biometeorology on the turn of the century. In de Dear, R., Oke, T., Kalma, J., and Auliciems, A. (Eds.), *Biometeorology and Urban Climatology at the Turn of the Millennium* (pp. 383-386). Geneva: World Meteorological Organization.

Jagtap, S.S. and Chan, A.K. (2000). Agrometeorological aspects of agriculture in the sub-humid and humid zones of Africa and Asia. *Agricultural and Forest Meteorology* 103: 59-72.

Lomas, J., Milford, J.R., and Mukhala, E. (2000). Education and training in agricultural meteorology: Current status and future needs. *Agricultural and Forest Meteorology* 103: 197-208.

Maracchi, G., Pérarnaud, V., and Kleschenko, A.D. (2000). Applications of geographical information systems and remote sensing in agrometeorology. *Agricultural and Forest Meteorology* 103: 119-136.

Mavi, H.S. (1994). *Introduction to Agrometeorology.* New Delhi: Oxford & IBH.

Molga, M. (1962). *Agricultural Meteorology.* Part II—*Outline of Agrometeorological Problems.* Warsaw: Polish Academy of Sciences.

Monteith, J.L. (2000). Agricultural meteorology: Evolution and application. *Agricultural and Forest Meteorology* 103: 5-9.

Newman, J.E. (1974). Applying meteorology to agriculture. *Agricultural Meteorology* 13: 1-3.

Ogallo, L.A., Boulahya, M.S., and Keane, T. (2000). Applications of seasonal to interannual climate prediction in agricultural planning and operations. *Agricultural and Forest Meteorology* 103: 159-166.

Olufayo, A.A., Stigter, C.J., and Baldy, C. (1998). On needs and deeds in agrometeorology in tropical Africa. *Agricultural and Forest Meteorology* 92: 227-240.

Overdieck, D. (1997). Statement of the field editor for plants. *International Journal of Biometeorology* 40: 2.

Paw U, K.T. (2000). Recent advances in agricultural, forest and plant biometeorology. In de Dear, R., Oke, T., Kalma, J., and Auliciems, A. (Eds.), *Biometeorology and Urban Climatology at the Turn of the Millennium* (pp. 217-222). Geneva: World Meteorological Organization.

Perry, K.B. (1994). Current and future agricultural meteorology and climatology education needs of the US Extension service. *Agricultural and Forest Meteorology* 69: 33-38.

Plant, S. (2000). The relevance of seasonal climate forecasting to a rural producer. In Hammer, G.L., Nicholls, N., and Mitchell, C. (Eds.), *Applications of Seasonal Climate Forecasting in Agricultural and Natural Ecosystems* (pp. 23-28). London: Kluwer Academic Publishers.

Rijks, D. and Baradas, M.W. (2000). The clients for agrometeorological information. *Agricultural and Forest Meteorology* 103: 27-42.

Salinger, M.J., Stigter, C.J., and Das, H.P. (2000). Agrometeorological adaptation strategies to increasing climate variability and climate change. *Agricultural and Forest Meteorology* 103: 167-184.

Seeley, M.W. (1994). The future of serving agriculture with weather/climate information and forecasting: Some indication and observation. *Agricultural and Forest Meteorology* 69: 47-49.

Serafin, R.J., Macdonald, A.E., and Gall, R.L. (2002). Transition of weather research to operations: Opportunities and challenges. *Bulletin of the American Meteorological Society* 83: 377-392.

Sivakumar, M.V.K., Stigter, C.J., and Rijks, D. (2000). Foreword. *Agricultural and Forest Meteorology* 103: 1-2.

Smith, L.P. (1970). Aims and extent of agricultural meteorology. *Agricultural Meteorology* 7: 193-196.

Stigter, C.J., Sivakumar, M.V.K., and Rijks, D.A. (2000). Agrometeorology in the 21st century: Workshop summary and recommendations on needs and perspectives. *Agricultural and Forest Meteorology* 103: 209-227.

Strand, J.F. (2000). Some agrometeorological aspects of pest and disease management for the 21st century. *Agricultural and Forest Meteorology* 103: 73-82.

Weiss, A., Van Crowder, L., and Bernardi, M. (2000). Communicating agrometeorological information to farming communities. *Agricultural and Forest Meteorology* 103: 185-196.

White, B. (2000). The importance of climate variability and seasonal forecasting to the Australian economy. In Hammer, G.L., Nicholls, N., and Mitchell, C. (Eds.), *Applications of Seasonal Climate Forecasting in Agricultural and Natural Ecosystems* (pp. 1-22). London: Kluwer Academic Publishers.

Chapter 2

Aikman, D.P. (1989). Potential increase in photosynthetic efficiency from the redistribution of solar radiation in a crop. *Journal of Experimental Botany* 40: 855-864.

Alados, I., Foyo-Moreno, I., Olmo, F.J., Alados-Arboledas, L., and Grupo de Física de la Atmósfera (2002). Improved estimation of diffuse photosynthetically active radiation using two spectral models. *Agricultural and Forest Meteorology* 111: 1-12.

Anisimov, O. and Fukshansky, L. (1997). Optics of vegetation: Implications for the radiation balance and photosynthetic performance. *Agricultural and Forest Meteorology* 85: 33-49.

Arkebauer, T.J., Weiss, A., Sinclair, T.R., and Blum, A. (1994). In defence of radiation use efficiency: A response to Demetriades Shah. (1992). *Agricultural and Forest Meteorology* 68: 221-227.

Barrett, E.C. (1992). *Introduction to Environmental Remote Sensing*. London: Chapman & Hall.

Baumgartner, A. (1973). Estimation of the radiation and thermal micro-environment from meteorological and plant parameters. In Slatyer, R.O. (Ed.), *Plant Response to Climatic Factors. Proceedings of the Uppsala Symposium* (pp. 313-323). Paris: UNESCO.

Berbigier, P. and Hassika, P. (1998). Annual cycle of photosynthetically active radiation in maritime pine forest. *Agricultural and Forest Meteorology* 90: 157-171.

Black, C. and Ong, C. (2000). Utilisation of light and water in tropical agriculture. *Agricultural and Forest Meteorology* 104: 25-47.

Boes, E. (1981). Fundamentals of solar radiation. In Krieder, J.F. and Krietb, F. (Eds.), *Solar Energy Hand Book* (pp. 2.1-2.76). New York: McGraw-Hill.

Bonhomme, R. (2000). Beware of comparing RUE values calculated from PAR vs solar radiation or absorbed vs intercepted radiation. *Field Crops Research* 68: 247-252.

Butler, W.L. and Roberts, J.D. (1966). Light and Plant Development. In Janick, J. (Ed.), *Plant Agriculture; Readings from Scientific American* (pp. 78-85). New York: W.H. Freeman & Co.

Caldwell, M.M. (1981). Plant response to solar ultra-violet radiation. In Lange, O.L. (Ed.), *Physiological Plant Ecology I. Response to Physical Environment* (pp. 169-198). Berlin: Springer-Verlag.

Campbell, C.S., Heilman, J.L., McInnes, K.J., Wilson, L.T., Medley, J.C., Wu, G., and Cobos, D.R. (2001). Seasonal variation in radiation use efficiency of irrigated rice. *Agricultural and Forest Meteorology* 110: 45-54.

Chang, J.H. (1968). *Climate and Agriculture. An Ecological Survey*. Chicago: Aldine Publishing Co.

Choudhury, B.J. (2000). A sensitivity analysis of the radiation use efficiency for gross photosynthesis and net carbon accumulation by wheat. *Agricultural and Forest Meteorology* 101: 217-234.

Cohen, S., Schaffer, A., Shen, S., Spiegelman, M., Ben-Moshe, Z., Cohen, S., and Sagi, M. (1999). Light distribution and canopy structure in greenhouse muskmelon. In Bar-Yosef, B. and Seginer, I. (Ed.), *Proceedings of the Third International Workshop on Models for Plant Growth and Control of the Shoot and Root Environments in Greenhouses,* Bet Dagan, Israel, February 21-25, 1999. *Acta Horticulturae* 507: 17-24.

Corlett, J.E., Black, C.R., Ong, C.K., and Monteith, J.L. (1992). Above and belowground interactions in a leucaena/mill ley cropping system. 2. Light interception and dry matter production. *Agricultural and Forest Meteorology* 60: 73-91.

Correia, C.M., Coutinho, J.F., Bjorn, L.O., and Torres Pereira, J.M.G. (2000). Ultraviolet-B radiation and nitrogen effects on growth and yield of maize under Mediterranean field conditions. *European Journal of Agronomy* 12: 117-125.

Courbaud, B., Coligny, F., and Cordonnier, T. (2003). Simulating radiation distribution in a heterogeneous Norway spruce forest on a slope. *Agricultural and Forest Meteorology* 116: 1-18.

Demetriades Shah, T.H., Fuchs, M., Kanemasu, E.T., and Flitcroft, I. D. (1992). A note of caution concerning the relationship between cumulated intercepted solar radiation and crop growth. *Agricultural and Forest Meteorology* 58: 193-207.

Demetriades Shah, T.H., Fuchs, M., Kanemasu, E.T., and Flitcroft, I.D. (1994). Further discussions on the relationship between cumulated intercepted solar radiation and crop growth. *Agricultural and Forest Meteorology* 68: 231-242.

Evans, L.T. (1973). The effect of light on plant growth, development and yield. In Slatyer, R.O. (Ed.), *Plant Response to Climatic Factors. Proceedings of the Uppsala Symposium* (pp. 21-31). Paris: UNESCO.

Flint, S.D. and Caldwell, M.M. (1998). Solar UV-B and visible radiation in tropical forest gaps: Measurements partitioning direct and diffuse radiation. *Global Change Biology* 4: 863-870.

Goody, R. M. and Yung, Y.L. (1989). *Atmospheric Radiation: Theoretical Basis*. London: Oxford University Press.

Hassika, P. and Berbigier, P. (1998). Annual cycle of photosynthetically active radiation in maritime pine forest. *Agricultural and Forest Meteorology* 90: 157-171.

Herbert, T.U. (1991). Variation in interception of direct solar beam by top canopy layers. *Ecology* 72: 17-22.

Iqbal, M. (1983). *An Introduction to Solar Radiation*. London: Academic Press.

Jin, X., Zhu, J., and Zeiger, E. (2001). The hypocotyl chloroplast plays a role in phototropic bending of Arabidopsis seedlings: Developmental and genetic evidence. *Journal of Experimental Botany* 52: 91-97.

Karsten, U., Bischof, K., Hanelt, D., Tug, H., and Wiencke, C. (1999). The effect of ultraviolet radiation on photosynthesis and ultraviolet-absorbing substances in the endemic artic macroalga *Devaleraea ramentacea* (Rhodophyta). *Physiologia Plantarium* 105: 58-66.

Kiehl, J.T. and Trenberth, K.E. (1997). Earth's annual global mean energy budget. *Bulletin of the American Meteorological Society* 78: 197-208.

Kiniry, J.R. (1994). A note of caution concerning the paper by Demetriades Shah. *Agricultural and Forest Meteorology* 68: 229-230.

Koller, D., Ritter, S., and Heller, E. (2001). Light-driven movements of the primary leaves of bean (*Phaseolus vulgaris* L.): A kinetic analysis. *Israel Journal of Plant Sciences* 49: 1-7.

Kull, O. and Kruijt, B. (1998). Leaf photosynthetic light response: A mechanistic model for scaling photosynthesis to leaves and canopies. *Functional Ecology* 12: 767-777.

Li, S., Kurata, K., and Takakura, T. (1998). Solar radiation enhancement in a lean-to greenhouse by use of reflection. *Journal of Agricultural Engineering Research* 71: 157-165.

Mariscal, M.J., Orgaz, F., and Villalobos, F.J. (2000). Modelling and measurement of radiation interception by olive canopies. *Agricultural and Forest Meteorology* 100: 183-197.

Marques Filho, A. de O. and Dallarosa, R.G. (2000). Interception of solar radiation and spatial distribution of leaf area in "terra firme" forest of the Central Amazonia, Brazil. *Acta Amazonica* 30: 453-470.

Mavi, H.S. (1994). *Introduction to Agrometeorology*. New Delhi: Oxford & IBH.

Monteith, J.L. (1990). Conservative behavior in the response of crops to water and light. In Rabbinge, R., Goudriaan, J., van Keulen, H., Penning de Vries, F.W.T., and van Laar, H.H. (Eds.), *Theoretical Production Ecology: Reflections and Prospects* (pp. 3-16). Wageningen, the Netherlands: PUDOC Scientific Publishers.

Monteith, J.L. (1994). Validity of the correlation between intercepted radiation and biomass. *Agricultural and Forest Meteorology* 68: 213-220.

Monteith, J.L. and Elston, J. (1983). Performance and productivity of foliage in the field. In Dale, J.E. and Milthorpe, L. (Eds.), *The Growth and Function of Leaves* (pp. 499-518). Cambridge, UK: Cambridge University Press.

Olesen, T. (2000). Is there a representative wavelength for photosynthetically active radiation in the terrestrial environment? *Austral-Ecology* 25: 626-630.

Prasad, R. and Sastry, C.V.S. (1994). A study on radiation interception, albedo, net radiation, crop growth and yield in wheat (*Triticum aestivum* L.). *Indian Journal of Ecology* 21: 112-116.

Predieri, S. and Gatti, E. (2000). Effects of gamma radiation on microcuttings of plum *(Prunus salicina). Advances in Horticultural Science* 14: 7-11.

Roberto, S., Alejandro, S., Ricardo, S., Saldaña, F., and Teodoro, G. (1999). Day-time net radiation parameterisation for Mexico City suburban areas. *Atmospheric Research* 50: 53-58.

Rodriguez, L.A., Orozco, V., Velasco, E., Medina, R., Verdecia, J., and Fonseca, I. (1999). Optimum levels of solar radiation and their relation to vegetative growth, leaf development and yield of coffee (*Coffea arabica* L.). *Cultivos Tropicales* 20: 45-49.

Roujean, J.L. (1999). Measurement of PAR transmittance within boreal forest stands during BOREAS. *Agricultural and Forest Meteorology* 93: 1-6.

Ruppel, N.J., Hangarter, R.P., and Kiss, J.Z. (2001). Red-light-induced positive phototropism in Arabidopsis roots. *Planta* 212: 424-430.

Russell, G., Jarvis, P.G., and Monteith, J.L. (1988). Absorption of radiation by canopies and stand growth. In Russell, G., Marshall, B., and Jarvis, P.G. (Eds.), *Plant Canopies: Their Growth, Form and Function* (pp. 21-40). Cambridge, UK: Cambridge University Press.

Sabins, F.F. (1997). *Remote Sensing: Principles and Interpretations,* Third Edition. New York: W.H. Freeman and Company.

Salisbury, F.B. (1981). Responses to photoperiod. In Lange, O.L. (Ed.), *Physiological Plant Ecology. 1. Response to Physical Environment* (pp. 135-168). Berlin: Springer-Verlag.

Sharp, J.L. and Polavarapu, P. (1999). Gamma doses radiation for preventing pupariation and adult emergence of *Rhagoletis mendax* (Diptera: Tephritidae). *Canadian Entomologist* 131: 549-555.

Skorska, E. (2000). The effect of ultraviolet-B radiation on triticale plants. *Folia Universitatis Agriculturae Stetinensis, Agricultura* 82: 249-254.

Squire, G.R. (1990). *The Physiology of Tropical Crop Production.* Wallingford, UK: CAB International.

Stirling, C.M., Williams, J.H., Black, C.R., and Ong, C.K. (1990). The effect of timing of shade on development, dry matter production and light use efficiency in groundnut (*Arachis hypogaea* L.) under field conditions. *Australian Journal of Agricultural Research* 41: 633-644.

Stowe-Evans, E.L., Luesse, D.R., and Liscum, E. (2001). The enhancement of phototropin-induced phototropic curvature in *Arabidopsis* occurs via a photoreversible phytochrome A-dependent modulation of auxin responsiveness. *Plant Physiology* 126: 826-834.

Sumit, M. and Kler, D.S. (2000). Solar radiation and its use efficiency in maize (*Zea mays* L.) canopy: A review. *Environment and Ecology* 18: 597-615.

Takaichi, M., Shimaji, H., Higashide, T., and Bodson, M. (2000). Effect of red/far-red photon flux ratio of solar radiation on growth of fruit vegetable seedlings. In Verhoyen, M.N.J. (Ed.), *Proceedings of the XXV International Horticultural Congress. Part 4. Culture Techniques with Special Emphasis on Environmental Implications: Chemical, Physical and Biological Means of Regulating Crop Growth in Vegetables and Fruits,* Brussels, Belgium, August 2-7, 1998. *Acta Horticulturae* 514: 147-156.

Thornley, J.H.M., Hand, D.W., and Wilson, J.W. (1992). Modelling light absorption and canopy net photosynthesis of glasshouse row crops and application to cucumber. *Journal of Experimental Botany* 43: 383-391.

Udo, S.O. and Aro, T.O. (1999). Global PAR related to solar radiation for central Nigeria. *Agricultural and Forest Meteorology* 97: 21-31.

Vijaya Kumar, P., Srivastava, N.N., Victor, U.S., Gangadhar Rao, D., Subba Rao, A.V.M., Ramakrishna, Y.S., and Ramana Rao, B.V. (1996). Radiation and water use efficiencies of rainfed castor beans (*Ricinus communis* L.) in relation to different weather parameters. *Agricultural and Forest Meteorology* 81: 241-253.

Vorasoot, N., Tienroj, U., and Apinakapong, K. (1996). Effects of leaf angle and leaf area index on utilization of solar radiation of two soybean cultivars grown under different water regimes. *Kaen Kaset Khon Kaen Agriculture Journal* 24: 158-164.

Chapter 3

Arachchi, D.H.M., Naylor, R.E.L., and Bingham, I.J. (1999). A thermal time analysis of ageing of maize (*Zea mays* L.) seed can account for reduced germination in hot moist soil. *Field Crops Research* 63: 159-167.

Atta-Aly, M.A. and Brecht, J.K. (1995). Effect of postharvest high temperature on tomato fruit ripening and quality. In Ait-Oubahou, A. and El-Otmani, M. (Eds.), *Postharvest Physiology, Pathology and Technologies for Horticultural Commodities: Recent Advances. Proceedings of the International Symposium Held at Agadir, Morocco* (pp. 250-256). Agadir, Moracco: Institute Agronomique.

Bierhuizen, J.H. (1973). The effect of temperature on plant growth, development and yield. In Slatyer, R.O. (Ed.), *Plant Response to Climatic Factors. Proceedings of Uppsala Symposium* (pp. 88-98). Paris: UNESCO.

Boer, R., Campbell, C.L., and Fletcher, D. (1993). Characteristics of frost in a major wheat-growing region of Australia. *Australian Journal of Agricultural Research* 44: 1731-1743.

Bonhomme, R. (2000). Bases and limits of using "degree day" units. *European Journal of Agronomy* 13: 1-10.

Chang, J.H. (1968). *Climate and Agriculture. An Ecological Survey.* Chicago: Aldine Publishing Co.

Chaurasia, R., Mahi, G.S., and Mavi, H.S. (1985). Effect of soil temperature on the mortality of cotton seedlings. *International Journal of Ecology and Environmental Science* 11: 119-123.

Chen, C., Lin, H.S., and Chang, L.R. (1994). Growing peaches under shelter in Taiwan. In Lin, H.S. and Chang, L.R., (Eds.), *Proceedings of a Symposium on the Practical Aspects of Some Economically Important Fruit Trees in Taiwan* (pp. 199-222). Special Publication No. 33. Changhua, Taiwan: Taichung District Agricultural Improvement Station.

Chowdhury, S., Kulshrestha, V.P., and Deshmukh, P.S. (1996). Thermo tolerance and yield parameters of late sown wheat. *Crop Improvement* 23: 263-267.

Degan, C. (1989). Plants tolerant of frost. Agdex 280/32. Sydney, Australia: NSW Agriculture and Fisheries.

Frank, A.B. and Bauer, A. (1996). Temperature, nitrogen, and carbon dioxide effects on spring wheat development and spikelet numbers. *Crop Science* 36: 659-665.

Gallasch, P.T. (1992). Guidelines for handling frost damaged citrus trees. *Riverland Newsletter* 113: 2-3.

Hanchinal, R.R., Tandon, J.P., Salimath, P.M., and Saunders, D.A. (1994). Variation and adaptation of wheat varieties for heat tolerance in Peninsular India. In Saunders, D.A. and Hettel, G.P. (Eds.), *Wheat in Heat-Stressed Environments: Irrigated, Dry Areas and Rice-Wheat Farming Systems. Proceedings of the In-*

ternational Conferences, held at Wad Medani, Sudan, 1-4 February, 1993 and Dinajpur, Bangladesh, 13-15 February 1993 (pp. 175-183). Mexico, DF: International Maize and Wheat Improvement Center (CIMMYT).

Hayasaka, M. and Imura, E. (1996). Relationship of climatic factors with root yield and quality of sugar beets grown in gravel culture. *Proceedings of the Japanese Society of Sugar Beet Technologists* 38: 72-78.

Helms, T.C., Deckard, E.L., and Gregoire, P.A. (1997). Corn, sunflower, and soybean emergence influenced by soil temperature and soil water content. *Agronomy Journal* 89: 59-63.

Hernandez, L.F. and Paoloni, P.J. (1998). Germination and seedling emergence of four sunflower (*Helianthus annuus* L.) hybrids differing in lipid content in relation to temperature. *Investigacion Agraria, Produccion y Proteccion Vegetales* 13: 345-358.

Hutton, R. (1998). Minimising frost damage to citrus. Frost report. Leeton, Australia: Leeton Growers Association.

Inaba, M. and Crandal, P.G. (1988). Electrolyte leakage as an indicator of high temperature injury to harvest mature green tomatoes. *Journal of the American Society for Horticultural Science* 113: 96-99.

Jamieson, G.I. (1986). Frost—Management in horticulture. Farm note. Brisbane, Australia: Queensland Department of Primary Industries.

Johns, R. (1986). Successful frost control in vineyards. *The Australian Grapegrowers and Winemakers* (Septmeber, No. 273): 14-16.

Judith, F.T. and Raper, C.D. (1981). Day and night temperature influence on carpel initiation and growth in soybeans. *Botanical Gazette* 142: 183-187.

Ki, W.K. and Warmund, M.R. (1992). Low temperature injury to strawberry floral organs at several stages of development. *Horticultural Science* 27: 1302-1304.

Kim, G.J., Kim, J.H., Woo, I.S., and Bae, J.H. (2001). Effect of root zone temperature on the yield and quality of sweet pepper (*Capsicum annum* L.) in hydroponics. *Journal of the Korean Society for Horticultural Science* 42: 48-52.

Kozlovskaya, Z.A. and Myalik, M.G. (1998). Low temperature injury to apple and pear seedlings following autumn planting in Belarus. In Utkhede, R. and Veghelyi, K. (Eds.), *Proceedings of the Fourth International Susposium on Replant Problems,* Budapest, Hungary. *Acta Horticulturae* 477: 173-178.

Kozlowski, T.T. (1983). Reduction in yield of forest and fruit trees by water and temperature stress. In Raper, C.D. and Kramer, P.J. (Eds.), *Proceedings of the Workshop on Crop Reactions to Water and Temperature Stresses in Humid and Temperate Climates* (pp. 50-67). Durham, NC: Duke University.

Kwon, Y.W., Kim, S.D., and Park, S.W. (1996). Effect of soil temperature on the emergence speed of rice and barnyard grasses under dry direct-seeding conditions. *Korean Journal of Weed Science* 16: 81-87.

Lad, B.L., Pujari, K.H., and Magdum, M.B. (1999). Effect of thermoperiodic changes on flowering behavior in mango under coastal climate of Maharashtra. *Annals of Plant Physiology* 13: 38-46.

Larcher, W. (1980). *Physiological Plant Ecology.* Berlin: Springer-Verlag.

Liakatas, A., Roussopoulos, D., and Whittington, W.J. (1998). Controlled-temperature effects on cotton yield and fiber properties. *Journal of Agricultural Science* 130: 463-471.

Lipman, A. and Duddy, N. (1999). Frost damage prevention and detection in navel oranges. *Australian Citrus News* (June): 7-8.

Mavi, H.S. (1994). *Introduction to Agrometeorology*. New Delhi: Oscord and IBH.

Mavi, H. (2000). Frost damage and control. Agnote DPI/231. Orange, Australia: NSW Agriculture.

McMaster, G.S. and Wilhelm, W.W. (1997). Growing degree-days: One equation, two interpretations. *Agricultural and Forest Meteorology* 87: 291-300.

Muthuvel, I., Thamburaj, S., Veeraragavathatham, D., and Kanthaswamy, V. (1999). Screening of tomato (*Lycopersicon esculentum* Mill.) genotypes for high temperature. *South Indian Horticulture* 47: 231-233.

Nieuwhof, M., Keizer, L.C.P., and Van Oeveren, J.C. (1997). Effects of temperature on growth and development of adult plants of genotypes of tomato (*Lycopersicon esculentum* Mill.). *Journal of Genetics and Breeding* 51: 185-193.

Ntare, B.R., Williams, J.H., and Ndunguru, B.J. (1998). Effects of seasonal variation in temperature and cultivar on yield and yield determination of irrigated groundnut *(Arachis hypogaea)* during the dry season in the Sahel of West Africa. *Journal of Agricultural Science* 131: 439-448.

Nykiforuk, C.L. and Flanagan, J.A. (1998). Low temperature emergence in crop plants: Biochemical and molecular aspects of germination and early seedling growth. *Journal of Crop Production* 1: 249-289.

Oda M., Thilakaratne, D.M., Li, Z.J., and Sasaki, H. (1994). Effects of abscisic acid on high temperature stress injury in cucumber. *Journal of the Japanese Society for Horticultural Science* 63: 393-399.

Oosterhuis, D.M. (1997). Effect of temperature extremes on cotton yields in Arkansas. In Oosterhuis, D.M. and Stewart, J.M. (Eds.), *Proceedings of the 1997 Cotton Research Meeting and 1997 Summaries of Cotton Research in Progress* (pp. 94-98). Fayetteville, Arkansas: Arkansas Agricultural Experiment Station.

Pallais, N. (1995). Storage factors control germination and seedling establishment of freshly harvested true potato seed. *American Potato Journal* 72: 427-436.

Pamplona, R.R., Bajita, J.B., Rebuelta, P.I., and Cruz, R.T. (1995). Effect of radiation and temperature on rice yield. In *Proceedings of the 8th National Rice R&D Review and Planning Workshop,* March 1-3, 1995, Philippines (pp. 37-44). Maligaya, Philippines: Philippine Rice Research Institute.

Perry, K.B. (1994). *Frost/Freeze Protection for Horticultural Crops.* North Carolina State University Horticulture Information Leaflet No. 705-A. Raleigh: North Carolina State University.

Perry, K.B., Wu, Y., Sanders, D.C., Garrett, J.T., Decoteau, D.R., Nagatta, R.T., Dufault, R.J., Batal, K.D., Granberry, D.M., and Mclaurin, W.J. (1997). Heat units to predict tomato harvest in the southeast USA. *Agricultural and Forest Meteorology* 84: 249-254.

Porter, J. and Gawith, M. (2000). Temperature and the growth and development of wheat: A review. *European Journal of Agronomy* 10: 23-36.

Powell, A.A. and Himelrick, D.G. (1998). *Principles of Freeze Protection for Fruit Crops.* Extension Leaflet. Raleigh: North Carolina State University.

Rogers, W.J. (1970). *Frost and the Prevention of Frost Damage.* Washington, DC: U.S. Government Printing Office.

Roltsch, W.J., Zalom, F.G., Strawn, A.J., Strand, J.F., and Pitcairn, M.J. (1999). Evaluation of several degree-day methods in California climates. *International Journal of Biometeorology* 42: 169-176.

Roundy, B.A. and Biedenbender, S.H. (1996). Germination of warm-season grasses under constant and dynamic temperatures. *Journal of Range Management* 49: 425-431.

Sakthivel, T. and Thamburaj, S. (1998). Temperatures regulate seed germination in tomato (*Lycopersicon esculentum* Mill) cv. Naveen. *South Indian Horticulture* 46: 198-199.

Singh, D., Singh, S., and Rao, V.U.M. (1998). Effect of soil temperature on seedling survival in *Brassica* spp. in different environments. *Annals of Biology* 14: 87-90.

Spieler, G.P. (1994). Microsprinklers and microclimate. *International Water and Irrigation Review* 14: 14-17.

Stolyarenko, V.S., Samoshkin, A. A., Bondar, P.S., and Chernousova, N.M. (1992). Effect of diurnal thermoperiodism in raising seedlings on the growth, development and yield of inbred maize lines in the phytotron. *Fiziologiya I Biokhimiya Kul'turnykh Rastenii* 24: 604-610.

Stone, P.J., Sorensen, I.B., and Jamieson, P.D. (1999). Effect of soil temperature on phenology, canopy development, biomass and yield of maize in a cool-temperature climate. *Field Crops Research* 63: 169-178.

Stone, R., Nicholls, N., and Hammer, G. (1996). Frost in northeast Australia: Trends and influences of phases of the southern oscillation. *Journal of Climate* 9: 1896-1909.

Szabo, Z., Buban, T., Soltesz, M., and Nyeki, J. (1995). Low winter temperature injury to apricot flower buds in Hungary. In Gulcan, R. and Askoy, U. (Eds.), *Tenth International Symposium on Apricot Culture,* Izmir, Turkey. *Acta Horticulturae* 384: 273-276.

Tibbitts,T.W., Bennet, S.M., and Cao, W.X. (1990).Control of continuous injury on potatoes with daily temperature cycling. *Plant Physiology* 93: 409-411.

Trione, S.O. and Cony, M.A. (1990). Thermoperiodism and other physiological traits of *Solanum elaeagnifolium* seeds in relation to germination. *Seed Science and Technology* 18: 525-539.

Ullio, L. (1986). Frost—Its nature and control. *Vegetable Growers News* 9: 5-6.

Ventskevich, O.Z. (1961). *Agrometeorology.* Jerusalem: Israel Program for Scientific Translation.

Vittum, M.T., Dethier, B.E., and Lesser, R.C. (1995). Estimating growing degree days. Proceedings of the American Society for Horticultural Science 87: 449-452.

Wickson, R.J. (1990). Frost control. *Proceedings of Irrigated Horticulture Workshop* (pp. 140-144). Orange, Australia: NSW Agriculture and Fisheries.

Woodruff, D., Douglas, N., and French, V. (1997). *Frost Damage in Winter Crops.* Information Series Q 987050. Queensland, Australia: Farming Systems Institute, Queensland Department of Primary Industries.

Xu, Q., Paulsen, A.Q., Guikema, J.A., and Paulsen, G.M. (1995). Functional and ultrastructural injury to photosynthesis in wheat by high temperature during maturation. *Environmental and Experimental Botany* 35: 43-54.

Yin, X., Kropff, M.J., McLaren, G., and Visperas, R.M. (1995). A nonlinear model for crop development as a function of temperature. *Agricultural and Forest Meteorology* 77: 1-16.

Youiang, Ho. and Ellison, F. (1996). Frost damage in wheat crops can be managed. *Farming Ahead with Kondinin Group* 54: 35-36.

Zalom, F.G., Goodell, P.B., Wilson, L.T., Barnett, W.W., and Bentley, W.J. (1993). *Degree-Day: The Calculation and Use of Heat Units in Pest Management.* University of California, Division of Agricultural Sciences Leaflet 21373.

Chapter 4

Ahmad, S., Aslam, M., and Shafiq, M. (1996). Reducing water seepage from earthen ponds. *Agricultural Water Management* 30: 69-76.

Alqarawi, A.A., Aldoss, A.A., and Assaeed, A.M. (1997). Effect of amount and distribution of rain on seedling growth characteristics of *Hammada elegans* (BGE) Botsch. *Arab Gulf Journal of Scientific Research* 15: 805-824.

Alqarawi, A.A., Aldoss, A.A., and Assaeed, A.M. (1998). Effect of amount and distribution of rain on seedling survival and establishment of *Hammada elegans* (BGE) Botsch. *Arab Gulf Journal of Scientific Research* 16: 207-222.

Anonymous (2000). Water savings of up to 50 percent with partial rootzone drying in horticultural crops. *Irrigation Australia* (Winter): 12-13.

Ashok Raj, P.C. (1979). *Onset of Effective Monsoon and Critical Dry Spells.* IARI Research Bulletin No. 11. New Delhi: Indian Agricultural Research Institute.

Benoit, P. (1977). The start of growing season in Northern Nigeria. *Agricultural Meteorology* 18: 91-99.

Boldt, A.L., Eisenhauer, D.E., Martin, D.L., and Wilmes, G.J. (1999). Water conservation practices for a river valley irrigated with groundwater. *Agricultural Water Management* 38: 235-256.

Bos, M.G., Murray-Rust, D.H., Merrey, D.J., Johnson, H.G., and Sneller, W.B. (1994). Methodologies for assessing performance of irrigation and drainage management. *Irrigation and Drainage Systems* 7(4): 231-261.

Bouwerg, H. (2000). Integrated water management: Emerging issues and challenges. *Agricultural Water Management* 45: 217-228.

Chiew, F.S.H., Kamaladasa, N.N., Malano, H.M., and Mcmahon, T.A. (1995). Penman-Monthieth, FAO-24 reference crop evapotranspiration and class-A pan data in Australia. *Agricultural Water Management* 28: 9-21.

Chin, K., Komamura, M., and Takasu, T. (1987). Studies on the estimation of effective rainfall and a model at the Taoyuan Irrigation Area in Taiwan. *Journal of Agricultural Science* 32: 151-162.

Collis, B. (2000). Underground dams bolster water resources. *Farming Ahead* 100: 68-69.

Dastane, N.G. (1974). *Effective Rainfall in Irrigated Agriculture.* FAO Irrigation and Drainage Paper 25. Rome: Food and Agriculture Organization of the United Nations.

Di Bella, C.M., Rebella, C.M., and Paruelo J.M. (2000). Evapotranspiration estimates using NOAA AVHRR imagery in the Pampa region of Argentina. *International Journal of Remote Sensing* 21: 791-797.

Doorenbos, J. and Pruitt, W.O. (1977). *Guidelines for Predicting Crop Water Requirements.* FAO Irrigation and Drainage Paper 24. Rome: Food and Agriculture Organization of the United Nations.

Food and Agriculture Organization (FAO) (1977). *Crop Water Requirements.* FAO Irrigation and Drainage Paper 24. Rome: Food and Agriculture Organization of the United Nations.

Food and Agriculture Organization (FAO) (1993). *AGROSTAT. PC, Computerized Information Series.* Rome: Food and Agriculture Organization of the United Nations.

Food and Agriculture Organization (FAO) (1994). *Water for Life.* World Food Day 1994. Rome: Food and Agriculture Organization of the United Nations.

Goussard, J. (1996). Interaction between water delivery and irrigation scheduling. In Smith, M. (Ed.), *Irrigation Scheduling. From Theory to Practice* (pp. 263-272). Rome: Food and Agriculture Organization of the United Nations.

Hargreaves, G.H. and Samani, Z.A. (1985). Reference crop evapotranspiration from temperature. *Journal of Applied Engineering in Agriculture* 1: 96-99.

Harrosh, J.H. (1992). Reducing evaporation losses from surface water reservoirs. *Water and Irrigation* 12: 7-9.

Heermann, D.F. (1996). Irrigation scheduling. In Pereira, L.S., Feddes, R.A., Gilley, J.R., and Lesaffre, B. (Eds.), *Sustainability of Irrigated Agriculture* (pp. 233-249). Dordrecht, The Netherlands: Kluwer Academic Publishers.

Jalota, S.K and Prihar, S.S. (1998). *Reducing Soil Water Evaporation with Tillage and Straw Mulching.* London: Eurospan Group.

Jensen, M.E., Burman, R.D., and Allen, R.G. (Eds.) (1990). *Evapotranspiration and Irrigation Water Requirements.* ASCE Manuals and Reports on Engineering Practice No.70. New York: American Society of Civil Engineers.

Jin M.G., Zhang R.Q., Sun, L.F., and Gao Y.F. (1999). Temporal and spatial soil water management: A case study in the Heilonggang region, PR China. *Agricultural Water Management* 42: 173-187.

Joshi, M.B., Murthy, J.S.R., and Shah, M.M. (1995). CROSOWAT: A decision tool for irrigation schedule. *Agricultural Water Management* 27: 203-223.

Kale, S.R., Raqmteke, J.R., Kadrekar, S.B., and Charpe, P.S. (1986). Effect of various sealant materials on seepage losses in tanks in the lateritic soil. *Indian Journal of Soil Conservation* 14: 58-59.

Kanber, R., Koksal, H., Bastug, R., and Eylen, M. (1991). Determining the effective rainfall by means of lysimeters for different soils and cultural practices. *Doga-Turk Tarim ve Ormancilik Dergisi* 15: 105-120.

Kashyap, P.S. and Panda, R.K. (2001). Evaluation of evapotranspiration estimation methods and development of crop-coefficients for potato crop in a sub-humid region. *Agricultural Water Management* 50: 2-25.

Kazinja, V.A. (1999). Promising technologies of water conservation in irrigation in arid and semiarid regions. Irrigation under conditions of water scarcity. In *Transactions: 17th Congress on Irrigation and Drainage,* Granada, Spain, Volume 1D (pp. 81-106). New Delhi: International Commission on Irrigation and Drainage.

Komamura, M. (1992). Fundamental studies on the effective rainfall of upland irrigation design water requirement. Part 1. Lower limit of effective rainfall for small rainfall of upland field. *Journal of Agricultural Science* 36: 233-242.

Kowal, J.M. and Krabe, D.T. (1972). *An Agroclimatalogical Atlas of the Northem States of Nigeria.* Zaria, Nigeria: Ahmadu Bello University Press.

Lamaddalena, N. (1997). Integrated simulation modelling for design and performance analysis of on-demand pressurised irrigation systems. PhD thesis, Instituto Superior de Agronomia, Technical University of Lisbon.

Liu, Y., Fernando, R.M., Li, Y., and Pereira, L.S. (1997). Irrigation scheduling strategies for wheat-maize cropping sequence in North China Plain. In de Lager, J.M., Vermes, L.P., and Ragab, R. (Eds.), *Sustainable Irrigation in Areas of Water Scarcity and Drought* (pp. 97-107). Oxford: British National Committee, International Commission on Irrigation and Drainage.

Malano, H.M., Turral, H.N., and Wood, M.L. (1996). Surface irrigation management in real time in South Eastern Australia: Irrigation scheduling and field application. In Smith, M. (Ed.), *Irrigation Scheduling: From Theory to Practice* (pp. 105-118). FAO Water Report 8. Rome: Food and Agriculture Organization of the United Nations.

Mando, A. and Stroosnijder, L. (1999). The biological and physical role of mulch in the rehabilitation of crusted soil in the Sahel. *Soil Use & Management* 15: 123-127.

McKenny, M.S. and Rosenberg, N.J. (1993). Sensitivity of some potential evapotranspiration methods to climate change. *Agricultural and Forest Meteorology* 64: 81-110.

Meyer, W.S., Smith, D., and Shell, H. (1995). *Estimating Reference Crop Evaporation and Crop Evapotranspiration from Weather Data and Crop Coefficients.* Technical Memo. Griffith, Australia: Division of Water Resources, CSIRO.

Mizutani, M., Rath, B., Mohanty, B.P., and Kalita, P.K. (1991). Estimation of effective rainfall in wet season paddy—Observational studies on water requirement of lowland rice in Thailand (II). *Journal of Irrigation Engineering and Rural Planning* 21: 15-28.

Mohan, S., Simhadrirao, B., and Arumugam, N. (1996). Comparative study of effective rainfall estimation methods for lowland rice. *Water Resource Management* 10: 35-44.

Muller, J.C. (1995). Contribution of lysimetric studies to the progress in agronomy: Historic representativeness of lysimeters and functioning of soil in situ. Paper presented at Lysimetry: Evaluation and Control of Nitrogen Transfers, Colloquium organized by COMIFER, Versailles, France.

Narayana, V.V.D. and Kamra, S.K. (1980). Studies on the use of alkali soil and sodium carbonate as sealants for controlling seepage losses. *Indian Journal of Soil Conservation* 8: 23-28.

Patwardhan, A.S. and Nieber, J. L. (1987). Effective rainfall in agriculture. No. 87-2018. St. Paul, MN. American Society of Agricultural Engineers.

Payten, I. (1999). Aquacaps put a lid on water evaporation for farmers. Resources. *The Weekend Australian* (September 18-19): 18.

Pereira, S.L. (1999). Higher performance through combined improvements in irrigation methods and scheduling: A discussion. *Agricultural Water Management* 40: 153-169.

Pitt, D., Peterson, K., Gilbert, G., and Fastenau, R. (1996). Field assessment of irrigation system performance. *Journal of Applied Engineering in Agriculture* 123: 307-313.

Ramalan, A.A. and Hill, R.W. (2000). Strategies for water management in gravity sprinkler irrigation system. *Agricultural Water Management* 43: 51-74.

Raman, C.R.V. (1974). *Analysis of Commencement of Monsoon Rains over Maharashtra State for Agricultural Planning.* Published Scientific Report No. 216. Poona, India: India Meteorological Department.

Raoa Mohan Rama, M.S., Adhikaria, R.N., Chittaranjana, S., and Chandrappaa, M. (1996). Influence of conservation measures on groundwater regime in a semi arid tract of South India. *Agricultural Water Management* 30: 301-312.

Sabrah, R.E.A. (1994). Water movement in a conditioner-treated sandy soil in Saudi Arabia. *Journal of Arid Environment* 27: 363-373.

Smajstrla A.G., Stanley C.D., and Clark, G.A. (1997). Estimating runoff and effective rainfall for high water-table soils in south-west Florida. *Soil & Crop Science Society of Florida Proceedings* 56: 94-98.

Smith, M. (1992). *CROPWAT, a Computer Program for Irrigation Planning and Management.* FAO Irrigation and Drainage Paper 26. Rome: Food and Agriculture Organization of the United Nations.

Smith, M. (2000). The application of climatic data for planning and management of sustainable rainfed and irrigated crop production. *Agricultural and Forest Meteorology* 103: 99-108.

Smith, M., Pereira, L.S., Beregena, J., Itier, B., Goussard, J., Ragab, R., Tollefson, L., and Van Hoffwegen, P. (Eds.) (1996). *Irrigation Scheduling: From Theory to Practice.* FAO Water Report 8. Rome: Food and Agriculture Organization of the United Nations.

Stern, R.D. and Coe, R. (1982). The use of rainfall models in agricultural planning. *Agricultural Meteorology* 26: 35-50.

Ventrella, D., Rinaldi, M., Rizzo, V., and Carlone, G. (1996). Soil water content and water use efficiency in nine cropping systems in conditions with a limited water supply. *Rivista di Agronomia* 30: 1-8.

Wheeler, R. (1994). *Water Spreading to Reduce Erosion and Increase Soil Moisture.* QDNR Water Facts. Rural Water Advisory Service. Queensland: Department of Natural Resources and Mines.

Chapter 5

Abdel-Samie, A.G., Gad, A., and Abdel-Rahman, M.A. (2000). Study on desertification of arable lands in Egypt. III. Drift sands and active dunes. *Egyptian Journal of Soil Science* 40: 385-398.

Agnew, C.T. (2000). Using the SPI to identify drought. *Drought Network News* 1: 6-11.

Alley, W.M. (1984). The Palmer Drought Severity Index: Limitations and assumptions. *Journal of Climate and Applied Meteorology* 23: 1100-1109.

Anonymous (2000). Conclusions and recommendations from the Central and Eastern European Workshop on Drought Mitigation. *Drought Network News* 12(2): 11-12.

Ayoub, A. (1998). Degradation of dryland ecosystems: Asessments and suggested actions to combat it. *Advances in Geoecology* 31: 457-463.

Bedo, D. (1997). Rainfall decile analysis for drought exceptional circumstances (DEC). In White, D.H. and Bordas, V. (Eds.), *Proceedings of the Workshop on Indicators of Drought Exceptional Circumstances* (pp. 19-20). Canberra, Australia: Bureau of Resource Sciences.

Bogardi, I., Matyasovszky, A., Bardossy, L., and Duckstein, L. (1994). Estimation of local and areal drought reflecting climate change. *Transactions of the American Society of Agricultural Engineers* 37: 1771-1781.

Brook, K. (1996). *Development of a National Drought Alert Strategic Information System.* Research Summary, Final Report on QPI 20 to Land and Water Resources Research and Development Corporation, Volume 1. Canberra, Australia: LWRRDC.

Carter, J. and Brook, K. (1996). Developing a national drought alert framework. In *Proceedings of the Managing Climate Variability Conference, "Of Droughts and Flooding Rains"* (pp. 53-60). LWRRDC Occasional Paper CV 03/96. Canberra, Australia: Land and Water Resources Research and Development Corporation.

Cheryl, A. (2000). Talking imperative for grieving farmers. *Drought Network News* 12: 12-14.

Clark, A., Brinkley T., Lamont, B., and Laughlin, G. (2000). Exceptional circumstances: A case study in the application of climate information to decision making. Paper presented at Cli-Manage 2000, Albury, NSW, October 23-25.

Cridland, S. (1997). Season quality in the rangelands as estimated from NDVI data by Agriculture Western Australia. In White, D.H. and Bordas, V. (Eds.), *Proceedings of the Workshop on Indicators of Drought Exceptional Circumstances* (pp. 59-61). Canberra, Australia: Bureau of Resource Sciences.

di Castri, F. (1990). Enrayer la progression de la désertification: An enjeu mondial. (Stopping the progression of desertification: A world stake.) *La Recherche* 221: 638-643.

Dixon, D. (1995). *Drought Exceptional Circumstances Explained.* Agnote ESU/7. Orange, Australia: NSW Agriculture.

Donnelly, J. and Freer, M. (1997). Analysis of drought incidence in the Wellington district over the period 1901-1995. In White, D.H. and Bordas, V. (Eds.), *Proceedings of the Workshop on Indicators of Drought Exceptional Circumstances* (pp. 33-35). Canberra, Australia: Bureau of Resource Sciences.

Gibbs, W.J. and Maher, J.V. (1967). *Rainfall Deciles As Drought Indicators*. Bureau of Meteorology Bulletin No. 48. Melbourne: Commonwealth of Australia.

Gommes, R. and Petrassi, F (1994). *Rainfall Variability and Drought in Sub-Saharan Africa Since 1960*. Agrometeorology Series Working Paper No. 9. Rome: Food and Agriculture Organization of the United Nations.

Gonzalez, P. (2001). Desertification and a shift of forest species in the West African Sahel. *Climate Research* 17: 217-228.

Graetz, D. (1997). Capturing and interpreting the behavior of the continental land surface, 1981 onward. In White, D.H. and Bordas, V. (Eds.), *Proceedings of the Workshop on Indicators of Drought Exceptional Circumstances* (p. 57). Canberra, Australia: Bureau of Resource Sciences.

Hare, K.F. (1993). *Climate Variations, Drought and Desertification*. WMO Publication No. 653. Geneva: World Meteorological Organization.

Hayes, M. (1996). *Comparison of Drought Indices*. A Bulletin. Lincoln: National Drought Mitigation Center, University of Nebraska.

Hayes, M. (2000). Revisiting the SPI: Clarifying the process. *Drought Network News* 12: 13-14.

Horstmann, B. (2001). Desertification—A global problem. *Entwicklung Landlicher Raum* 35: 4-6.

Hoven, I.G. (2001). The convention to combat desertification—Objectives and activities. *Entwicklung Landlicher Raum* 35: 7-11.

Karl, T.R. and Knight, R.W. (1985). *Atlas of Monthly Palmer Hydrological Drought Indices (1931-1983) for the Contiguous United States*. Historical Climatology Series 3-7. Asheville, NC: National Climatic Data Center.

Keating, B.A., Meinke, H., and Dimes, J.P. (1997). Prospectus of using a crop simulator to assess an exceptional drought: A case study for grain production in North-East Australia. In White, D.H. and Bordas, V. (Eds.), *Proceedings of the Workshop on Indicators of Drought Exceptional Circumstances* (pp. 43-46). Canberra, Australia: Bureau of Resource Sciences.

Kerley, G.H. and Whitford, W.G. (2000). Impact of grazing and desertification in the Chichuahuan desert: Plant communities, granivores and granivory. *American Midland Naturalist* 144: 78-91.

Le Houerou, H.N., Popov, G.F., and See, L. (1993). *Agro-Bioclimatic Classification of Africa*. Agrometeorology Series Working Paper No. 6. Rome: Food and Agriculture Organization of the United Nations.

Lembit, M. (1995). Commonwealth drought policies. In *Coping with Drought—Report* (pp. 57-65). Occasional Publication No. 87. Australian Institute of Agricultural Science.

McKee, T.B., Doesken, N.J., and Kleist, J. (1993). The relationship of drought frequency and duration to time scales. *Pre-prints, 8th Conference on Applied Climatology*, January 17-22 (pp. 179-184). Anaheim, CA: American Meteorological Society.

McKee, T.B., Doesken, N.J., and Kleist, J. (1995). Drought monitoring with multiple time scales. *Pre-prints, 9th Conference on Applied Climatology,* January 15-20 (pp. 233-236). Dallas, TX: American Meteorological Society.

McKeon, G. (1997). Development of a national drought alert strategic information system. In White, D.H. and Bordas, V. (Eds.), *Proceedings of the Workshop on Indicators of Drought Exceptional Circumstances* (pp. 63-66). Canberra, Australia: Bureau of Resource Sciences.

McVicar, T.R. (1997). *The Use of Remote Sensing to Aid Decisions on Drought Exceptional Circumstances.* Report to Bureau of Resource Science. Department of Primary Industries and Energy, Consultancy Report 97-14. CSIRO, Division of Water Resources.

McVicar, T.R., Jupp, D., Yi, Q., and Guoliang, T. (1997). Remote sensing and drought exceptional circumstances: An underutilised resource. In White, D.H. and Bordas, V. (Eds.), *Proceedings of the Workshop on Indicators of Drought Exceptional Circumstances* (pp. 53-55). Canberra, Australia: Bureau of Resource Sciences.

Murty, V.V.N. and Takeuchi, K. (1996). Assessment and mitigation of droughts in the Asia-Pacific region. In Murty, V.V.N. and Takeuchi, K. (Eds.), *Land and Water Development for Agriculture in the Asia-Pacific Region* (pp. 98-119). Barking, UK: Science Publishers, Inc.

National Drought Mitigation Center (1996a). *Basics of Drought Planning.* A Bulletin. Lincoln, NE: Author.

National Drought Mitigation Center (1996b). *The Devastation of Drought.* A Bulletin. Lincoln, NE: Author.

National Drought Mitigation Center (1996c). *What Is Drought?* A Bulletin. Lincoln, NE: Author.

National Drought Mitigation Center (1996d). *What We Can Do About Drought.* A Bulletin. Lincoln, NE: Author.

National Drought Mitigation Center (1996e). *Why Drought Has Been Hard To Prepare For.* A Bulletin. Lincoln, NE: Author.

Oba, G., Post, E., and Stenseth, N.C. (2001). Sub-saharan desertification and productivity are linked to hemispheric climate variability. *Global Change Biology* 7: 241-246.

Olulumazo, A.K. (2000). *Improving Tenure Security in Northern Togo: A Means to Address Desertification.* Issue Paper Dryland Programme No. 20. Dakoui, Senegal: International Institute for Environment and Development.

Palmer, W.C. (1965). *Meteorological Drought.* Research Paper No. 45. Washington, DC: U.S. Weather Bureau, Department of Commerce.

Palmer, W.C. (1968). Keeping track of crop moisture conditions, nation-wide: The new Crop Moisture Index. *Weatherwise* 21: 156-161.

Pandey, A.N., Pamar, T.D., and Tanna, S.R. (1999). Desertification: A case study from Saurashtra region of Gujarat state of India. *Tropical-Ecology* 40: 213-220.

Queensland Department of Primary Industries (1995). *Conditions and Procedures for Drought Declaring Individual Properties.* Information Sheet. Brinsbane, Australia: Author.

Quiring, S.M. and Papakryiakou, T.N. (2003). An evaluation of agricultural drought indices for Canadian prairies. *Agricultural and Forest Meteorology* 118: 49-62.

Rural Industry Business Services (1997). *Drought Bulletin.* Queensland Department of Primary Industries. October. No. 10.

Saiko. T.A. and Zonn. I.S. (2000). Irrigation expansion and dynamics of desertification in the Circum-Aral region of central Asia. *Applied Geography* 20: 349-367.

Shafer, B.A. and Dezman, L.E. (1982). Development of a Surface Water Supply Index (SWSI) to assess the severity of drought conditions in snowpack runoff areas. In *Proceedings of the Western Snow Conference* (pp. 164-175). Fort Collins: Colorado State University.

Sivakumar, M.V.K. (1991). *Drought Spells and Drought Frequencies in West Africa.* ICRISAT Research Bulletin No. 13. Patancheru, India: International Crops Research Institute for the Semi-Arid Tropics.

Sivakumar, M.V.K.. Stigter, C.J., and Rijks, D. (2000). Foreword. *Agricultural and Forest Meteorology* 103: 1-2.

Smith, R. (1996). Managing climate variability from satellite mesures of vegetation growth. In *Proceedings of the Managing Climate Variability Conference, "Of Droughts and Flooding Rains"* (pp. 48-51). LWRRDC Occasional Paper CV 03/96, Canberra, Australia: Land and Water Resources Research and Development Corporation.

Smith, S.M. and McKeon, G. (1997). Assessing the frequency of drought events in rangeland grazing properties. In White, D.H. and Bordas. V. (Eds.), *Proceedings of the Workshop on Indicators of Drought Exceptional Circumstances* (pp. 27-28). Canberra, Australia: Bureau of Resource Sciences.

Stehlik, D., Gray. I., and Lawrence, G. (1999). *Drought in the 1990s. Australian Farm Families Experience.* RIRDC Publication No 99/14. Canberra, Australia: Rural Industries Research and Development Corporation.

Stephens, D.J. (1996). *Objective Criteria for Drought Exceptional Circumstances Declaration—Cropping Area.* Final Report. Perth: Agriculture West Australia.

Stephens. D.J. (1997). Objective criteria for Drought Exceptional Circumstances Declaration—Cropping areas. In White, D.H. and Bordas. V. (Eds.), *Proceedings of the Workshop on Indicators of Drought Exceptional Circumstances* (pp. 37-39). Canberra, Australia: Bureau of Resource Sciences.

Stigter, C.J., Sivakumar, M.V.K., and Rijks, D. A. (2000). Agrometeorology in the 21st century: Workshop summary and recommendations on needs and perspectives. *Agricultural and Forest Meteorology* 103: 209-227.

Wang, C.Y. (1990). Desertification monitoring with remote sensing technology and GIS (study of Yilin County, Shaanxi Province, as an example). *Chinese Journal of Arid Land Research* 3: 193-198.

Wang, S., Zheng R., and Yang, Y.L. (2000). Combating desertification: The Asian experience. *International Forestry Review* 2: 112-117.

White D.H. (1997). Providing scientific advice on drought exceptional circumstances. In White, D.H. and Bordas, V. (Eds.), *Proceedings of the Workshop on Indicators of Drought Exceptional Circumstances* (pp. 1-11). Canberra, Australia: Bureau of Resource Sciences.

Wilhite, D.A. (2001). Creating a network of regional drought preparedness networks: A call for action. *Drought Network News* 13: 1-4.

Zhang, D.F. and Bian, J.M. (2000). Analysis of the eco-environmental vulnerable mechanisms of land desertification in the farming-pastoral ecotone in northern China. *Arid Land Geography* 2: 133-137.

Chapter 6

Agarwala, B.K. and Bhattacharya, S. (1994). Anholocyclicity tropical aphids: Population trends and influence of temperature on development, reproduction, and survival of three aphid species (Homoptera: Aphidoidea). *Phytophaga Palermo* 6: 17-27.

Arundel, J.H. and Sutherland, A.K. (1988). *Ectoparasitic Diseases of Sheep, Cattle, Goats and Horses. Animal Health in Australia,* Volume 10. Canberra: Australian Government Publishing Service.

Atkins, M.D. (1978). *Insects in Perspective.* New York: Macmillan Publishing Co.

Aylor, D.E. (1999). Biophysical scaling and the passive dispersal of fungus spores: Relationship to integrated pest management strategies. *Agricultural and Forest Meteorology* 97: 275-292.

Baker, R.H.A., Guillet, P., Seketeli, A., Poudiougo, P., Boakye, D., Wilson, M.D., and Bissan,Y. (1990). Progress in controlling the reinvasion of windborne vectors into the western area of the Onchocerciasis Control Programme in West Africa. *Philosophical Transactions of the Royal Society of London* Series B, Biological Sciences 328: 731-750.

Bateman, M.A. (1972). Determinants of abundance in a population of the Queensland fruit fly. *Symposia of the Royal Entomological Society of London* 4: 119-131.

Baylis, M., Bouayoune, H., Tou, J., and El Hasnaoui, H. (1998). Use of climatic data and satellite imagery to model the abundance of *Culicoides imicola,* the vector of African horse sickness virus, in Morocco. *Medical and Veterinary Entomology* 12: 255-266.

Baylis, M., Mellor, P.S., and Meiswinkel, R. (1999). Horse sickness and ENSO in South Africa. *Nature* 397: 574.

Bishnoi, O.P., Singh, M., Rao, V.U.M., Niwas, R., and Sharma, P.D. (1996). Population dynamics of cotton pests in relation to weather parameters. *Indian Journal of Entomology* 58: 103-107.

Blood, D.C. and Radostits, O.M. (Eds.) (1989). *Veterinary Medicine—A Textbook of the Diseases of Cattle, Sheep, Pigs, Goats and Horses.* London: Bailliere Tindall.

Bouma, M.J. and van der Kaay, H.J. (1996). The El-Nino Southern Oscillation and the historic malaria epidemics on the Indian subcontinent and Sri Lanka: An early warning system for future epidemics? *Tropical Medicine and International Health* 1: 86-96.

Byrne, D.N. (1999). Migration and dispersal by the sweet potato whitefly, *Bemisia tabaci. Agricultural and Forest Meteorology* 97: 309-316.

Capinera, J.L. and Horton, D.R. (1989). Geographic variation in effects of weather on grasshopper infestation. *Environmental Entomology* 18: 8-14.

Chandra, S. (1993). A long-term study on major changes in desert locust (*Schistocerca gregaria* Forsk) activity in relation to rainfall in Indo-Pak region. *Plant Protection Bulletin* 45: 42-49.

Chattopadhyay, N., Samui, R.P., Wadekar, S.N., Satpute, U.S., and Sarode, S.V. (1996). Sensitivity of aphid infestation to meteorological parameters at Akola, Maharashtra, India. *Indian Journal of Entomology* 58: 291-301.

Cole, V.G. (1986). *Helminth Parasites of Sheep and Cattle. Animal Health in Australia,* Volume 8. Canberra: Australian Government Publishing Service.

Crawford, S., James, P.J., and Maddocks, S. (2001). Survival away from sheep and alternative methods of transmission of sheep lice *(Bovicola ovis). Veterinary Parasitology* 94: 205-216.

De Chaneet, G.C. and Dunsmore, J.D. (1988). Climate and the distribution of intestinal *Trichostrongylus* spp. of sheep. *Veterinary Parasitology* 26: 273-283.

Diss, A.L., Kunkel, J.G., Montgomery, M.E., and Leonard, D.E. (1996). Effect of maternal nutrient and egg provisioning on parameters of larval hatch, survival and dispersal in the gypsy moth, *Lymantria dispar* L. *Oecologia* 106: 470-477.

Drake, V.A. (1994). The influence of weather and climate on agriculturally important insects: An Australian perspective. *Australian Journal of Agricultural Research* 45: 487-509.

Drake, V.A. and Farrow, R.A. (1988). The influence of atmospheric structure and motions on insect migration. *Annual Review of Entomology* 33: 183-210.

El Bashir, S. (1996). Strategies of adaptation and survival of the desert locust in an area of recession and multiplication. *Secheresse* 7: 115-118.

El Sadaany, G.B., Hossain, A.M., El Fateh, R.S.M., and Romeilah, M.A. (1999). The simultaneous effect of physical environmental factors governing the population activity of cotton bollworm moths. *Egyptian Journal of Agricultural Research* 77: 591-609.

Estrada, P.A. and Genchi, C. (2001). Forecasting habitat suitability for ticks and prevention of tick-borne diseases. Special issue. *Veterinary Parasitology* 98: 111-132.

Fay, H.A.C. and Meats, A. (1987). The sterile insect release method and the importance of thermal conditioning before release: Field-cage experiments with *Dacus tryoni* in spring weather. *Australian Journal of Zoology* 35: 197-204.

Fensham, R.J. (1994). Phytophagous insect-woody sprout interactions in tropical eucalypt forest. 1. Insect herbivory. *Australian Journal of Ecology* 19: 178-188.

Freier, B. and Triltsch, H. (1996). Climate chamber experiments and computer simulations on the influence of increasing temperature on wheat-aphid-predator interactions. Implications of "global environmental change" for crops in Europe. *Aspects of Applied Biology* 45: 293-298.

Gergis, M.F., Moftah, E.A., Soliman, M.A., and Khidr, A.A. (1990). Temperature-dependant development and functional responses of pink bollworm *Pectinophora gossypiella* (Saund.). *Australian Journal of Agricultural Sciences* 21: 119-128.

Gerozisis, J.J. and Hadlington, P.W. (1995). *Urban Pest Control in Australia,* Third Edition. Sydney: University of New South Wales Press.

Gregg, P. (1983). Development of the Australian plague locust, *Chortoicetes terminifera,* in relation to weather. 1. Effects of constant temperature and humidity. *Journal of the Australian Entomological Society* 22: 247-251.

Gu, H. and Danthanarayana, W. (1992). Influence of larval rearing conditions on the body size and flight capacity of *Epiphyas postvittana* moths. *Australian Journal of Zoology* 40: 573-581.

Gupta, M.P., Gupta, D.P., and Shrivastava, K.K. (1996). Population dynamics of cotton bollworms in Madhya Pradesh. *Annals of Entomology* 14: 61-66.

Haggis, M.J. (1996). Forecasting the severity of seasonal outbreaks of African armyworm, *Spodoptera exempta* (Lepidoptera: Noctuidae), in Kenya from the previous year's rainfall. *Bulletin of Entomological Research* 86: 129-136.

Hansen, L.O. and Somme, L. (1994). Cold hardiness of the elm bark beetle *Scolytus laevis* Chapuis, 1873 (Col., Scolytide) and its potential as Dutch elm disease vector in the northernmost elm forests of Europe. *Journal of Applied Entomology* 117: 444-450.

Hernandez, E.L., Prats,V.V., and Ruiz, A.C. (1992). Determination and estimation of the infective larvae of gastrointestinal nematodes on pasture during two periods of the year in a humid tropical climate. *Tecnica Pecuariaen Mexico* 30: 31-36.

Hill, J.K. and Gatehouse, A.G. (1992). Effects of temperature and photoperiod on development and pre-reproductive period of the silver Y moth *Autographa gamma* (Lepidoptera: Noctuidae). *Bulletin of Entomological Research* 82: 335-341.

Hughes, R.D. (1981). The Australian bushfly: A climate-dominated nuisance pest of man. In Kitching, R.L. and Jones, R.E. (Eds.), *The Ecology of Pests. Some Australian Case Histories* (pp. 177-191). Collingwood, Victoria: CSIRO Publishing.

Hunter, A.F. (1993). Gypsy moth population sizes and the window of opportunity in spring. *Oikos* 68: 531-538.

Hunter, D.M. (1981). Mass take-off after sunset in the Australian plague locust. In *Australian Plague Locust Commission Annual Report Research Supplement 1979-1980* (pp. 68-72). Canberra, Australia: Australian Plague Locust Commission.

Hunter, D.M., McCulloch, L., and Wright, D.E. (1981). Lipid accumulation and migratory flight in the Australian plague locust, *Chortoicetes terminifera* (Walker) (Orthoptera: Acrididae). *Bulletin of Entomological Research* 71: 543-546.

Johnson, H.D. (1987). *Bioclimatology and the Adaptation of the Livestock.* Amsterdam: Elsevier.

Kaakeh, W. and Dutcher, J.D. (1993). Survival of yellow pecan aphids and black pecan aphids (Homoptera: Aphididae) at different temperature regimes. *Environmental Entomology* 22: 810-817.

Kemp, W.P. and Cigliano, M.M. (1994). Drought and rangeland grasshopper species diversity. *Canadian Entomologist* 126: 1075-1092.

Leimar, O. (1996). Life history plasticity: Influence of photoperiod on growth and development in the common blue butterfly. *Oikos* 76: 228-234.

Li, D.Y., Xu, D.W., Qian, D.J., and Cheng, C.L. (1996). A study on the relation between bollworm incidence and precipitation in Fuyang District. *China Cottons* 23: 17-18.

Limpert, E., Godet, F., and Müller, K. (1999). Dispersal of cereal mildews across Europe. *Agricultural and Forest Meteorology* 97: 293-308.

Lockwood, J.A. and Lockwood, D.R. (1991). Rangeland grasshopper (Orthoptera: Acrididae) population dynamics: Insights from catastrophe theory. *Environmental Entomology* 20: 970-980.

Mackenzie, J., Lindsay, M., and Daniel, P. (2000). The effect of climate on the incidence of vector-borne viral diseases in Australia: The potential value of seasonal forecasting. In Hammer, G.L., Nicholls, N., and Mitchell, C. (Eds.), *Applications of Seasonal Climate Forecasting in Agricultural and Natural Ecosystems* (pp. 429-452). London: Kluwer Academic Publishers.

Mavi, H.S. and Dominiak, B.C. (2000). Impact of climate variability on the population of Queensland fruit fly in southwestern New South Wales. Paper presented at 13th Australia New Zealand Climate Forum, Hobart Australia, April 10-12.

Mazanec, Z. (1989). Jarrah leafminer, an insect pest of jarrah. In Dell, B., Havel, J.J., and Malajczuk, N. (Eds.), *The Jarrah Forest: A Complex Mediterranean Ecosystem* (pp. 123-131). Dordrecht, The Netherlands: Kluwer.

Meats, A. (1981). The bioclimatic potential of the Queensland fruit fly, *Dacus tryconi*, in Australia. *Proceedings of the Ecological Society, Australia* 11: 151-161.

Montealegre, F.A., Boshell, F., and Leon, G.A. (1998). Influence of the climatic factors on the development and establishment of the locust *Rhammatocerus schistocercoides* (Orthoptera: Acrididae) in the Colombian Orinoquia. *Revista Colombiana de Entomologia* 24: 83-88.

Murray, M.D. (1987). Akabane epizootics in New South Wales: Evidence for long-distance dispersal of the biting midge *Culicaides brevitarsis. Australian Veterinary Journal* 64: 305-308.

Nikitenko, V.G. (1995). Locusts in the Stavropol region. *Zashchita Rastenii Moskva* 6: 9.

Pair, S.D. and Westbrook, J.K. (1995). Agro-ecological and climatological factors potentially influencing armyworm populations and their movement in the Southeastern United States. *Southwestern Entomologist Supplement* 18: 101-118.

Pedgley, D.E. (1990). Concentration of flying insects by the wind. *Philosophical Transactions of the Royal Society of London.* Series B, Biological Sciences 328: 113-135.

Pegram, R.G. and Banda, D.S. (1990). Ecology and phenology of cattle ticks in Zambia development and survival of freeliving stages. *Experimental and Applied Acarology* 8: 291-301.

Pegram, R.G., Lemche, J., Chizyuka, H.G.B., Sutherst, R.W., Floyd, R.B., Kerr, J.D., and McCosker, P.J. (1989). Ecological aspects of cattle tick control in central Zambia. *Medical and Veterinary Entomology* 3: 307-312.

Pons, X., Comas, J., and Albajes, R. (1993). Overwintering of cereal aphids (Homoptera: Aphididae) on durum wheat in a Mediterranean climate. *Environmental Entomology* 22: 381-387.

Prange, H.D. and Pinshow, B. (1994). Thermoregulation of an unusual grasshopper in a desert environment: The importance of food source and body size. *Joural of Thermal Biology* 19: 75-78.

Reid, D.G., Wardhaugh, K.G., and Roffey, J. (1979). *Radar Studies of Insect Flight at Benalla, Victoria, in February 1974.* Australia Division of Entomolgy Technical Paper No. 16. Melbourne, Australia: CSIRO.

Ridsdill, S.T.J., Scott, J.K., and Nieto, N.J.M. (1998). Aphids of annual pastures in south-western Australia. Aphids in natural and managed ecosystems. In Dixon, A.F.G. (Ed.), *Proceedings of the Fifth International Symposium on Aphids,* Leon, Spain, September 15-19, 1997 (pp. 457-462). Léon, Spain: Universidad de Léon.

Risch, S.J. (1987). Agricultural ecology and insect outbreaks. In Barbosa, P. and Schtuz, J.C. (Eds.), *Insect Outbreaks* (pp. 217-238). San Diego: Academic Press.

Robinson, T., Rogers, D., and Williams, B. (1997a). Mapping tsetse habitat suitability in the common fly belt of southern Africa using mulvariate analysis of climate and remotely sensed vegetation data. *Medical and Veterinary Entomology* 11: 235-245.

Robinson, T., Rogers, D., and Williams, B. (1997b). Univariate analysis of tsetse habitat in the common fly belt of southern Africa using climate and remotely sensed vegetation data. *Medical and Veterinary Entomology* 11: 223-234.

Romero, C.G., Valcarcel, F., and Vazquez, F.A.R. (1997). Influence of climate on pasture infectivity of ovine trichostrongyles in dry pastures. *Journal of Veterinary Medicine.* Series B. 44: 437-443.

Rose, D.J.W., Dewhurst, C.F., and Page, W.W. (1995). The bionomics of the African armyworm *Spodoptera exempta* in relation to its status as a migrant pest. *Integrated Pest Management Reviews* 1: 49-64.

Rutter, J.F., Mills, A.P., and Rosenberg, L.J. (1998). Weather associated with autumn and winter migrations of rice pests and other insects in south-eastern and eastern Asia. *Bulletin of Entomological Research* 88: 189-197.

Sellers, R.F. and Walton, T.E. (1992). Weather, Culicoides, and the distribution and spread of bluetongue and African horse sickness viruses. In Osburn, B.I. (Ed.), *Bluetongue, African Horse Sickness, and Related Orbiviruses. Proceedings of Second International Symposium* (pp. 284-290). Boca Raton, FL: CRC Press.

Shah, F.M. and Ralph, S.R. (1989). *Manual of Tropical Veterinary Parasitology.* Wallingford, UK: CAB International.

Shields, J.E. and Testa, A.M. (1999). Fall migratory flight initiation of the potato leafhopper, *Empoasca fabae* (Homoptera: Cicadellidae): Observations in the lower atmosphere using remote piloted vehicles. *Agricultural and Forest Meteorology* 97: 317-330.

Skirvin, D.J., Perry, J.N., and Harrington, R. (1996). A model to describe the effect of climate change on aphid and coccinellid population dynamics. *Bulletin OILS CROP* 19: 30-40.

Smeal, M.G. (1995). *Parasites of Cattle*. Veterinary Review No. 32. Sydney: The University of Sydney.

Speight, M.R., Hunter, M.D., and Watt, A.D. (1999). *Ecology of Insects: Concepts and Applications*. Oxford, UK: Blackwell Science Ltd.

Stewart, S.D., Layton, M.B. Jr., and Williams, M.R. (1996). Occurrence and control of beet armyworm outbreaks in the cotton belt. In Richter, D.A. and Armour, J. (Eds.), *Proceedings of the Beltwide Cotton Production Conference*, Nashville, TN, January 9-12, Volume 2 (pp. 846-848). Memphis, TN: National Cotton Council of America.

Sutherst. R.W., Collyer, B.S., and Yonow, T. (2000). The vulnerability of Australian horticulture to the Queensland fruit fly *Bactrocera tryoni* (Froggatt), under climate change. *Australian Journal of Agricultural Research* 51: 467-480.

Telfer, M.G. and Hassall. M. (1999). Ecotypic differentiation in the grasshopper *Chorthippus brunneus:* Life history varies in relation to climate. *Oecologia* 121: 245-254.

Thackray. D. (1999). *Forecasting Aphid Outbreaks and Virus Epidemics in Lupins*. The 1999 Australian Grain Field Research Manual. Toowoomba, Australia: Grain Research and Development Corporation.

Thireau, J.C. and Regniere, J. (1995). Development. reproduction, voltinism and host synchrony of *Meteorus trchynotus* with it hosts *Chorisoneura fumeferana* and *C. rosaceana. Entomologia Expermentalis et Applicata* 76: 67-82.

Torr, S.J. and Hargrove, J.W. (1999). Behavior of tsetse *(Diptera glossinidae)* during the hot seasons in Zimbabwe: The interaction of microclimate and reproductive status. *Bulletin of Entomological Research* 89: 365-379.

Tucker, M.R. (1994). Inter- and intra-seasonal variation in outbreak distribution of the armyworm. *Spodoptera exempta* (Lepidoptera: Noctuidae), in eastern Africa. *Bulletin of Entomological Research* 84: 275-287.

Waghmare, A.G., Varshneya, M.C., Khandge, S.V., Thankur, S.S., and Jadhav, A.S. (1995). Effects of meteorological parameters on the incidence of aphids on sorghum. *Journal of Maharashtra Agricultural Universities* 20: 307-308.

Waller, P.J., Mahon, R.J., and Wardhaugh, K.G. (1993). Regional and temporal use of avermectins in Australia. *Veterinary Parasitology for Ruminants* 48: 29-47.

Ward, M.P. (1996a). Climatic factors associated with the infection of herds of cattle with bluetongue viruses. *Veterinary Research Communications* 20: 273-283.

Ward, M.P. (1996b). Seasonality of infection of cattle with bluetongue viruses. *Preventive Veterinary Medicine* 26: 133-141.

Ward, M.P. and Johnson, S.J. (1996). Bluetongue virus and the southern oscillation index—Evidence of an association. *Preventive Veterinary Medicine* 28: 57-68.

Wardhaugh, K.G. and Morton, R. (1990). The incidence of flystrike in sheep in relation to weather conditions, sheep husbandry, and the abundance of the Australian sheep blowfly, *Lucilia cuprina* (Wiedemann) (Diptera: Calliphoridae). *Australian Journal of Agricultural Research* 41: 1155-1167.

Willott, S.J. and Hassall, M. (1998). Life-history responses of British grasshoppers (Orthoptera: Acrididae) to temperature change. *Functional Ecology* 12: 232-241.

Wu, K.J., Chen Y.P., and Li, M.H. (1993). Performances of the cotton bollworm, *Heliothis armigera* (Hubner), at different temperatures and relative humidities. *Journal of Environmental Sciences* 5: 158-168.

Yonow, T. and Sutherst, R.W. (1998). The geographical distribution of the Queensland fruit fly, *Bactrocera* (Dacus) *tryoni,* in relation to climate. *Australian Journal of Agricultural Research* 49: 935-953.

Zhou, X., Harrington, R., Woiwood, I.P., Perry, J.N., Clark, S.J., and Zhou, X.L. (1996). Impact of climate change of aphid flight phenology. Implications of "global environmental change" for crops in Europe. *Aspects of Applied Biology* 45: 299-305.

Chapter 7

Ballestra, G., Bertozzi, R., Buscaroli, A., Gherardi, M., and Vianello, G. (1996). *Applicazioni dei Sistemi Informativi Geografici nella Valutazione delle Modificazioni Ambientali e Territoriali.* (Application of geographic information system for evaluating environmental and infrastructural modifications.) Milano, Italy: Franco Angeli.

Bierwirth, P.N. and McVicar, T.R. (1998). *Rapid Monitoring and Assessment of Drought in Papua New Guinea Using Satellite Imagery.* Port Moresby, Papua New Guinea: Consultancy Report to United Nations Development Program.

Brook, K.D. and Carter, J.B. (1994). Integrating satellite data and pasture growth models to produce feed deficit and land condition alerts. *Agricultural Systems & Information Technology* 6: 54-56.

Burrough, P.A. (1990). *Principles of Geographical Information Systems for Land Resources Assessment.* London: Oxford University Press.

Carter, J., Flood, N., Danaher, T., Hugman, P., Young, R., Duncalfe, F., Barber, D., Flavel, R., Beeston, G., Mlodawski, G., et al. (1996). *Development of a National Drought Alert Strategic Information System,* Volume 3: *Development of Data Rasters for Model Inputs.* Final report on QPI20. Canberra, Australia: Land and Water Resources Research and Development Corporation.

Carter, J.O., Hall, W.B., Brook, K.D., McKeon, G.M., Day, K.A., and Paul, C.J. (2000). Aussie GRASS: Australian Grassland and Rangeland Assessment by Spatial Simulation. In Hammer, G.L., Nicholls, N., and Mitchell, C. (Eds.), *Applications of Seasonal Climate Forecasting in Agricultural and Natural Ecosystems* (pp. 329-350). London: Kluwer Academic Publishers.

Coulson, R.N., Lovelady, C.N., Flamm, R.O., Spradling, S.L., and Saunders, M.C. (1991). Intelligent geographic information systems for natural resource management. In Turner, M.G. and Gardner, R.H. (Eds.), *Quantitative Methods in Landscape Ecology* (pp. 153-172). Berlin: Springer-Verlag.

Cridland, S.W., Burnside, D.G., and Smith, R.C.G. (1994). Use by managers in rangeland environments of near real-time satellite measurements of seasonal vegetation response. In *Mapping Resources, Monitoring the Environment and Managing the Future. Proceedings of the 7th Australasian Remote Sensing Con-*

ference, Volume 2, March 1-4, Melbourne (pp. 1134-1141). Floreat, Australia: Remote Sensing and Photogrammetry Association.

Diak, G.R., Anderson, M.C., Bland, W.L., Norman, J.M., Mecikalski, J.M., and Aune, R.M. (1998). Agricultural management decision aids driven by real-time satellite data. *Bulletin of the American Meteorological Society* 79: 1345-1355.

Diak, G.R., Bland, W.L., and Mecikalski, J.R. (1996). Preparing for the AMSU. *Bulletin of the American Meteorological Society* 82: 219-226.

Diak, G.R. and Gautier, C. (1983). Improvements to a simple physical model for estimating insolation from GOES data. *Journal of Climate and Applied Meteorology* 22: 505-508.

Di Chiara, C. and Maracchi, G. (1994). *Guide au S.I.S.P, Version 1.0.* (Guide to SISP). Technical Manual No. 14. Firenze, Italy: Centre for the Application of Computer Science Agriculture.

Eva, H.D. and Lambin, E.F. (1998). Remote sensing of biomass burning in tropical regions: Sampling issues and multisensor approach. *Remote Sensing of Environment* 64: 292-315.

Hutchinson, M.F. (1995). *ANUSPLIN VERSION 3.2.*, <http://cres.anu.edu.au/outputs/anusplin.html>.

Jupp, D.L.B., Tian, G., McVicar, T.R., Qin, Y., and Fuqin, L. (1998). *Soil Moisture and Drought Monitoring Using Remote Sensing I: Theoretical Background and Methods.* Canberra, Australia: CSIRO Earth Observation Centre.

Longley, P.A., Goodchild, M.F., Maguire, D.J., and Rhind, D.W. (2001). *Geographic Information Systems and Science.* New York: John Wiley and Sons.

Maracchi, G., Pérarnaud, V., and Kleschenko, A.D. (2000). Applications of geographical information systems and remote sensing in agrometeorology. *Agricultural and Forest Meteorology* 103: 119-136.

McKeon, G.M., Carter, J.O., Day, K.A., Hall, W.B., and Howden, S.M. (1998). *Evaluation of the Impact of Climate Change on Northern Australian Grazing Industries.* Final Report for the Rural Industries Research and Development Corporation (DAQ139A). Canberra, Australia: Rural Industries Research and Development Corporation.

McVicar, T.R. and Jupp, D.L.B. (1999). One time of day interpolation of meteorological data from daily data as inputs to remote sensing based estimates of energy balance components. *Agriculture and Forest Meteorology* 96: 219-238.

McVicar, T.R., Jupp, D.L.B., Billings, S.D., Tian, G., and Qin, Y. (1996). Monitoring drought using AVHRR. In *Proceedings of the 8th Australasian Remote Sensing Conference,* March 25-29, Canberra (pp. 254-261). Canberra, Australia: Remote Sensing and Photogrammetry Association of Australia.

McVicar, T.R., Jupp, D.L.B., Reece, P.H., and Williams, N.A. (1996). *Relating LANDSAT TM Vegetation Indices to in Situ Leaf Area Index Measurements.* Canberra, ACT: CSIRO, Division of Water Resources.

McVicar, T.R., Jupp, D.L.B., and Williams, N.A. (1996). *Relating AVHRR Vegetation Indices to LANDSAT TM Leaf Area Index Estimates.* Canberra, ACT: CSIRO, Division of Water Resources.

McVicar, T.R., Walker, J., Jupp, D.L.B., Pierce, L.L., Byrne, G.T., and Dallwitz, R. (1996). *Relating AVHRR Vegetation Indices to in Situ Leaf Area Index.* Canberra, ACT: CSIRO, Division of Water Resources.

Menzel, W.P. and Purdum, J.F.W. (1994). Introducing GOES-1: The first of a new generation of geostationary operational environmental satellites. *Bulletin of the American Meteorological Society* 75: 757-781.

Moore, I.D., Grayson, R.B., and Ladson, A.R. (1991). Digital terrain modelling: A review of hydrological, geomorphological and biological applications. *Hydrological Processes* 5: 3-30.

Postel, S. (1993). Water and agriculture. In Gleick, P.H. (Ed.), *Water in Crisis: A Guide to the World's Fresh Water Resources* (pp. 56-66). London: Oxford University Press.

Rijks, D., Terres, J.M., and Vossen, P. (1998). *Agrometeorological Applications for Regional Crop Monitoring and Production Assessment.* Account of the EU Support Group on Agrometeorology (SUGRAM) 1991-1996. EUR 17735 EN. Ispra, Italy: Joint Research Centre of the European Union.

Roderick, M., Smith, R.C.G., and Ludwick, G. (1996). Calibrating long term AVHRR-derived NDVI imagery. *Remote Sensing of the Environment* 58: 1-12.

Romanelli, S., Bottai, L., and Maselli, F. (1998). *Studio Preliminare per la Stima del Rischio d'Incendio Boschivo a Scala Regionale per Mezzo dei Dati Satellitari e Ausiliari.* (Preliminary study into an estimate of brush fire risk on a regional scale using satellite and auxiliary data.) Firenze, Italy: Tuscany Region—Laboratory for Meteorology and Environmental Modelling (LaMMA).

Sabins, F.F. (1997). *Remote Sensing: Principles and Interpretation.* New York: W.H. Freeman and Co.

Setzer, A.W. and Verstraete, M.M. (1994). Fire and glint in AVHRRs channel 3: A possible reason for the non-saturation mystery. *International Journal of Remote Sensing* 15: 711-718.

Smith, R.B. (2002). *Introduction to Remote Sensing of Environment (RSE) with TNTmips® TNTview®.* Lincoln, NE: Microimages, Inc.

Smith, R.C.G. (1994). *Australian Vegetation Watch.* Final report to the Rural Industries Research and Development Corporation. RIRDC Reference No. DOL-1A. Canberra, Australia: RIRDC.

Stevenson, W.R. (1993). Management of early blight and late blight. In Rowe, R.C. (Ed.), *Potato Health Management* (pp. 141-147). St. Paul, MN: APS Press.

Tuddenham, W.G. and Le Marshall, J.F. (1996). The interpretation of "NDVI" data and the potential use of a differential technique for monitoring sequential changes in vegetative cover. In *Proceedings of the Second Australian Conference on Agricultural Meteorology*, October 1-4, The University of Queensland (pp. 57-61). Melbourne, Australia: Bureau of Meteorology.

Tuddenham, W.G., Le Marshall, J.F., Rouse, B.J., and Ebert, E.E. (1994). The real time generation and processing of NDVI from NOAA-11: A perspective view from the Bureau of Meteorology. In *Proceedings of the 7th Australasian Remote Sensing Conference,* March, Melbourne (pp. 495-502). Canberra, Australia: Remote Sensing and Photogrammetry Association of Australia.

Tupper, G.J., Worsley, P.M., Bowler, J.K., Freckelton, D.A., McGowen, I.J., Pradhan, U.C., Roger, R.E., and Worsley, M.A. (2000). The use of remote sensing and GIS technology by NSW Agriculture for emergency management. In *Proceedings of the IEEE 2000 International Geoscience and Remote Sensing Symposium*, July, Waikiki, HI (CDROM). New York: IEEE.

White, D.H., Tupper, G.J., and Mavi, H.S. (1999). *Climate Variability and Drought Research in Relation to Australian Agriculture.* Occasional Paper CV01/99. Canberra, Australia: Land and Water Resources Research and Development Corporation.

Wood, H., Hassett, R., Carter, J., and Danaher, T. (1996). *Development of a National Drought Alert Strategic Information System*, Volume 2: *Field Validation of Pasture Biomass and Tree Cover.* Final Report on QPI20 to Land and Water Resources Research and Development Corporation. Brisbane, Australia: Queensland Department of Primary Industries.

Chapter 8

Abrecht, D.G. and Robinson, S. (1996). TACT: A tactical decision aid using a CERES based wheat simulation model. *Ecological Modeling* 86: 241-244.

Ascough, J.C. II, Hoag, D.L., Marshall Frasier, W., and McMaster, G.S. (1999). Computer use in agriculture: An analysis of Great Plains producers. *Computers and Electronics in Agricultures* 23: 189-204.

Balston, J.M. and Egan, J.P. (1998). *Development and Evaluation of SOWHAT, A Decision Support Tool for Sowing Decisions Based on Early Season Rainfall.* South Australian Research and Development Institute Research Report. Adelaide, Australia: SARDI.

Boote, K.J., Jones, J.W., and Pickering, N.B. (1996). Potential uses and limitations of crop models. *Agronomy Journal, American Society of Agronomy* 88: 704-716.

Bowen, W.T., Thornton, P.K., and Hoogenboom, G. (1998). The simulation of cropping sequences using DSSAT. In Tsuji, G.Y., Hoogenboom, G., and Thornton, P.K. (Eds.), *Understanding Options for Agricultural Production* (pp. 313-327). Dordrecht, The Netherlands: Kluwer Academic Publishers.

Bowman, P.J., Cottle, D.J., White, D.H., and Bywater, A.C. (1993). Simulation of wool growth rate and fleece characteristics of Merino sheep in southern Australia. Part 1. Model description. *Agricultural Systems* 43: 287-299.

Bowman, P.J., White, D.H., Cottle, D.J., and Bywater, A.C. (1993). Simulation of wool growth rate and fleece characteristics of Merino sheep in southern Australia. Part 2. Assessment of biological components of the model. *Agricultural Systems* 43: 301-321.

Brook, R.D. and Hearn, A.B. (1990). *The "STRATAC" Pest Management Computer Program: Program Content* (July 1988). Divisional Technical Paper. Canberra, Australia: CSIRO Division of Plant Industry.

Cain, J.D., Jinapala, K., Makin, I.W., Sumaratna, P.G., Ariyaratna, B.R., and Perera, L.R. (2003). Participatory decison support for agricultural management: A case study from Sri Lanka. *Agricultural Systems* 76: 457-482.

Carter, J.O., Hall, W.B., Brook, K.D., MeKeon, G.M., Day, K.A., and Paul, C.J. (2000). Aussie GRASS: Australian Grassland and Rangeland Assessment by Spatial Simulation. In Hammer, G.L., Nicholls, N., and Mitchell, C. (Eds.), *Applications of Seasonal Climate Forecasting in Agricultural and Natural Ecosystems* (pp. 329-350). London: Kluwer Academic Publishers.

Clewett, J.F., Smith, P.G., Partridge, I.J., George, D.A., and Peacock, A. (1999). *Australian Rainman Version 3: An Integrated Software Package of Rainfall Information for Better Management.* Q198071. Queensland: Department of Primary Industries.

Curtis, F.M.S., Bowden, J.W., and Fels, H.E. (1987). PGAP—A pasture growth model. In White, D.H. and Weber, K.M. (Eds.), *Proceedings of a Workshop on Computer-Assisted Management of Agricultural Production Systems* (pp. 45-51). Melbourne, Australia: Royal Melbourne Institute of Technology.

Decker, W.L. (1994). Developments in agricultural meteorology as a guide to its potential for the twenty-first century. *Agricultural and Forest Meteorology* 69: 9-25.

Dimes, J.P., Freebairn, D.M., and Glanville, S.F. (1993). HOWWET?—A tool for predicting fallow water storage. In *Proceedings of the 8th Australian Agronomy Conference 1993,* Toowoomba, Australia, January 30-February 2 (pp. 207-210). Toowoomba, Australia: Australian Society of Agronomy.

Donnelly, J.R., Freer, M., and Moore, A.D. (1994). Evaluating pasture breeding objectives using computer models. *New Zealand Journal of Agricultural Research* 37: 269-75.

Donnelly, J.R., Moore, A.D., and Freer, M. (1997). GrazPlan: Decision support systems for Australian grazing enterprises. I. Overview of the GrazPlan project, and a description of the MetAccess and LambAlive DSS. *Agricultural Systems* 54: 57-76.

Dorward, P., Galpin, M., and Shepherd, D. (2003). Participatory farm management methods for assessing the suitability of portential innovations: A case study of green manuring options for tomato growers in Ghana. *Agricultural Systems* 75: 97-117.

Duncan, W.G., McCloud, D.W., McGraw, R.Z., and Boote, K.J. (1978). Physiological aspects of peanut yield improvement. *Crop Science* 18: 1015-1020.

Freer, M. and Moore, A.D. (1990). *GrazFeed: A Nutritional Management System for Grazing Animals.* User and Technical Manuals. Sydney: Horizon Technology Pty Ltd.

Grains Research and Development Corporation (1999). *The Australian Grains Field Research Manual and CD-ROM.* Kingston, Australia: Author.

Grains Research and Development Corporation (2001). On-farm computers: Adults take over but complain about Internet. *Groundcover* (Winter, No. 35): 26.

Guardrian, J. (1977). *Crop Micrometeorology: A Simulation Study.* Wageningen, Netherlands: Centre for Agricultural Publishing and Documentation.

Hackett, C. and Harris, G. (1996). *PLANTGRO: A Software Package for Prediction of Plant Growth.* Brisbane, Australia: Griffith University.

Hammer, G., Carberry, P., and Stone, R. (2000). Comparing the value of seasonal forecasting systems in managing cropping systems. In Hammer, G.L., Nicholls,

N., and Mitchell, C. (Eds.), *Applications of Seasonal Climate Forecasting in Agricultural and Natural Ecosystems* (pp. 183-195). London: Kluwer Academic Publishers.

Hammer, G.L., Holzworth, D.P., and Stone, R. (1996). The value of skill in seasonal climate forecasting to wheat crop management in a region with high climatic variability. *Australian Journal of Agricultural Research* 47: 717-737.

Hartkamp, A.D., White, J.W., and Hoogenboom, G. (1999). Interfacing geographic information systems with agronomic Modeling: A review. *Agronomy Journal* 91: 762-772.

Hearn, A.B. (1994). OZCOT: A simulation model for cotton crop management. *Agricultural Systems* 44: 257-299.

Hodges, T. (1998). Water and nitrogen applications for potato: Commercial and experimental rates compared to a simulation model. *Journal of Sustainable Agriculture* 13: 79-90.

Hoogenboom, G. (2000). Contribution of agrometeorology to the simulation of crop production and its applications. *Agricultural and Forest Meteorology* 103: 137-157.

Hook, A.R. (Ed.) (1997). *Predicting Farm Production and Catchment Processes. A Directory of Australian Modeling Groups and Models:* Collingwood, VIC, Australia: CSIRO Publishing.

Hume, C.J. and Callander, B.A. (1990). Agrometeorology and model building. *Outlook on Agriculture* 19: 25-30.

James, Y.W. and Cutforth, H.W. (1996). Crop growth models for decision support systems. *Canadian Journal of Plant Science* 76: 9-19.

Jones, C.A., Wegener, M.X., Russell, J.S., McLeod, I.M., and Williams, J.R. (1989). *AUSCANE: Simulation of Australian Sugarcane with EPIC.* GSIRO Division of Tropical Crops and Pastures Technical Paper No. 29. Brisbane: CSIRO Division of Tropical Crops and Pastures.

Jorgensen, S.E. (1999). State-of-the-art of ecological modeling with emphasis on development of structural models. *Ecological Modeling* 120: 75-96.

Kamp, J.A.L.M. (1999). Knowledge based systems: From research to practical application: Pitfalls and critical success factors. *Computers and Electronics in Agriculture* 22: 243-250.

Kingwell, R.S., Morrison, D.A., and Bathgate, A.D. (1991). *MUDAS, Model of an Uncertain Dryland Agricultural System: A Description.* Miscellaneous publication 17/91. Perth, Australia: Western Australia Department of Agriculture.

Kirschbaum, M.U.F. (1999). CenW, a forest growth model with linked carbon, energy, nutrient and water cycles. *Ecological Modelling* 118: 17-59.

Kuhlmann, F. and Brodersen, C. (2001). Information technology and farm management: Developments and perspectives. *Computers and Electronics in Agriculture* 30: 71-83.

Larsen, D. (2001). EntomoLOGIC. <http://cotton.pi.csiro.au/Tools/>.

Lewis, T. (1998). Evolution of farm management systems. *Computers and Electronics in Agricultures* 19: 233-248.

Lynch, T., Gregor, S., and Midmore, D. (2000). Intelligent support systems in agriculture: How can we do better. *Australian Journal of Experimental Agriculture* 40: 609-620.

Maas, S.J. (1993). Within-season calibration of modelled wheat growth using remote sensing and field sampling. *Agronomy Journal* 85: 669-672.

Maracchi, G., Pérarnaud, V., and Kleschenko, A.D. (2000). Applications of geographical information systems and remote sensing in agrometeorology. *Agricultural and Forest Meteorology* 103: 119-136.

Marshall, G.R., Parton, K.A., and Hammer, G.L. (1996). Risk attitude, planting conditions and the value of seasonal forecasts to a dryland wheat grower. *Australian Journal of Agricultural Economics* 40: 211-233.

McCown, R.L., Hammer, G.L., Hargreaves, J.N.G., Holzworth, D.P., and Freebairn, D.M. (1996). APSIM: A novel software system for model development, model testing, and simulation in agricultural systems research. *Agricultural Systems* 50: 255-271.

McKeon, G.M., Day, K.A., Howden, S.M., Mott, J.J., Orr, D.M., and Scattini, W.J. (1990). Management of pastoral production in northern Australian savannas. *Journal of Biogeography* 17: 355-372.

McLeod, C.R. and Bowman, P.J. (1992). *SheepO Users Guide,* Version 3. East Melbourne: Victorian Department of Food and Agriculture.

McMurtrie, R.E. and Landsberg, J.J. (1991). BIOMASS—A mechanistic model of the growth of tree stands. *Agricultural Systems and Information Technology Newsletter* 3: 33-35.

Meinke, H. (2000). Anticipating climate variability to improve agricultural decision making. Paper presented at Cli-Manage 2000. Conference on Climate Variability, Albury, NSW, October 23-25.

Moore, A.D., Donnelly, J.R., and Freer, M. (1997). GRAZPLAN: Decision support systems for Australian grazing enterprises. III. Pasture growth and soil moisture submodels, and the GrassGro DSS. *Agricultural Systems* 55: 535-582.

Norman, J.M. (1979). Modeling the complete crop canopy. In Barfield, B.J. and Gerber, J.F. (Eds.), *Modification of the Aerial Environment of Plants.* St. Josephs, MI: American Society of Agricultural Engineers.

Parker, C. (1999). Decision support systems: Lessons from past failures. *Farm Management* 10: 273-289.

Parker, C.G., Campion, S., and Kure, H. (1997). Improving the uptake of decision support systems in agriculture. In Thysen, I. and Kristensen, A.R. (Eds.), *Proceedings of the First European Conference for Information Technology in Agriculture.* Copenhagen, Denmark, June 15-18 (pp.129-134). Copenhagen: EFITA, The Royal Veterinary and Agricultural University.

Paull, C.J. and Peacock, A. (1999) *Australian Climate/Weather Services and Use of the Information.* Indooroopilly: Queensland Centre for Climate Applications.

Penning de Vries, F.W.T., Jansen, D.M., Berge, H.F.M., and Bakenta, A. (1989). *Simulation of Crop Physiological Processes of Growth in Several Annual Crops.* Simulation Monograph. Wageningen, Netherlands: PUDOC Scientific Publishers.

Pusey, P.L. (1997). Crab apple blossoms as model for research on biological control of fire blight. *Phytopathology* 87: 1096-1102.

Rickert, K.G., Thompson, P.J.M., Pritchard, J.R., and Scattini, W.J. (1996). *FEEDMAN—A Feed-to-Dollar Beef Management Package.* Brisbane, Queensland: Department of Primary Industries.

Ritche, J.T., Kniry, J.R., Jones, C.A., and Dyke, P.T. (1986). Model inputs. In Jones, C.A., Kiniry, J.R., and Dyke, P.T. (Eds.), CERES—*Maize: A Simulation Model for Maize Growth and Development* (CD-ROM). Temple, TX: Texas A&M University Press.

Skarrat, D.B., Sutherst, R.W., and Maywald, G.F. (1995). *CLIMEX for Windows Version 1.0. User's Guide. Computer Software for predicting the effect of climate on plants and animals.* Brisbane, Australia: CSIRO and CRC for Tropical Pest Management.

Stafford Smith, D.M., Clewett, J.F., Moore, A.D., McKeon, G.M., and Clark, R. (1996). *DroughtPlan.* Full Project Report. DroughtPlan Working Paper No. 10, CSIRO Alice Springs/LWRRDC Occasional Paper Series. Canberra, Australia: Land and Water Resources Research and Development Corporation.

Stafford Smith, D.M. and Foran, B.D. (1989). Strategic decisions in pastoral management. *Australian Rangelands Journal* 8: 110-117.

Stapper, M. (1984). *SIMTAG: A Simulation Model of Wheat Genotypes.* Aleppo, Syria: International Center for Agricultural Research in the Dry Areas (ICARDA).

Stephens, D.J., Lyons, T.J., and Lamond, M.H. (1989). A simple model to forecast wheat yield in Western Australia. *Journal of the Royal Society of Western Australia* 71: 77-81.

Stephens, D.J., Walker, G.X., and Lyons, T.J. (1994). Forecasting Australian wheat yields with a weighted rainfall index. *Agricultural and Forest Meteorology* 71: 247-63.

Stubbs, A.K., Markham, N.K., and Straw, W.M. (1998). *Personal Computers for Farmers.* Report No. 98/33. Kingston, ACT: Rural Industries Research and Development Corporation.

Thornton, P.K., Hoogenboom, G., Wilkens, P.W., and Bowen, W.T. (1995). A computer program to analyse multiple-season crop model outputs. *Agronomy Journal* 87: 131-136.

Thornton, P.K. and Wilkens, P.W. (1998). Risk assessment and food security. In Tsuji, G.Y., Hoogenboom, G., and Thornton, P.K. (Eds.), *Understanding Options for Agricultural Production* (pp. 329-345). Dordrecht, Netherlands: Kluwer Academic Publishers.

Tsuji, G.Y., Hoogenboom, G., and Thornton, P.K. (Eds.) (1998). *Understanding Options for Agricultural Production. Systems Approaches for Sustainable Agricultural Development,* Volume 7. Dordrecht, Netherlands: Kluwer Academic Publishers.

Wang, K.M. and Orsini, P.G. (1992). SUMMER-PAK a user-friendly simulation software for the management of sheep grazing dry pastures or stubbles. *Agricultural Systems and Information Technology Newsletter* 4: 15-16.

Wang, Y.P. and Gifford, R.M. (1995). A model of wheat grain growth and its applications to different temperature and carbon dioxide levels. *Australian Journal of Plant Physiology* 22: 843-55.

Whelan, M.B., Bowman, P.J., White, D.H., and McLeod, C.R. (1987). SheepO: A sheep management optimization package for sheep industry specialists. *Proceedings of the International Conference on Veterinary Preventive Medicine and Animal Production. Australian Veterinary Journal* (special issue): 142-143.

Whisler, F.D., Acock, B., Baker, D.N., Fye, R.F., Hodges, H.F., Lambert, J.R., Lemmain, H.E., Mckinion, J.M., and Reddy, V.R. (1986). Crop simulation models in agronomic systems. *Advances in Agronomy*. 40: 141-208.

Woodruff, D.R. (1992). "WHEATMAN," a decision support system for wheat management in subtropical Australia. *Australian Journal of Agricultural Research* 43: 1483-1499.

Chapter 9

Adams, R.M., Houston, L.L., McCarl, B.A., Tiscareno, L.M., Matus, G.J., and Weiher, R.F. (2003). The benefits to Mexican agriculture of an El Niño-southern oscillation (ESNO) early warning system. *Agricultural and Forest Meteorology* 115: 183-194.

Albrecht, G.A. and Gow, J. (1997). *Rural Needs and Climate Variability*. Final Report to NSW Agriculture. The Rural Needs Research Unit, The University of Newcastle.

Anonymous (1996). Weather forecasting: Hail, rain or shine? *Australian Farm Journal CROPS* (April): 28-29.

Balston, J. (2000). Climate tools and services provided by the Queensland Center for Climate Applications (QDPI/ODNR). Paper presented at Cli-Manage 2000. Conference on Climate Variability, Albury, NSW, October 23-25.

Beard, G. (2000). Climate services provided by the Bureau of Meteorology. Paper presented at Cli-Manage 2000. Conference on Climate Variability, Albury, NSW, October 23-25.

Blad, B.L. (1994). Future directions and needs for academic education in agricultural meteorology. *Agricultural and Forest Meteorology* 69: 27-32.

Bowman, P.J., McKeon, G.M., and White, D.H. (1995). The impact of long range climate forecasting on the performance of sheep flocks in Victoria. *Australian Journal of Agricultural Research* 46: 687-702.

Buckley, D. (2002). Forecast accuracy limits on-farm use. *Farming Ahead* 124: 34-46.

Bureau of Meteorology (1997a). *Climate Activities in Australia 1997*. Melbourne, Australia: Author.

Bureau of Meteorology (1997b). *Directory of Bureau of Meteorology Products.* Melbourne, Australia: Author.

Bureau of Meteorology (1997c). RIVERWATCH, River States as at 9 am Friday, 6/5/97. A joint service for which information is provided by Bureau of Meteorology, Department of Land and Water Conservation, and Government Astronomer. Sydney Office.

Bureau of Meteorology (2001). Climate services and applications. In *Climate Activities in Australia 2001* (pp. 50-72). Melbourne, Australia: Author.

Cane, M.A., Eshel, G., and Buckland, R.W. (1994). Forecasting Zimbabwean maize yield using eastern equatorial Pacific sea surface temperature. *Nature* 370: 204-205.

Carberry, P., Hammer, G., Meinke, H., and Bange, M. (2000). The potential value of seasonal climate forecasting in managing cropping systems. In Hammer, G.L., Nicholls, N., and Mitchell, C. (Eds.), *Applications of Seasonal Climate Forecasting in Agricultural and Natural Ecosystems* (pp. 167-181). London: Kluwer Academic Publishers.

Carlson, D.J. (1998). Seasonal forecasting. *Quarterly Journal of the Royal Meteorological Society* 124: 1-26.

Clewett, J.F. and Drosdowsky, L. (1996). Study 1: Use of the Southern Oscillation Index as a drought management tool. In Stafford Smith, D.M., Clewett, J.F., Moore, A.D., McKeon, G.M., and Clark, R. *Drought Plan: Building on Participation.* Full project report, Working Paper No. 10 (pp. 120-124). Alice Spring, Australia: CSIRD/LWRDC Occasional Paper Series.

Clewett, J.F., Howden, S.M., McKeon, G.M., and Rose, C.W. (1991). Optimising farm dam irrigation in response to climatic risk. In Muchow, R.C. and Bellamy, J.A. (Eds.), *Climatic Risk in Crop Production: Models and Management for the Semiarid Tropics and Subtropics* (pp. 307-328). Wallingford, UK: CAB International.

Crichton, J., Mavi, H., Tupper, G., and McGufficke, A. (1999). *A Survey of the Assessment of Seasonal Conditions in Pastoral Australia Part 2: New South Wales. Benchmarking in the Aussie GRASS Project.* Queensland Department of Primary Industries Report Series QO99015. Brisbane: Department of Primary Industries.

Doraiswamy, P.C., Pasteris, P.A., Jones, K.C., Motha R.P., and Nejedlik, P. (2000). Techniques for methods of collection, database management and distribution of agrometeorological data. *Agricultural and Forest Meteorology* 103: 83-97.

Dudley, N.J. and Hearn, A.B. (1993). El Niño effects hurt Naomi irrigated cotton growers, but they can do little to ease the pain. *Agricultural Systems* 42: 103-126.

Elliott, G. and Foster, I. (1994). *Requirements for Seasonal Climate Forecasts by Agricultural Producers in Western Australia.* Milestone Report to LWRRDC, National Climate Variability Program Funded Project BOM1. Perth: Department of Agriculture, Western Australia.

Freebairn, J. (1996). Some economic considerations in the provision of meteorological services to agriculture. In *Proceedings of the Second Australian Conference on Agricultural Meteorology,* October 1-4, The University of Queensland (pp. 1-5). Brisbane: The Institute of Continuing and TESOL Education, University of Queensland.

Glantz, M.H. (1994). Forecasting El Niño: Science's gift to the 21st century. *Ecodecision* 4: 78-81.

Gommes, R., Snijders, F., and Rijks J.Q. (1996). FAO crop forecasting philosophy in national food warning systems. In Dunkel, Z. (Ed.), *The Use of Remote Sensing Technique in Agricultural Meteorology Practice,* COST 77 Workshop, Budapest, September 19-20 (pp. 146-147). Luxembourg: European Commission.

Hammer, G., Carberry, P., and Stone, R. (2000). Comparing the value of seasonal forecasting systems in managing cropping systems. In Hammer, G.L., Nicholls, N., and Mitchell, C. (Eds.), *Applications of Seasonal Climate Forecasting in Agricultural and Natural Ecosystems* (pp. 183-195). London: Kluwer Academic Publishers.

Hammer, G. L., Hansen, J.W., Phillips, J.G., Mjelde, J.W., Hill, H., Love A., and Potgieter A. (2001). Advances in application of climate prediction in agriculture. *Agricultural Systems* 70: 515-553.

Hammer, G.L., Holzworth, D.P., and Stone, R. (1996). The value of skill in seasonal climate forecasting to wheat crop management in a region with high climatic variability. *Australian Journal of Agricultural Research* 47: 717-737.

Hastings, P. and O'Sullivan, D.B. (1998). Impacts of seasonal climate forecasting on primary producers in regional Queensland. Paper presented at the 12th ANZ Climate Forum, November 30 to December 2, Perth, Western Australia.

Hatfield, J.L. (1994). Future needs in agricultural meteorology: Basic and applied research. *Agricultural and Forest Meteorology* 69: 39-45.

Hollinger, S.E. (1994). Future directions and needs in agricultural meteorology/climatology and modeling. *Agricultural and Forest Meteorology* 69: 1-7.

Hughes, J.P. and Guttorp, P. (1994). A class of stochastic models for relating atmospheric patterns to regional hydrologic phenomena. *Water Resources Research.* 30: 1535-1546.

Jessop, R. (1977). The influence of time of sowing and plant density on the yield and oil content of dryland sunflowers. *Australian Journal of Experimental Agriculture and Animal Husbandry* 17: 664.

Jones, J. (1997). What weather forecasting services are available. *Australian Farm Journal CROPS* (April): 17.

Katz, R.W. and Murphy, A.H., (Eds.) (1997). *Economic Value of Weather and Climate Forecasts.* Cambridge, UK: Cambridge University Press.

Keplinger, K. and Mjelde, J.W. (1995). The influence of Southern Oscillation on sorghum yield in selected regions of the world, Abstr. *Journal of Agricultural and Applied Economics* 26: 324.

Kidson, J.W. and Thomson, C.S. (1998). A comparison of statistical and model-based downscaling techniques for estimating local climate variations. *Journal of Climate* 11: 735-753.

Lyon, N. (1997). Weather forecasting help in managing farm risk. *Australian Farm Journal CROPS* (April): 6-7.

Marshall, G.R., Parton, K.A., and Hammer, G.L. (1996). Risk attitude, planting conditions and the value of seasonal forecasts to a dryland wheat grower. *Australian Journal of Agricultural Economics* 40: 211-233.

Mavi, H.S. (1994). *Introduction to Agrometeorology.* New Delhi: Oxford and IBH.

McKeon, G.M. and White, D.H. (1992). El Niño and better land management. *Search* 23: 197-200.

Meinke, H. (2000). Anticipating climate variability to improve agricultural decision making. Paper presented at Cli-Manage 2000. Conference on Climate Variability, Albury, NSW. October 23-25.

Mjelde, J.W. and Keplinger, K. (1998). Use of Southern Oscillation to forecast Texas winter wheat and sorghum crop yields. *Journal of Climate* 11: 54-60.

Mjelde, J.W., Thompson, T.N., Nixon, C.J., and Lamb, P.J. (1997). Utilising a farm-level decision model to help prioritise future climate prediction research needs. *Meteorological Applications* 4: 161-170.

National Research Council (NRC) (1996). *Learning to predict with El Nino and Southern Oscillation.* Washington, DC: National Academy Press.

Nicholls, N. (1985). Impact of the Southern Oscillation on the Australian crops. *International Journal of Climatology* 5: 553-560.

Nicholls, N. (2000). Opportunities to improve the use of seasonal climate forecasts. In Hammer, G.L., Nicholls, N., and Mitchell, C. (Eds.), *Applications of Seasonal Climate Forecasting in Agricultural and Natural Ecosystems* (pp. 309-327). London: Kluwer Academic Publishers.

NSW Agriculture (1997). *Weather, Irrigation and Disease Report,* May 5. Alstonville, Australia: Tropical Fruit Research Station.

Ogallo, L.A., Boulahyab, M.S., and Keane, T. (2000). Applications of seasonal to interannual climate prediction in agricultural planning and operations. *Agricultural Meteorology* 103: 159-166.

Palmer, T.N. and Anderson, D.L.T. (1993). *Scientific Assessment of the Prospects for Seasonal Forecasting a European Perspective.* European Center for Medium Range Weather Forecasts (ECMWF) Technical Report No. 70. Reading, UK: ECMWF.

Paul, C., Cliffe, N., and Hall, W. (2001). *A Survey of the Assessment of Seasonal Conditions in Pastoral Australia.* Aussie GRASS Extension Sub-project Final Report. Queensland: Department of Natural Resources and Mines.

Perry, K.B. (1994). Current and future agricultural meteorology and climatology education needs of the US Extension service. *Agricultural and Forest Meteorology* 69: 33-38.

Petersen, E.H. and Fraser, R.W. (2001). An assessment of the value of seasonal forecasting technology for Western Australia. *Agricultural Systems* 70: 259-274.

Rijks, D. and Baradas, M.W. (2000). The clients for agrometeorological information. *Agricultural and Forest Meteorology* 103: 27-42.

Seeley, M.W. (1994). The future of serving agriculture with weather/climate information and forecasting: Some indications and observations. *Agricultural and Forest Meteorology* 69: 47-49.

Serafin, R.J., Macdonald, A.E., and Gall, R.L. (2002). Transition of weather research to operations: Opportunities and challenges. *Bulletin of the American Meteorological Society* 83: 377-392.

Stafford Smith, M., McKeon, G., Ash, A., Buxton, R., and Breen, J. (1996). *Evaluating the Use of SOI Forecasts in North Queensland Using the Herd-Econ/*

GRASP Linked Model. DroughtPlan Working Paper No. 9/LUCNA Working Paper No. 1. Alice Springs: CSIRO.

Stigter, C.J., Sivakumar, M.V.K., and Rijks, D.A. (2000). Agrometeorology in the 21st century: Workshop summary and recommendations on needs and perspectives. *Agricultural and Forest Meteorology* 103: 209-227.

Stone, R.C. and Marcussen, T. (1994). *Queensland Producer Requirements from Seasonal Climate Forecast Systems.* Milestone Report to LWRRDC, NCVP Funded Project BOM1. Queensland: Department of Primary Industries.

Tennant, D. and Stephens, D. (2000). Climate services provided by Agriculture Western Australia. Paper presented at Cli-Manage 2000. Conference on Climate Variability, Albury, NSW. October 23-25.

Truscott, M. and Egan, J. (2000). Climate services provided by SARDI. Paper presented at Cli-Manage 2000. Conference on Climate Variability, Albury, NSW. October 23-25.

Von Storch, H., Zorita, E., and Cubasch, U. (1993). Down scaling of climate change estimates to regional scales: An application to winter rainfall in the Iberian Peninsula. *Journal of Climate* 6: 1161-1171.

Weiss, A., Van Crowder, L., and Bernardi, M. (2000). Communicating agrometeorological information to farming communities. *Agricultural and Forest Meteorology* 103: 185-196.

White, D.H., Tupper, G., and Mavi, H. (1999). *Agricultural Climate Research and Services in Australia.* Occasional Paper CV02/99. Canberra, Australia: LWRRDC.

World Meteorological Organization (WMO) (1997a). *CLIPS: Stepping Forward Implementation of the WMO CLIPS Project.* WMO No. 864. Geneva, Switzerland: Author.

World Meteorological Organization (WMO) (1997b). *CLIVAR—A Research Programme on Climate Variability and Predicability for the 21st Century.* Report on World Climate Research Programme. WCRP No. 101, WMO/TD No. 853, ICPO No. 10. Southampton, UK: CLIVAR.

Zorita, E., Hughes, J.P., Lettermaier, D.P., and Von Storch, H. (1995). Stochastic characterization of regional circulation patterns for climate model diagnosis and estimation of local precipitation. *Journal of Climate* 8: 1023-1042.

Chapter 10

Australian Bureau of Statistics (2000). *Use of Information in Farms, Australia, 1999-2000.* Catalog No. 8150.0. Canberra, Australia: Bureau of Statistics.

Bayley, D. (2000). *Weather and Climate in Farming, Managing Risk for Profit.* Paterson, NSW, Australia: NSW Agriculture.

Burke, S.J.A. (1991). Effects of shelterbelts on crop yields at Rutherglen. Paper presented at the National Conference on the Role of Trees in Sustainable Agriculture, Rural Industry Research and Development Corporation, Canberra, Australia.

Buttel, F., Larson, O., and Gillespie, G. (1990). *The Sociology of Agriculture*. New York: Greenwood Press.

Clewett, J., Cliffe, N., Drosdowsky, L., George, D., O'Sullivan, D., Paull, C., Partridge, I., and Saal, R. (2000). Building knowledge and skills to use seasonal climate forecasts in property management planning. In Hammer, G.L., Nicholls, N., and Mitchell, C. (Eds.), *Applications of Seasonal Climate Forecasting in Agricultural and Natural Ecosystems* (pp. 291-308). London: Kluwer Academic Publishers.

Clewett, J.F., Smith, P.G., Partridge, I.J., George, D.A., and Peacock, A. (1999). *Australian Rainman,* Version 3. Queensland: Department of Primary Industries.

Cornish, P.S., Ridge, P., Hammer, G., Butler, D., Moll, J., and Macrow, I. (1998). Wheat yield trends in the northern grains region—A sustainable industry? In *Proceedings of the 9th Australian Agronomy Conference,* July 20-23 (pp. 649-652). Wagga Wagga: Australian Society of Agronomy.

Cornwall, A., Guijt, I., and Welbourn, A. (1994). Acknowledging process: Challenges for agricultural research and extension methodology. In Scoones, I. and Thompson, J. (Eds.), *Beyond Farmer First, Rural People's Knowledge, Agricultural Research and Extension Practice* (pp. 98-117). London: Intermediate Technology Publications.

Cousens, R. and Mortimer, M. (1995). *Dynamics of Weed Populations*. Cambridge, UK: Cambridge University Press.

Davidson, T.M., Silver, B.A., Lisle, A.T., and Orr, W.N. (1988). The influence of shade on milk production of Holstein-Friesian cows in a tropical upland environment. *Australian Journal of Experimental Agriculture* 28: 149-154.

Egan, J. and Hammer, G. (1995). Managing climate risks in grain production. In *Proceedings of Managing with Climate Variability Conference* (pp. 98-105). Occasional Paper CV03/95. Canberra, Australia: LWRRDC.

Fawcett, R.G. (1968). Factors influencing the development and grain yield of spring wheats grown on the north-west plains of New South Wales, with special reference to soil-water plant relationships. PhD Thesis, University of Sydney.

Fitzgerald, D. (1994). *Money Trees on Your Farm*. Melbourne: Inkata Press.

French, R.J. (1987). Future productivity on our farmlands. In *Proceedings of the 4th Australian Agronomy Conference,* August 24-27 (pp. 140-149). Melbourne: Australian Society of Agronomy.

Greggery, I. (2000). Managing climate through flexible cropping programs. *Australian Farm Journal* (December): 53-54.

Hammer, G. (2000). Applying seasonal climate forecasts in agricultural and natural ecosystems—A synthesis. In Hammer, G.L., Nicholls, N., and Mitchell, C. (Eds.), *Applications of Seasonal Forecasting in Agricultural and Natural Ecosystems* (pp. 453-462). London: Kluwer Academic Publishers.

Hammer, G.L., Holzworth, D.P., and Stone, R. (1996). The value of skill in seasonal climate forecasting to wheat crop management in a region with high climatic variability. *Australian Journal of Agricultural Research* 47: 717-737.

Hammer, G.L., Nicholls, N., and Mitchell, C. (2000). *Applications of Seasonal Climate Forecasting in Agricultural and Natural Ecosystems*. London: Kluwer Academic Publishers.

Hayman, P., Cox, P., and Huda, S. (1996). Will more climate information assist farmers better to assess and manage the risks of opportunity cropping on the Liverpool Plains. In *Proceedings of the Second Australian Conference on Agricultural Meteorology,* October 1-4, The University of Queensland (pp. 67-71). National Committee on Agrometeorology.

Hayman, P. and de Vries, J. (1995). *Managing Crop Sequences.* Orange: NSW Agriculture.

Hayman, P. and Pollock, K. (2000). Communicating probabilities to farmers: Some experiences with charts and chocolate wheels. Paper presented at the 13th Australian New Zealand Climate Forum, Hobart, Australia, April 10-12.

Hayman, P.T. and Turpin, J.E. (1998). Nitrogen fertiliser decisions for wheat on the Liverpool Plains, NSW. II. Should farmers consider stored soil water and climate forecasts? In *Proceedings of the 9th Australian Agronomy Conference,* July 20-23 (pp. 653-656). Wagga Wagga: Australian Society of Agronomy.

Jones, R. (2000). Greenhouse warming will lower cow's milk production. *Land and Water News* 4(8) (September 5).

Keogh, D. (2000). Seasonal forecasts helping northern Murray Darling irrigators. *CLIMAG* 4: 7.

Kilpatrick, S. (1996). *Change, Training and Farm Profitability.* National Farmer's Federation Research Paper 10. Canberra, Australia: National Farmer's Federation.

Lynch, J.J. and Donnelly, J.B. (1980). Changes in pastures and animal production resulting from the use of windbreaks. *Australian Journal of Agricultural Research* 31: 967-79.

Mavi, H.S. (1994). *Introduction to Agrometeorology.* New Delhi: Oxford and IBH.

Nicholls, N. (2000). Opportunities to improve the use of seasonal climate forecasts. In Hammer, G.L., Nicholls, N., and Mitchell, C. (Eds.), *Applications of Seasonal Forecasting in Agricultural and Natural Ecosystems* (pp. 309-327). London: Kluwer Academic Publishers.

Nicholson, S.W. and Albert, R.D. (1988). *Clearing Practices for Central West New South Wales.* Agfact P1.1.2. Sydney: Department of Agriculture, NSW.

Perry, M.W. (1987). Water use efficiency of non-irrigated field crops. In *Proceedings of the 4th Australian Agronomy Conference,* August 24-27 (pp. 83-89). Melbourne: Australian Society of Agronomy.

Pratley, J.E. (1987). Soil, water and weed management—The key to productivity in southern Australia. *Plant Protection Quarterly* 2: 21-30.

Ridge, P. and Wylie, P. (1996a). Farmers' training needs and learning for improved management of climatic risk. Paper presented at the Conference on Managing with Climate Variability, Canberra, Australia, November 16-17.

Ridge, P. and Wylie, P. (1996b). *Improved Management of Climate Risk.* Occasional Paper No. CV01/96. Canberra: Land and Water Resources Research and Development Corporation.

Roling, N. (1988). *Extension Science—Information Systems in Agricultural Development.* Cambridge, UK: Cambridge University Press.

Russell, D., Ison, R., Gamble, D., and Williams, R. (1989). *A Critical Review of Rural Extension Theory and Practice.* Richmond: Australian Wool Corporation/ University of Western Sydney, Hawkesbury.

Shrapnel, M. and Davie, J. (2000). The influence of personality in determining farmer responsiveness to risk. Paper presented at the International Workshop on Farm Management Decisions with Climatic Risk, April 17-19, Toowoomba, Australia: Queensland Department of Primary Industries.

Simpson, P. (1999). Recognising and managing landscape and pasture diversity. *Proceedings of Fourteenth Annual Conference of the Grasslands Society* NSW, 18-24.

Stafford Smith, M., Buxton, R., McKeon, G., and Ash, A. (2000). Seasonal climate forecasts and the management of rangelands: Do production benefits translate into enterprise profits? In Hammer, G.L., Nicholls, N., and Mitchell, C. (Eds.), *Applications of Seasonal Forecasting in Agricultural and Natural Ecosystems* (pp. 271-291). London: Kluwer Academic Publishers.

Standing Committee on Agriculture and Resource Management (SCARM) (1998). *Sustainable Agriculture Assessing Australia's Recent Performance.* Technical Report 70. Collingwood, Victoria: CSIRO Publishing.

Tow, P.G. (1991). Factors in the development and classification of dryland farming systems. In Squires, V. and Tow, P. (Eds.), *Dryland Farming—A Systems Approach.* Sydney: Sydney University Press.

Tow, P.G. and Schultz, J.E. (1991). Crop and crop-pasture sequences. In Squires, V. and Tow, P. (Eds.), *Dryland Farming—A Systems Approach.* Sydney: Sydney University Press.

Vanclay, F. (1992). The barriers to adoption often have a rational basis. Paper presented at the 7th ISCO Conference—People Protecting Their Land, September 27-30, Sydney, Department of Conservation and Land Management (CALM).

Vanclay, F. and Lawrence, G. (1995). Agricultural extension as social welfare. *Rural Society* 5: 20-33.

Woodhill, J. and Robins, L. (1998). *Participatory Evaluation for Landcare and Catchment Groups. A Guide for Facilitators.* Yarralumla, Australia: Greening Australia.

Chapter 11

Abdalati, W. and Steffen, K. (1997). Snow melt on the Greenland ice sheet as derived from passive microwave satellite data. *Journal of Climate* 10: 165-175.

Adams, R.M. (1999). The economic effects of climate change on U.S. In Mendolsohn, R. and Neumann, J. (Eds.), *The Economics of Climate Change.* Cambridge, UK: Cambridge University Press.

Albritton, D. and Kuijpers, L. (Eds.) (1999). *Synthesis of the Reports of the Scientific, Environmental Effects, and Technology and Economic Assessment Panels of the Montreal Protocol: A Decade of Assessments for Decision Makers Regarding the Protection of the Ozone Layer, 1988-1999.* Nairobi: United Nations Environment Programme, Ozone Secretariat.

Appendini, K. and Liverman, D. (1993). Agricultural policy and climate change in Mexico. In Downing, T.E. (Ed.), *Climate Change and World Food Security* (pp. 525-547). NATO, ASI Series No. 37. Berlin: Springer-Verlag.

Arnell, N.W. (2000). *Impact of Climate Change on Global Water Resources*, Volume 2: *Unmitigated Emissions*. Southampton, UK: University of Southampton, Report of the Department of the Environment, Transport and the Regions.

Baethgen, W.E. (1997). Vulnerability of the agricultural sector of Latin America to climate change. *Climate Research* 9: 1-7.

Banks, L. (1996). The effect of greenhouse on eastern Australian agriculture. In Jarman, P. and Johnson, K. (Eds.), *Land Use in a Changing Climate* (pp. 33-42). Armidale, Australia: University of New England.

Beniston, M. (1997). Variation in snow depth and duration in the Swiss Alps over the last 50 years. Link to changes in large scale climatic forcing. *Climatic Change* 36: 281-300.

Bindi, M., Ferrini, F., and Miglietta, F. (1992). Climatic change and the shift in the cultivated area of olive trees. *Journal of Agriculture Mediterranea* 22: 41-44.

Brazdil, R. (1996). Trends in maximum and minimum temperatures in central and southeastern Europe. *International Journal of Climatology* 16: 765-782.

Brklacich, M., Bryant, C., Veenhof, B., and Beauchesne, A. (1997). Implications of global climate change for Canadian agriculture: A review and appraisal of research from 1994-1997. In Koshida, G. and Avis, W. (Eds.), *Canada Country Study: Climate Impacts and Adaptations* (pp. 220-256). Ottawa: Environment Canada.

Cambula, P. (1999). *Impacts of Climate Change on Water Resources of Mozambique*. The Final Report of the Mozambique/U.S. Country Study Program Project on the Climate Change Vulnerability and Adaptation Assessments in Mozambique. Maputo, Mozambique.

Campbell, B.D., McKeon, G.M., Gifford, R.M., Clark, H., Stafford Smith, M.S., Newton, P.C.D., and Lutze, J. L. (1996). Impacts of atmospheric composition and climate change on temperate and tropical pastoral agriculture. In Bouma, W.J., Pearman, G.I., and Manning, M.R. (Eds.), *Greenhouse: Coping with Climate Change* (pp. 171-189). Collingwood, Victoria: CSIRO Publishing.

Carter, T.R. (1998). Changes in the thermal growing season in Nordic countries during the past century and prospects for the future. *Agricultural and Food Science in Finland* 7: 161-179.

Carter, T.R., Saarikko, R.A., and Niemi, K.J. (1996). Assessing the risks and uncertainties of regional crop potential under a changing climate in Finland. *Agricultural and Food Science in Finland* 5: 324-350.

Commonwealth Scientific and Industrial Research Organisation (CSIRO) (1996). *Climate Change Scenarios for the Australian Region*. Aspendale, Victoria: Commonwealth Scientific and Industrial Research Organisation, Atmospheric Research.

Commonwealth Scientific and Industrial Research Organisation (CSIRO) (2001). *Climate Change—Impacts for Australia*. CSIRO Impacts and Adaptation Group. Hobart, Australia: CSIRO.

Conde, C. (1997). Vulnerability of rainfed maize crop in Mexico to climate change. *Climate Research* 9: 17-23.

Cooper, K., Parsons, D.L., and Dernmers, T. (1998). A thermal balance model for livestock buildings for use in climate change studies. *Journal of Agricultural Engineering Research* 69: 43-52.

Crick, H.Q.P., Dudley, C., Glue, D.E., and Thomson D.L. (1997). UK birds are laying eggs earlier. *Nature* 388: 526.

Cubasch. U., Waszkewitz, G., Hegrerl, G., and Perlwitz, J. (1995). Regional climate changes as simulated in time-slice experiments. *Climate Change* 31: 273-304.

Davies, A., Jenkins, T., Pike, A., Carson, I., Pollock, C. J., and Parry, M.L. (1998). Modelling the predicted geographic and economic response of UK cropping systems to climate change scenarios: The case of sugar beet. *Annals of Applied Biology* 133: 135-148.

Davison, T., McGowan, M., Mayer, D., Young, B., Jonsson, N., Hall, A., Matschoss, A., Goodwin, P., Goughan, J., and Lake, M. (1996). *Managing Hot Cows in Australia*. Queensland: Department of Primary Industries.

Eddy, J.A. (1976). The Maunder minimum. *Science* 192: 1189-1202.

Ellery, W., Scholes, M.C., and Scholes, R.J. (1996). The distribution of sweetveld and sourveld in South Africa's grassland biome in relation to environmental factors. *African Journal of Range and Forage Science* 12: 38-45.

Fearnside, P.M. (1999). Plantation forestry in Brazil: The potential impacts of climatic change. *Biomass and Bioenergy* 16: 91-102.

Furquay, L.W. (1989). Heat stress as it affects animal production. *Journal of Animal Science* 52: 164-174.

Fushimi, H. (1999). Recent changes in glacier phenomena in the Nepalese Himalayas. In *IUCN Report on Climate Change and Biodiversity* (pp. 131-137). San Jose, CA: International Union for Conservation of Nature.

Gates, W.L., Mitchell, J.F.B., Boer, G.J., Cubasch, U., and Meleshko, V.P. (1992). Climate modelling, climate prediction and model validation. In Houghton, L.T., Cailander, B.A., and Varney S.K. (Eds.), *Climate Change 1992: The Supplementary Report to the IPCC Scientific Assessment* (pp. 97-134). Cambridge, UK: Cambridge University Press.

Gleick, P.H. (Ed.) (1993). *Water in Crisis: A Guide to the World's Water Resources*. London: Oxford University Press.

Gleick, P.H. (1998). *The World's Water: The Biennial Report on Freshwater Resources, 1998-1999*. Washington, DC: Island Press.

Grabberr, G., Gottfried, M., and Pauli, H. (1995). Pattern and current changes in alpine plant diversity. In Korner, C. (Ed.), *Arctic and Alpine Biodiversity: Patterns, Causes and Ecosystem Consequences* (pp. 176-181). Heidelberg: Springer-Verlag.

Groisman, P.Y., Karl, T.R., and Knight, R.W. (1994). Observed impact of snow cover on the heat balance and the rise of continental spring temperatures. *Science* 263: 198-200.

Gruza, G.V., Rankova, E., Bardin, M., Korvkina, L., Rocheva, E., Semenjuk, E., and Platova, T. (1997). Modern state of the global climate system. In Laverov,

N.P. (Ed.), *Global Changes of Environment and Climate: Collection of Selected Scientific Papers* (pp. 194-216). Moscow: Russian Academy of Sciences.

Hall, A.J. and McPherson, H.G. (1997). Modelling the influence of temperature on the timing of budbreak in kiwifruit. *Acta Horticulturae* 444: 401-406.

Hall, W.B., McKeon, G.M., Carter, J.O, Day, K.A., Howden, S.M., Scanlan, J.C., Johnston, P.W., and Burrows, W.H. (1998). Climate change and Queensland's grazing lands. II. An assessment of impact on animal production from native pastures. *Rangeland Journal* 20: 177-205.

Harrison, P.A. and Butterfield, R.E. (1996). The effect of climate change on Europe-wide winter wheat and sunflower productivity. *Climate Research* 7: 225-241.

Hassall and Associates, NSW Department of Land and Water Conservation, NSW National Parks & Wildlife Service, and CSIRO Atmospheric Research (1998). *Climatic Change Scenarios and Managing the Scarce Water Resources of the Macquarie River.* Canberra: Australian Greenhouse Office.

Hennessy, K.J., Suppiah, R., and Page, C.M. (1999). Australian rainfall changes—1910-1995. *Australian Meteorological Magazine* 48: 1-13.

Holbrook, S.J., Schmitt, R.J., and Stephens, J.S., Jr. (1997). Changes in an assemblage of temperate reef fishes associated with a climate shift. *Ecological Applications* 7: 1299-1310.

Horie, T., Matsui, T., Nakagawa, H., and Omasa, K. (1996). Effect of elevated CO_2 and global climate change on rice yield in Japan. In Omasa, K. (Ed.), *Climate Change and Plants in East Asia* (pp. 39-56). Berlin: Springer-Verlag.

Howden, S.M., Hall, W.B., and Bruget, D. (1999). Heat stress and beef cattle in Australian rangelands. Recent trends and climate change. In Eldridge, D. and Freudenberger, D. (Eds.), *People and Rangelands: Building the Future* (pp. 43-45). Proceedings of the VI International Rangelands Congress, Townsville, July 1999. Wallingford, UK: CABI Publishing.

Hulme, M. (1996). *Climate Change in Southern Africa—An Exploration of Some Potential Impacts and Implications in the SADC Region.* Norwich, UK: Climatic Research Unit, University of East Anglia, UK.

Hulme, M. and Carter, T.R. (2000). The changing climate of Europe. In Parry, M.L. (Ed.), *Assessment of Potential Effects of Climate Change in Europe.* The ACACIA Report. Jackson Environment Institute, University of East Anglea.

Hurrell, J.W. (1995). Decadal trends in the North Atlantic Oscillation and relationships to regional temperature and precipitation. *Science* 269: 676-679.

Intergovernmental Panel on Climate Change (IPCC) (1994). *IPCC Technical Guidelines for Assessing Climate Change Impacts and Adaptations.* Prepared by Working Group II. Carter, T.R., Parry, M.L., Harasawa, H., and Nishioka, S. (Eds.), World Meteorological Organization/United Nations Environment Program. CGER-I015-94. University College London, UK and Center for Global Environmental Research, National Institute for Environmental Studies, Tsukuba, Japan.

Intergovernmental Panel on Climate Change (IPCC) (1996). *Climate Change 1995—The Science of Climate Change.* Contribution of Group I to the Second

Assessment Report on the Intergovernmental Panel on Climate Change. Cambridge: Cambridge University Press.

Intergovernmental Panel on Climate Change (IPCC) (1998). *The Regional Impacts of Climate Change: An Assessment of Vulnerability.* A Special Report of IPCC Working Group II. Watson, R.T., Zinyowera, M.C., and Moss, R.H. (Eds.), WMO/UNEP. Cambridge, UK: Cambridge University Press.

Izrael, Y., Anokhin, Y., and Eliseev, A.D. (1997). Adaptation of water management to climate change. In Laverov, N.P. (Ed.), *Global Changes of Environment and Climate: Collection of Selected Scientific Papers* (pp. 373-392). Moscow: Russian Academy of. Sciences.

Jallow, B.P., Toure, S., Barrow, M.M.K., and Mathieu, A.A. (1999). Coastal zone of the Gambia and the Abidjan region in Cote d'Ivoire. Sea level rise vulnerability, response strategies, and adaptation options. In Mimura, N. (Ed.), *National Assessment Results of Climate Change: Impacts and Responses* (pp. 129-136). Oldendorf, Germany: Luhe, Inter-Research.

Jones, R.N. and Hennessy, K.J. (2000). *Climate Change Impacts in the Hunter Valley: A Risk assessment of Heat Stress Affecting Dairy Cattle.* CSIRO Atmospheric Research. Aspendale, Australia: CSIRO.

Kenitzer, A. (2000). NASA scientist predicts less climate cooling from clouds. Release: 00-151, October 3. Goddard Space Flight Center, Greenbelt, MD.

Kimball, B.A., Pinter, P., Gracia, R.L., LaMorte, R.L., Wall, G.W., Hunsker, D.J., Wechsung, G., Wechsung, F., and Karschall, T. (1995). Productivity and water use of wheat under free-air CO_2 enrichment. *Global Change Biology* 1: 429-442.

Kothavala, Z. (1999). The duration and severity of drought over eastern Australia simulated by a coupled ocean-atmosphere GCM with a transient increase in CO_2. *Environmental Modelling & Software* 14: 243-252.

Kothyari, U.C. and Singh, V.P. (1996). Rainfall and temperature trends in India. *Hydrological Processes* 10: 357-372.

Krabill, W., Frenerick, E., Manizade, S., Martin, C., Sonntag, J., Swift, R., Thomas, R., Wright, W., and Jungel, J. (1999). Rapid thinning of parts of the southern Greenland ice sheet. *Science* 283: 1522-1524.

Lal, M., Singh, K.K., Rathore, L.S., Srinivasan G., and Saseendran, S.A. (1998). Vulnerability of rice and wheat yields in north-west India to future changes in climate. *Agricultural and Forest Meteorology* 89: 101-114.

Lou, Q. and Lin, E. (1999). Agricultural vulnerability and adaptation in developing countries: The Asia Pacific region. *Climatic Change* 43: 729-743.

Magana, V., Conde, C., Sanchez, O., and Gay, C. (1997). Assessment of current and future regional climate change scenarios for Mexico. *Climate Research* 9: 107-114.

Magrin, G.O., Grondona, M., Travasso, M.I., Boullon, D., Rodriguez, G., and Messina, C. (1999). ENSO Impacts on Crop Production in the Argentina's Pampas region. In *Tenth Symposium on Global Change Studies* (pp. 65-66). Dallas, TX, Preprint Volume. Boston, MA: American Meteorological Society (AMS).

Marengo, J. (1995). Variation and change in South American stream flow. *Climate Change* 31: 99-117.

Matalas, N.C. (1998). Note on the assumptions of hydrologic stationarity. *Water Resources Update* 112: 64-72.

Matthews, R.B., Horie, T., Kropff, M.J., Bachelet, D., Centeno, H.G., Shin, J.C., Mohandass, S., Singh, S., Defeng, Z., and Moon, H.L. (1995). A regional evaluation of the effect of future climate change on rice production in Asia. In Matthews, R.B. (Ed.), *Modelling the Impact of Climate Change on Rice Production in Asia* (pp. 95-139). Wallingford, UK: CAB International.

McKeon, G.M., Hall, W.B., Crimp, S.J., Howden, S.M., Stone, R.C., and Jones, D.A. (1998). Climate change in Queensland's grazing lands. 1. Approaches and climatic trends. *Rangeland Journal* 20: 147-173.

Mills, P.F. (1994). The agricultural potential of northwestern Canada and Alaska and the impact of climatic change. *Arctic* 47: 115-123.

Muchena, P. and Iglesias, A. (1995). Vulnerability of maize yields to climate change in different farming sectors in Zimbabwe. In Rosenzweig, C. (Ed.), *Climate change and agriculture: Analysis of potential international impacts* (pp. 229-239). Madison, WI: American Society of Agronomy.

Munk,W., Dzieciuch, M., and Jayne, S. (2002). Millennial climate variability: Is there a tidal connection? *Journal of Climate* 15: 370-385.

Murray-Darling Basin Commission (1999). *The Salinity Audit of the Murray-Darling Basin: A 100-Year Perspective.* Canberra, Australia: Author.

Nash, L.L. and Gleick, P.H. (1993). *The Colorado River Basin and Climatic Change: The Sensitivity of Streamflow and Water Supply to Variations in Temperature and Precipitation.* EPA 230-R-93-009. Oakland, CA: Report prepared for the U.S. Environmental Protection Agency, Office of Policy, Planning and Evaluation—Climate Change Division by Pacific Institute for Studies in Development, Environment, and Security.

Newman, P.A. (2000). What's happening to stratospheric ozone over the arctic, and why? U.S. Global Change Research Program, Seminar Series, Washington, DC, July 14.

Nicholls, N. (1997). Increased Australian wheat yield due to recent climate trends. *Nature* 387: 484-485.

O'Brien, B.J. (1998). Scientific vigour amidst political correctness. Keynote address for the Twelfth Australia-New Zealand Climate Forum, Perth, Australia, November 29-30.

Odingo, R.S. (1990). Implications for African agriculture of the greenhouse effect. *Developments in Soil Science* 20: 231-248.

Olesen, J.E. and Grevsen, K. (1993). Simulated effects of climate change on summer cauliflower production in Europe. *European Journal of Agronomy* 2: 313-323.

Parmesan, C., Ryrholm, N., Stefanescu, C.. Hill, J.K., Thomas, C.D., Descimon, H., Huntley, B., Kaila, L., Kullberg, J., Tammaru, T., et al. (1999). Poleward shifts in geographical ranges of butterfly species associated with regional warming. *Nature* 399: 579-583.

Peiris, D.R., Crawford, J.W., Grassoff, C., Jefferies, R.A., Porter, J.R., and Marshall, B. (1996). A simulation study of crop growth and development under climate change. *Agricultural and Forest Meteorology* 79: 271-287.

Pimentel, D. (1993). Climate changes and food supply. *Forum for Applied Research and Public Policy* 8: 54-60.

Pinter, L. (1996). Impact of climate change on sustainability of native lifestyles in the Mackenzie River Basin. In Cohen, S.J. (Ed.), *The Mackenzie Basin Impact Study (MBIS) Final Report* (pp. 62-65).

Pounds, J.A., Fogden, M.P.L., and Campbell, J.H. (1999). Biological response to climate change on a tropical mountain. *Nature* 398: 611-615.

Quintana-Gomez, R.A. (1999). Trends of maximum and minimum temperatures in Northern South America. *Journal of Climate* 12: 2104-2112.

Rankova, E. (1998). Climate change during the 20th century in for Russian Federation. In Abstract book of Seventh International Meeting on Statistical Climatology, Whistler, B.C., Canada, May 25-29, Abstract 102, p. 98.

Robinson, P.J. (2000). Temporal trends in United States dew point temperatures. *International Journal of Climatology* 20: 985-1002.

Rosenzweig, C. and Hillel, D. (1998). *Climate Change and the Global Harvest: Potential Impacts of the Greenhouse Effect on Agriculture.* London: Oxford University Press.

Rosenzweig, C., Parry, M.L., and Fischer, G. (1995). World food supplies. In Strzepek, K.M. and Smith, J.B. (Eds.), *Climate Changes: International Impacts and Implications* (pp. 27-56). Cambridge, UK: Cambridge University Press.

Rosenzweig, C. and Tubiello, F.N. (1997). Impacts of future climate change on Mediterranean agriculture. Current methodologies and future directions. *Mitig. Adaptive Strategies Global Change* 1: 219-232.

Rupakumar, K., Krishna Kumar, K., and Pant, G.B. (1994). Diurnal asymmetry of surface temperature trends over India. *Geophysical Research Letters* 21: 677-680.

Ryaboshapko, A.G., Gallardo, L., Klellstrorn, E., Gromov, S., Paramonov, S., Afinogenova, O., and Rodhe, H. (1998). Balance of oxidized sulfur and nitrogen over the former Soviet Union territory. *Atmospheric Environment* 32: 647-658.

Salinger M.J., Stigter, C.J., and Dasc, H.P. (2000). Agrometeorological adaptation strategies to increasing climate variability and climate change. *Agricultural and Forest Meteorology* 103: 167-184.

Samarakoon, A.B. and Gifford, R.M. (1995). Soil water content under plants at high CO_2 concentration and interactions with the direct CO_2 effects: A species comparison. *Journal of Biogeography* 22: 193-202.

Schreider, S.Y., Jakeman, A.J., Pittock, A.B., and Whetton, P.H. (1996). Estimation of possible climate change impacts on water availability, extreme flow events and soil moisture in the Goulburn and Ovens basins, Victoria. *Climatic Change* 34: 513-546.

Semenov, S.M., Kounina, I.M., and Koukhta, A. (1999). *Tropospheric Ozone and Plant Growth in Europe: Meteorology and Hydrology.* Moscow: Moscow Publishing Center.

Singh, P. (1998). Effect of global warming on the streamflow of high-altitude Spiti River. In Chalise, S. R., Herrmann, A., Khanal, N.R., Lang, H., Molnar, L., and Pokhrel, A.P. (Eds.), *Ecohydrology of High Mountain Areas* (pp. 103-114). Kathmandu: International Centre for Integrated Mountain Development.

Sirotenko, O. (1999). The global greenhouse effect, agroecosystems and future of agriculture. Paper presented at the International Workshop Agrometeorology in the 21st Century, Needs and Perspectives, Accra, Ghana, February 15-17.

Smith, I.N., Budd, W.F., and Reid, P. (1998). Model estimates of Antarctic accumulation rates and relationship to temperature changes. *Annals of Glaciology* 27: 246-250.

Sutherst, R.W., Collyer, B.S., and Yonow, T. (2000). The vulnerability of Australian horticulture to the Queensland fruit fly, *Bactrocera (Dacus) tryoni,* under climate change. *Australian Journal of Agricultural Research* 51: 467-480.

Topp, C.F.E. and Doyle, C.J. (1996). Simulating the impact of global warming on milk and forage production in Scotland. 2. The effects on milk yields and grazing management of dairy herds. *Agricultural Systems* 52: 243-270.

United States National Assessment (2000). Changing climate and changing agriculture, Draft Report of the Agricultural Sector Assessment Team. Washington, DC: U.S. Climate Change Program/U.S. Global Change Research Program.

Vaughan, D.G., Bamber, J.L., Giovinetto, M., Russal, J., and Coopar, A.P.R. (1999). Reassessment of net surface mass balance in Antractica. *Journal of Climate* 12: 933-946.

Visser, M.E., van Noordwijk, A.J., Tinbergen, J.M., and Lessells, C.M. (1998). Warmer springs lead to mistimed reproduction in great tits *(Pares major). Proceedings of the Royal Society of London* 265: 1867-1870.

Walsh, J.E., Kattsov, V., Portis, D., and Meleshko, V. (1998). Arctic precipitation and evaporation: Model results and observational estimates. *Journal of Climate* 11: 72-87.

Wang, F. (1996). Climate change and crop production in China. *Economic of Chinese Country* 11: 19-23.

Wheeler, T.R., Ellis, R.H., Hadley, P., Morison, J.I.L., Batts, G.R., and Daymond, A.J. (1996). Assessing the effects of climate change on field crop production. *Aspects of Applied Biology* 45: 49-54.

Whetton, P. (2001). Will climate variability change in the future *CLIMAG* 5: 1-2.

Wolf, I. and van Diepen, C.A. (1995). Effects of climate change on grain maize yield potential in the European Community. *Climate Change* 29: 299-331.

Wolf, J. (1999). Modelling climate change impacts at the site scale on soybean. In Harrison, P.A., Butterfield, R.E., and Downing, T.E. (Eds.), *Climate Change, Climate Variability and Agriculture in Europe: An Integrated Assessment* (pp. 217-237). Environmental Change Unit Research Report 21. Oxford, UK: University of Oxford.

World Meteorological Organization (WMO) (1998). *The Global Climate System Review, December 1993-May 1996.* WMO No. 856. Geneva: WMO.

World Meteorological Organization/Global Atmosphere Watch (WMO/GAW) 116 (1998). Expert Meeting on Chemistry of Aerosols, Clouds and Atmospheric Precipitation in the former USSR, St. Petersburg, Russia, November 13-15.

Author Index

Page numbers followed by the letter "f" indicate figures; those followed by the letter "t" indicate tables.

Abdalati, W. 281
Abdel-Rahman, M.A. 120
Abdel-Samie, A.G. 120
Abrecht, D.G. 191t
Adam, H. 9
Adams, R.M. 218, 289
Agarwala, B.K. 130
Agnew, C.T. 105
Ahmad, S. 93
Aikman, D.P. 38
Alados, I. 36
Albajes, R. 131
Albert, R.D. 239
Albrecht, G.A. 217
Albritton, D. 269
Aldoss, A.A. 74
Allen, R.G. 83
Alley, W.M. 102
Alqarawi, A.A. 74
Anderson, D.L.T. 222
Anisimov, O. 35, 35f, 36f
Anokhin, Y. 279
Apinakapong, K. 31
Appendini, K. 286
Arachchi, D.H.M. 45
Arkebauer, T.J. 41
Armitage, M. 249, 250
Armitage, S. 249, 250
Arnell, N.W. 280, 281
Aro, T.O. 37
Arumugam, N. 75
Arundel, J.H. 140, 141
Ascough, J.C. II 206, 207
Ashok Raj, P.C. 72, 73
Aslam, M. 93
Assaeed, A.M. 74
Atkins, M.D.I. 123
Atta-Aly, M.A. 52
Australian Bureau of Statistics 259
Aylor, D.E. 128
Ayoub, A. 119, 121

Baethgen, W.E. 288
Baker, R.H.A. 123
Baldy, C. 10
Ballestra, G. 159
Balston, J. 215
Balston, J.M. 192t
Banda, D.S. 142
Banks, L. 284, 288
Baradas, M.W. 2, 223
Barrett, E.C. 22, 24t
Bateman, M.A. 135
Bathgate, A.D. 191t
Bauer, A. 48
Baumgartner, A. 28t, 34
Bayley, D. 252
Baylis, M. 141
Beard, G. 214
Bedo, D. 109
Beniston, M. 269
Bennet, S.M. 65
Benoit, P. 72
Berbigier, P. 36, 37, 38f
Bernardi, M. 11, 223
Bhattacharya, S. 130
Bian, J.M. 120
Biedenbender, S.H. 44
Bierhuizen, J.H. 47t
Bierwirth, P.N. 162
Billings, S.D. 164
Bindi, M. 285
Bingham, I.J. 45
Bishnoi, O.P. 135

Black, C. 38
Blad, B.L. 11, 224
Bland, W.L. 170, 171
Blood, D.C. 136, 144
Boer, R. 55, 58
Boes, E. 16
Bogardi, I. 96, 102
Boldt, A.L. 88, 90
Bonhomme, R. 40, 66, 67
Boote, K.J. 179
Bos, M.G. 88
Boshell, F. 134
Bottai, L. 160
Boulahya, M.S. 4, 220, 221
Bouma, M.J. 140
Bourke, P.M.A. 2
Bouwerg, H. 69
Bowden, J.W. 189t
Bowen, W.T. 5, 183
Bowman, P.J. 189t, 193t, 220
Brazdil, R. 271
Brecht, J.K. 52
Brklacich, M. 286
Brodersen, C. 204
Brook, K. 111
Brook, K.D. 161, 169
Brook, R.D. 192t
Bruget, D. 283, 288
Buckland, R.W. 222
Buckley, D. 218
Budd, W.F. 272
Bureau of Meteorology 214, 215
Burke, S.J.A. 239
Burman, R.D. 83
Burnside, D.G. 165
Burrough, P.A. 159
Butler, W.L. 29
Buttel, F. 258
Butterfield, R.E. 285
Byrne, D.N. 127

Cain, J.D. 207
Caldwell, M.M. 29, 34
Callander, B.A. 181
Cambula, P. 279
Campbell, B.D. 288
Campbell, C.L. 55, 58
Campbell, C.S. 41
Campbell, J.H. 274
Campion, S. 205
Cane, M.A. 222
Cao, W.X. 65
Capinera, J.L. 134
Carberry, P. 184, 218, 219
Carlson, D.J. 222
Carter, J. 111, 169
Carter, J.B. 161, 169
Carter, J.O. 169, 194t
Carter, T.R. 272, 274, 285
Chan, A.K. 11
Chandra, S. 134
Chang, J.H. 32, 52
Chang, L.R. 52
Chattopadhyay, N. 130
Chaurasia, R. 44, 52
Chen, C. 52
Chen, Y.P. 134
Cheryl, A. 95
Chiew, F.S.H. 82, 83
Chin, K. 74
Choudhury, B.J. 40
Chowdhury, S. 48
Cigliano, M.M. 133
Clark, A. 108
Clark, G.A. 75
Clewett, J. 252
Clewett, J.F. 191t, 219, 220, 250, 259
Cliffe, N. 217
Coe, R. 73
Cohen, S. 31
Cole, V.G. 136, 137
Coligny, F. 31
Collis, B. 92
Collyer, B.S. 136, 284
Comas, J. 131
Commonwealth Scientific and Industrial Research Organisation (CSIRO) 280, 284
Conde, C. 286
Cony, M.A. 65
Cooper, K. 288
Cordonnier, T. 31
Corlett, J.E. 38, 39t
Cornish, P.S. 247
Cornwall, A. 258
Correia, C.M. 30
Cottle, D.J. 189t, 193t

Coulson, R.N. 159
Courbaud, B. 31
Cousens, R. 256
Cox, P. 252, 258
Crandal, P.G. 52
Crawford, S. 142
Crichton, J. 217, 223
Crick, H.Q.P. 274
Cridland, S. 113
Cridland, S.W. 165
CSIRO 280, 284
Cubasch, U. 222, 271
Cummins, T. 84
Curtis, F.M.S. 189t
Cutforth, H.W. 184

Dallarosa, R.G. 33
Daniel, P. 128
Danthanarayana, W. 125
Das, H.P. 10
Dasc, H.P. 263, 272
Dastane, N.G. 72
Davidson, T.M. 240
Davie, J. 257
Davies, A. 285
Davison, T. 288
De Chaneet, G.C. 138
De Pauw, E. 9
de Vries, J. 246
Deckard, E.L. 44
Decker, W.L. 5, 8, 179
Degan, C. 55
Demetriades Shah, T.H. 40, 41
Department of Agriculture (NSW) 215
Dernmers, T. 288
Deshmukh, P.S. 48
Dethier, B.E. 66
Dewhurst, C.F. 132
Dezman, L.E. 103
Di Bella, C.M. 83
di Castri, F. 119
Di Chiara, C. 160
Diak, G.R. 170, 171, 172
Dimes, J.P. 112, 191t
Diss, A.L. 129
Dixon, D. 108
Doesken, N.J. 104, 105
Dominiak, B.C. 135
Donnelly. J. 112
Donnelly, J.B. 239
Donnelly. J.R. 191t, 193t
Doorenbos, J. 78, 81
Doraiswamy, P.C. 220
Dorward, P. 207
Douglas, N. 56
Doyle, C.J. 288
Drake, V.A. 123, 124, 125, 126, 128, 129
Drosdowsky, L. 220
Duddy, N. 61
Dudley, N.J. 219
Duncan, W.G. 182
Dunsmore, J.D. 138
Dutcher, J.D. 124
Dzieciuch, M. 264

Easdown, W.J. 206
Eddy, J.A. 264
Egan, J. 216, 247, 248
Egan, J.P. 192t
El Bashir, S. 134
El Sadaany, G.B. 134
Eliseev, A.D. 279
Ellery, W. 287
Elliott, G. 217
Ellison, F. 56
Elston, J. 33, 40
Eshel, G. 222
Estrada, P.A. 142
Eva, H.D. 167
Evans, L.T. 25

FAO 69, 70, 72, 80, 84
Farrow, R.A. 123, 129
Fawcett, R.G. 248f
Fay, H.A.C. 125
Fearnside, P.M. 286
Fels, H.E. 189t
Fensham, R.J. 127
Ferrini, F. 285
Fischer, G. 286
Fitzgerald, D. 239
Flanagan, J.A. 44
Fleming, J. 5
Fletcher, D. 55, 58
Flint, S.D. 34
Fogden, M.P.L. 274

Food and Agriculture Organization (FAO) 69, 70, 72, 80, 84
Foran, B.D. 194t
Foster, I. 217
Frank, A.B. 48
Fraser, R.W. 219
Freebairn, D.M. 191t
Freebairn, J. 218
Freer, M. 112, 191t, 193t
Freier, B. 130
French, R.J. 246
French, V. 56
Fukshansky, L. 35, 35f, 36f
Furquay, L.W. 288
Fushimi, H. 273

Gad, A. 120
Gall, R.L. 9, 222
Gallasch, P.T. 60
Galpin, M. 207
Gatehouse, A.G. 130
Gates, W.L. 286
Gatti, E. 29
Gautier, C. 171
Gawith, M. 46
Genchi, C. 142
Gergis, M.F. 134
Gerozisis, J.J. 124
Gibbs, W.J. 99
Gifford, R.M. 188t, 282
Gillespie, G. 258
Glantz, M.H. 222
Glanville, S.F. 191t
Gleick, P.H. 279, 281
Göbel, W. 9
Godet, F. 128
Gommes, R. 100, 221
Gonzalez, P. 120
Goody, R, M. 13, 14, 16, 19t
Gottfried, M. 274
Goussard, J. 85
Gow, J. 217
Grabberr, G. 274
Graetz, D. 113
Grains Research and Development Corporation (GRDC) 190, 192t
Gray, I. 95
Grayson, R.B. 159
GRDC 190, 192t
Gregg, P. 134
Greggery, I. 251
Gregoire, P.A. 44
Gregor, S. 206, 207
Grevsen, K. 285
Groisman, P.Y. 273
Gruza, G.V. 269, 279
Gu, H. 125
Guardrian, J. 182
Guijt, I. 258
Gupta, D.P. 135
Gupta, M.P. 135
Guttorp, P. 222

Hackett, C. 191t
Hadlington, P.W. 124
Haggis, M.J. 127
Hall, A.J. 284
Hall, W. 217
Hall, W.B. 283, 287, 288
Hammer, G. 62, 184, 218, 247, 251
Hammer, G.L. 184, 219, 222, 251, 252
Hanchinal, R.R. 48
Hand, D.W. 37
Hangarter, R.P. 29
Hansen, L.O. 124
Happ, E. 241
Hare, K.F. 119
Hargreaves, G.H. 81
Hargrove, J.W. 139
Harrington, R. 130
Harris, G. 191t
Harrison, P.A. 285
Harrosh, J.H. 91
Hartkamp, A.D. 186
Hassall and Associates 280
Hassall, M. 132
Hassika, P. 36, 37, 38f
Hastings, P. 217
Hatfield, J.L. 2
Hayasaka, M. 46
Hayes, M. 97, 105
Hayman, P. 246, 250f, 252, 258
Hayman, P.T. 206, 248
Hearn, A.B. 188t, 192t, 219
Heermann, D.F. 84
Heller, E. 29
Helms, T.C. 44

Hennessy, K.J. 272, 288
Herbert, T.U. 37
Hernandez, E.L. 138
Hernandez, L.F. 44
Hill, J.K. 129
Hill, R.W. 84
Hillel, D. 283
Himelrick, D.G. 56, 60
Hodges, T. 186
Holbrook, S.J. 274
Hollinger, S.E. 1, 223
Holzworth, D.P. 184, 219, 222, 251
Hoogenboom, G. 7, 182, 183-184, 186
Hook, A.R. 188t
Hoppe, P. 1
Horie, T. 282, 283
Horstmann, B. 119
Horton, D.R. 134
Hoven, I.G. 121
Howden, S.M. 283, 288
Huda, S. 252, 258
Hughes, J.P. 222
Hughes, R.D. 125
Hulme, M. 272, 287
Hume, C.J. 181
Hunter, A.F. 124, 125
Hunter, D.M. 125, 134
Hunter, M.D. 140
Hurrell, J.W. 266
Hutchinson, M.F. 163
Hutton, R. 58

Iglesias, A. 282
Imura, E. 46
Inaba, M. 52
International Panel on Climate Change (IPCC) 263, 264, 265, 269, 275, 276, 279
Iqbal, M. 13, 18, 24t
Izrael, Y. 279

Jagtap, S.S. 11
Jallow, B.P. 279
Jalota, S.K. 88
James, P.J. 142
James, Y.W. 184
Jamieson, G.I. 56
Jamieson, P.D. 46
Jarvis, P.G. 40
Jayne, S. 264
Jensen, M.E. 83
Jessop, R. 219
Jin, M.G. 89
Jin, X. 29
Johns, R. 59
Johnson, H.D. 139
Johnson, S.J. 140, 141
Jones, C.A. 188t
Jones, J. 216
Jones, J.W. 179
Jones, R. 240
Jones, R.N. 288
Jorgensen, S.E. 179
Joshi, M.B. 86
Judith, F.T. 64
Jupp, D.L.B. 161, 162, 163, 164

Kaakeh, W. 124
Kale, S.R. 93
Kamp, J.A.L.M. 206
Kamra, S.K. 92
Kanber, R. 74
Karl, T.R. 101, 102, 273
Karsten, U. 30
Kashyap, P.S. 83
Katz, R.W. 218
Kazinja, V.A. 88
Keane, T. 4, 220, 221
Keating, B.A. 112
Keizer, L.C.P. 46
Kemp, W.P. 133
Kenitzer, A. 276
Keogh, D. 259
Keplinger, K. 222
Kerley, G.H. 120
Ki, W.K. 54
Kidson, J.W. 222
Kiehl, J.T. 16, 20, 21t, 25t
Kilpatrick, S. 259
Kim, G.J. 46
Kim, S.D. 44
Kimball, B.A. 282
Kingwell, R.S. 191t
Kiniry, J.R. 41
Kirschbaum, M.U.F. 188t
Kiss, J.Z. 29

Kleist, J. 104, 105
Kler, D.S. 40
Kleschenko, A.D. 7, 159, 160, 186
Knight, R.W. 101, 102, 273
Koller, D. 29
Komamura, M. 74
Kothavala, Z. 280
Kothyari, U.C. 272
Koukhta, A. 268, 282
Kounina, I.M. 268, 282
Kowal, J.M. 71
Kozlovskaya, Z.A. 54
Kozlowski, T.T. 54
Krabe, D.T. 71
Krabill, W. 271, 282
Krishna Kumar, K. 269
Kruijt, B. 33
Kuhlmann, F. 204
Kuijpers, L. 269
Kull, O. 33
Kulshrestha, V.P. 48
Kurata, K. 39
Kure, H. 205
Kwon, Y.W. 44

Lad, B.L. 65
Ladson, A.R. 159
Lal, M. 283
Lamaddalena, N. 84
Lambin, E.F. 167
Lamond, M.H. 192t
Landsberg, J.J. 188t
Larcher, W. 51t
Larsen, D. 193t
Larson, O. 258
Lawrence, G. 95, 257, 258
Layton, M.B, Jr. 132
Le Houerou, H.N. 100
Le Marshall, J.F. 162
Leimar, O. 129
Lembit, M. 108
Leon, G.A. 134
Lesser, R.C. 66
Lewis, T. 207
Li, D.Y. 134
Li, M.H. 134
Li, S. 39
Liakatas, A. 48
Limpert, E. 128
Lin, E. 283
Lin, H.S. 52
Lindsay, M. 128
Lipman, A. 61
Liscum, E. 29
Liu, Y. 85
Liverman, D. 286
Lockwood, D.R. 133
Lockwood, J.A. 133
Lomas, J. 10
Longley, P.A. 160
Lou, Q. 283
Ludwick, G. 168
Luesse, D.R. 29
Lymn, A. 247, 248
Lynch, J.J. 239
Lynch, T. 206, 207
Lyon, N. 216
Lyons, T.J. 185, 192t

Maas, S.J. 186
Macdonald, A.E. 9, 222
Mackenzie, J. 128
Maddocks, S. 142
Magana, V. 272
Magdum, M.B. 65
Magrin, G.O. 272, 285, 289
Maher, J.V. 99
Mahi, G.S. 44, 52
Mahon, R.J. 127
Malano, H.M. 85
Mando, A. 88
Maracchi, G. 7, 159, 160, 186
Marcussen, T. 217
Marengo, J. 281
Mariscal, M.J. 33
Markham, N.K. 206
Marques Filho, A. de O. 33
Marshall, G.R. 184, 219
Maselli, F. 160
Matalas, N.C. 281
Matthews, R.B. 283
Mavi, H.S. 4, 38, 44, 49f, 50f, 52, 56, 135, 161, 209, 212, 217, 239, 240f
Maywald, G.F. 193t
Mazanec, Z. 125, 127
McCown, R.L. 183, 188t, 192t

McCulloch, L. 125
McKee, T.B. 104, 105
McKenny, M.S. 83
McKeon, G. 111, 112
McKeon, G.M. 169, 188t, 220, 287
McLeod, C.R. 193t
McMaster, G.S. 68
McMurtrie, R.E. 188t
McPherson, H.G. 284
McVicar, T.R. 112, 113, 161, 162, 163, 164
MDBC 280
Meats, A. 125, 135
Mecikalski, J.R. 170, 171
Meinke, H. 112, 184, 217
Meiswinkel, R. 141
Mellor, P.S. 141
Menzel, W.P. 171
Meyer, W.S. 81
Midmore, D. 206, 207
Miglietta, F. 285
Milford, J.R. 10
Mills, A.P. 128
Mills, P.F. 286
Mitchell, C. 252
Mizutani, M. 74
Mjelde, J.W. 209, 222
Mohan, S. 75
Molga, M. 1
Montealegre, F.A. 134
Monteith, J.L. 5, 33, 39, 40, 41
Moore, A.D. 191t, 193t
Moore, I.D. 159
Morrison, D.A. 191t
Mortimer, M. 256
Morton, R. 127
Muchena, P. 282
Mukhala, E. 10
Muller, J.C. 80
Müller, K. 128
Munk, W. 264
Murphy, A.H. 218
Murray-Darling Basin Commission (MDBC) 280
Murray, M.D. 127
Murthy, J.S.R. 86
Murty, V.V.N. 98, 99
Muthuvel, I. 52
Myalik, M.G. 54

Narayana, V.V.D. 92
Nash, L.L. 281
National Drought Mitigation Center 95, 96, 115
National Research Council 222
Naylor, R.E.L. 45
Ndunguru, B.J. 49
Newman, J.E. 4
Newman, P.A. 268
Nicholls, N. 62, 217, 222, 251, 252, 254, 274
Nicholson, S.W. 239
Nieber, J. L. 73
Niemi, K.J. 285
Nieto, N.J.M. 131
Nieuwhof, M. 46
Nikitenko, V.G. 134
Norman, J.M. 180
NSW Agriculture 215
Ntare, B.R. 49
Nykiforuk, C.L. 44

Oba, G. 120
O'Brien, B.J. 275
Oda, M. 52
Odingo, R.S. 282
Ogallo, L.A. 4, 220, 221
Olesen, J.E. 285
Olesen, T. 37
Olufayo, A.A. 10
Olulumazo, A.K. 121
Ong, C. 38
Oosterhuis, D.M. 48
Orgaz, F. 33
Orsini, P.G. 193t
O'Sullivan, D.B. 212, 217
Overdieck, D. 8

Page, C.M. 272
Page, W.W. 132
Pair, S.D. 132
Pallais, N. 45
Palmer, T.N. 222
Palmer, W.C. 101, 105
Pamplona, R.R. 48
Panda, R.K. 83
Pandey, A.N. 121

Pant, G.B. 269
Paoloni, P.J. 44
Papakryiakou, T.N. 97
Park, S.W. 44
Parker, C. 205
Parker, C.G. 205
Parmar, T.D. 121
Parmesan, C. 274
Parry, M.L. 286
Parsons, D.L. 288
Parton, K.A. 184, 219
Paruelo, J.M. 83
Patwardhan, A.S. 73
Paul, C.J. 217
Pauli, H. 274
Paull, C.J. 191t
Paw U, K.T. 7
Payten, I. 91
Peacock, A. 191t
Pedgley, D.E. 123
Pegram, R.G. 142, 143
Peiris, D.R. 285
Penning de Vries, F.W.T. 181
Pérarnaud, V. 7, 159, 160, 186
Pereira, S.L. 84
Perry, J.N. 130
Perry, K.B. 11, 59, 60, 66, 67, 223
Perry, M.W. 244
Petersen, E.H. 219
Petrassi, F. 100
Pickering, N.B. 179
Pimentel, D. 282
Pinshow, B. 133
Pinter, L. 282
Pitt, D. 84
Plant, S. 4
Polavarapu, P. 29
Pollock, K. 250f
Pons, X. 131
Popov, G.F. 100
Porter, J. 46
Post, E. 120
Postel, S. 170
Pounds, J.A. 274
Powell, A.A. 56, 60
Prange, H.D. 133
Prasad, R. 37
Pratley, J.E. 244
Prats, V.V. 138
Predieri, S. 29
Prihar, S.S. 88
Primary Industries Department (QLD) 107, 109
Pruitt, W.O. 78, 81
Pujari, K.H. 65
Purdum, J.F.W. 171
Pusey, P.L. 185

Queensland Department of Primary Industries 107, 109
Quintana-Gomez, R.A. 271
Quiring, S.M. 97

Radostits, O.M. 136, 144
Ralph, S.R. 142
Ramalan, A.A. 84
Raman, C.R.V. 71
Rankova, E. 272
Rao, V.U.M. 44
Raoa Mohan Rama, M.S. 89
Raper, C.D. 64
Rebella, C.M. 83
Reece, P.H. 163
Regniere, J. 124
Reid, D.G. 125
Reid, P. 272
Rickert, K.G. 194t
Ridge, P. 253, 258
Ridsdill, S.T.J. 131
Rijks, D. 2, 9, 121, 159, 223
Rijks, D.A. 224
Rijks, J.Q. 221
Risch, S.J. 127
Ritche, J.T. 181
Ritter, S. 29
Roberto, S. 16, 23
Roberts, J.D. 29
Robins, L. 260f
Robinson, P.J. 271
Robinson, S. 191t
Robinson, T. 139
Roderick, M. 168
Rodriguez, L.A. 26
Roffey, J. 125
Rogers, D. 139
Rogers, W.J. 57t
Roling, N. 257
Roltsch, W.J. 67

Romanelli, S. 160
Romero, C.G. 138
Rose, D.J.W. 132
Rosenberg, L.J. 128
Rosenberg, N.J. 83
Rosenzweig, C. 283, 286
Roujean, J.L. 33, 34f
Roundy, B.A. 44
Roussopoulos, D. 48
Ruiz, A.C. 138
Rupakumar, K. 269
Ruppel, N.J. 29
Rural Industry Business Services 107
Russell, D. 257
Russell, G. 40
Rutter, J.F. 128
Ryaboshapko, A.G. 267

Saarikko, R.A. 285
Sabins, F.F. 22, 27, 166, 167
Sabrah, R.E.A. 88
Saiko, T.A. 121
Sakthivel, T. 46
Salinger, M.J. 10, 263, 272
Salisbury, F.B. 26
Samani, Z.A. 81
Samarakoon, A.B. 282
Sastry, C.V.S. 37
SCARM 242
Schmitt, R.J. 274
Scholes, M.C. 287
Scholes, R.J. 287
Schreider, S.Y. 280
Schultz, J.E. 241
Scott, J.K. 131
See, L. 100
Seeley, M.W. 11, 223
Sellers, R.F. 140
Semenov, S.M. 268, 282
Serafin, R.J. 9, 222
Setzer, A.W. 167
Shafer, B.A. 103
Shafiq, M. 93
Shah, F.M. 142
Shah, M.M. 86
Sharp, J.L. 29
Shell, H. 81
Shepherd, D. 207
Shields, J.E. 127
Shrapnel, M. 257
Shrivastava, K.K. 135
Simhadrirao, B. 75
Simpson, P. 238, 239
Singh, D. 44
Singh, P. 280
Singh, S. 44
Singh, V.P. 272
Sirotenko, O. 265
Sivakumar, M.V.K. 9, 96, 121, 224
Skarrat, D.B. 193t
Skirvin, D.J. 130
Skorska, E. 29
Smajstrla, A.G. 75
Smeal, M.G. 139, 140, 143
Smith, D. 81
Smith, I.N. 272
Smith, L.P. 1
Smith, M. 70, 78, 81, 83, 85, 85f, 86, 86f, 87, 88
Smith, R. 112
Smith, R.B. 155t, 156t, 173f, 174f, 175f, 176f
Smith, R.C.G. 164-165, 165, 168
Smith. S.M. 112
Snijders, F. 221
Somme, L. 124
Sorensen, I.B. 46
Speight, M.R. 140
Spieler, G.P. 61
Squire, G.R. 31, 40
Stafford Smith, D.M. 194t
Stafford Smith, M. 220, 251
Standing Committee on Agriculture and Resource Management (SCARM) 242
Stanley C.D. 75
Stapper, M. 188t
Steffen, K. 281
Stehlik, D. 95
Stenseth, N.C. 120
Stephens, D. 216
Stephens, D.J. 110, 185, 192t
Stephens, J.S. Jr 274
Stern, R.D. 73
Stevenson, W.R. 172
Stewart, S.D. 132
Stigter, C.J. 9, 10, 121, 224, 263, 272
Stirling, C.M. 39, 40
Stolyarenko, V.S. 65

Stone, P.J. 46
Stone, R. 62, 184, 218, 219, 222, 251
Stone, R.C. 217
Stowe-Evans, E.L. 29
Strand, J.F. 10
Straw, W.M. 206
Stroosnijder, L. 88
Stubbs, A.K. 206
Sumit, M. 40
Suppiah, R. 272
Sutherland, A.K. 140, 141
Sutherst, R.W. 135, 136, 193t, 284
Szabo, Z. 54

Takaichi, M. 29
Takakura, T. 39
Takasu, T. 74
Takeuchi, K. 98, 99
Tanna, S.R. 121
Telfer, M.G. 132
Tennant, D. 216
Terres, J.M. 159
Testa, A.M. 127
Thackray, D. 131
Thamburaj, S. 46
Thireau, J.C. 124
Thomson, C.S. 222
Thornley, J.H.M 37
Thornton, P.K. 183, 183-184, 184
Tibbitts, T.W. 65
Tienroj, U. 31
Topp, C.F.E. 288
Torr, S.J. 139
Tow, P.G. 241, 242, 242f, 243f, 244f
Trenberth, K.E. 16, 20, 21t, 25t
Triltsch, H. 130
Trione, S.O. 65
Truscott, M. 216
Tsuji, G.Y. 183-184
Tubiello, F.N. 286
Tucker, M.R. 132
Tuddenham, W.G. 162
Tupper, G.J. 161, 165, 217
Turpin, J.E. 248
Turral, H.N. 85

Udo, S.O. 37
Ullio, L. 60
U.S. National Assessment 271, 274, 286

Valcarcel, F. 138
Van Crowder, L. 11, 223
van der Kaay, H.J. 140
van Diepen, C.A. 285
Van Oeveren, J.C. 46
Vanclay, F. 257, 258
Vaughan, D.G. 272, 273, 282
Vazquez, F.A.R. 138
Ventrella, D. 88
Ventskevich, O.Z. 57t
Verstraete, M.M. 167
Vijaya Kumar, P. 40
Villalobos, F.J. 33
Visser, M.E. 274
Vittum, M.T. 66
Von Storch, H. 222
Vorasoot, N. 31
Vossen, P. 159

Waghmare, A.G. 130
Walker, G.X. 185
Walker, J. 163
Waller, P.J. 127
Walsh, J.E. 271
Walton, T.E. 140
Wang, C.Y. 121
Wang, F. 283
Wang, K.M. 193t
Wang, S. 121
Wang, Y.P. 188t
Ward, M.P. 140, 141
Wardhaugh, K.G. 125, 127
Warmund, M.R. 54
Watt, A.D. 140
Weiss, A. 11, 223
Welbourn, A. 258
Westbrook, J.K. 132
Wheeler, R. 89
Wheeler, T.R. 285
Whelan, M.B. 193t
Whetton, P, 269
Whisler, F.D. 181, 182, 183
White, B. 4
White, D.H. 109, 161, 189t, 193t, 217, 220
White, J.W. 186
Whitford, W.G. 120
Whittington, W.J. 48
Wickson, R.J. 55, 61

Wilhelm, W.W. 68
Wilhite, DA. 115
Wilkens, P.W. 184
Williams, B. 139
Williams, J.H. 49
Williams, M.R. 132
Williams, N.A. 163
Willott, S.J. 132
Wilson, J.W. 37
WMO 222, 263, 265
Wolf, I. 285
Wolf, J. 285
Wood, H. 161, 169
Wood, M.L. 85
Woodhill, J. 260f
Woodruff, D. 56
Woodruff, D.R. 192t
World Meteorological Organization (WMO) 222, 263, 265
World Meteorological Organization/Global Atmosphere Watch (WMO/GAW) 267, 268
Wright, D.E. 125
Wright, T. 254
Wu, K.J. 134
Wylie, P. 253, 258

Xu, Q. 52

Yang, Y.L. 121
Yin, X. 44, 45f
Yonow, T. 135, 136, 284
Youiang, Ho. 56
Yung, Y.L. 13, 14, 16, 19t

Zalom, F.G. 66, 67
Zeiger, E. 29
Zhang, D.F. 120
Zheng, R. 121
Zhou, X. 131
Zhu, J. 29
Zonn, I.S. 121
Zorita, E. 222

Subject Index

Page numbers followed by the letter "f" indicate figures; those followed by the letter "t" indicate tables.

absolute potential for efficient water use 246
absorption of EMR 147
  by leaves 26-30, 27f, 28t, 157
acidifying atmospheric compounds 267-268
acoustic energy 146
active instruments 150
active protection against frost 56
adaptive research 10
adult learning for agricultural producers 258-261
Advanced Very High Resolution Radiometer (AVHRR) 162, 164-167
advection frost 55
aerial photography 7. *See also* remote sensing
Africa
  armyworms 132
  global warming effects 278-279, 282, 287
  locusts 133-134
  National Rainfall Index 100
  Sahelian region 119-120
  tsetse flies 139
African horse sickness 140-141
agricultural drought 96
agricultural producers
  adapting to climate 237-261
  attitudes to decision support 205-207
  effective support for 256-258
  forecasting needs 211-212, 213f
  modelling needs 11
  tailored products for 223-224
  with computers 190
agricultural production. *See also* crop production; livestock
  adapting to seasons 247-249
  farm layout planning 238-240
  forecasts for 7-8
  impact of climate change 263-289
  income from 247
  on-farm conditions 10
  sowing dates 58
  water use efficiency 88, 244-246
Agricultural Production Systems Simulation Model (APSIM) 183-184
agroclimatological services 1-12, 179-207, 209-235. *See also* forecasts
agrometeorological databases 220-221
air drainage 58
air temperature 47-50. *See also* temperature
Airborne Visual and Infrared Imaging Spectrometer (AVIRIS) 158
Akabane disease 127-128
albedo
  of the earth 16, 23
  of shortwave radiation 24t
ALEX 171
algal bloom, reduced by aquacaps 91
America. *See* North America; South America
ammonia, accumulation of 267
animal simulation models. *See* simulation models
animals, global warming and 274
Antarctica
  global warming effects 282
  hydrological changes 273
  ozone loss over 268
  precipitation changes 272

aphids 130-131
apples, freezing injuries 54
apricots, freezing injuries 54
APSIM 183-184
aquacaps 91
Arabidopsis thaliana 29
Arctic region 271-273
  global warming effects 281-282
  ozone loss over 268-269
armyworms 127, 128, 132
arthropod parasites 138-144
Asia
  global warming effects 269, 279-280, 283
  hydrological changes 273
aspect, soil temperature and 43
ASTER Spectral Library 157-158, 176f
astronomical periodicities 263-264
atmosphere
  air temperature 47-50
  convergence zones 128
  gases in 22, 267-272
  radiation balance in 25
  radiation scattering 20-22, 173f
  scattering of EMR 148-149
Atmosphere-Land Exchange system (ALEX) 171
atmosphere-ocean models. *See* general circulation models
atmospheric window 23
Aussie GRASS project 161, 169, 215
Australia
  agroclimatological services 214-217
  aphids 131
  cane toads 203f
  decision support systems in 206
  drought monitoring 105-108
  frost losses 55
  global warming effects 269, 280, 283-284, 287-288
  modelling in 187, 188t-189t
  precipitation changes 272
  Queensland fruit fly 136
  rainfall records 113
  rainfall use 242-252
  remote sensing used in 161-169
  water use efficiency 84
  wind-borne pests and diseases 128
Australian Association of Agricultural Consultants (AAAC) 216-217
Australian Bureau of Meteorology. *See* Bureau of Meteorology (AUS)
Australian Farm Journal 251-252
Australian paralysis tick 143
Australian RAINMAN model 190-197, 250
available water content of soil (AWC) 101
AVHRR 162, 164-167
AVIRIS 158

Bactrocera tryoni 125, 135-136
barn itch 143-144
bean weevils 129
Beer's law 33
Bhalme and Mooley Drought Index (BMDI) 102, 102t
biodiversity 119-120, 239, 253
biological control of seepage 93
biomass production. *See* crop production
biometeorology 1
bioplastic 93
black body radiation 16
black cone-headed grasshopper 133
Blaney and Criddle method for irrigation scheduling 6
blowflies 139
blue light, effect on plants 29
bluetongue infection 128, 140-141
BMDI 102, 102t
Bovicola ovis 141-142
breeding crops, simulation models 182-183
Britain, decision support systems in 205
brushing for frost control 59
Bureau of Meteorology (AUS) 214, 252
  long-term forecasts 210-211
  rainfall rankings 99
  Research Centre 162
bushfires 165-168

Canada, early sowing in 44. *See also* North America
canal irrigation 84
cane toads, modelling spread of 201-204
canopy cover. *See* screening
carbon dioxide 22, 264, 267
cardinal temperatures 46-47
cassava, soil temperature and germination 44, 45f
castor beans, conversion coefficient 41
cattle. *See* livestock
Ceratopogonidae 140
Ceratovacuna silvestrii 130
cereal crops 86f, 131
CFCs 268
Chabertina ovina 137
checker-spot butterfly 274
chemicals
  frost control with 60
  petrochemicals 88
  plumes attract insect pests 129
  to reduce seepage 92
  remote sensing of 145
chilling injury to plants 53-54
China
  cotton bollworms in 134
  global warming effects 283
  soil water management 89
chlorofluorocarbons (CFCs) 268
chlorophyll 26-27, 157
Chorthippus brunneus 132-133
Chortoicetes terminifera 125, 128
chromosphere 14
CIMSS 171
Cinara atrotibialis 130
Class A pans for measuring evaporation 79-81
Clermont (QLD), crop rotation 242f
climate change 209. *See also* global warming
  effect on aphids 131
  future scenarios 275-276
  impact assessment 8, 10, 185, 263-289
  modelling effects of 201-204
  risk management workshops 216
  seasonal forecasts 222
climate forecasts. *See* forecasts
climatological management
  data archives for 8-9
  of pests and diseases 123-144, 135-136
  of water use 69-93, 85-90
CLIMEX model 201-204
clouds
  block remote sensors 149
  effect on aphids 131
  light scattered by 148
  radiation reflected by 20
CMI 105, 106t
Cobar (NSW), MetAccess model 197-199, 198f, 199f
cold. *See* chilling injury; frost damage and control
colloidal materials 93
color 19t, 148. *See also* spectral bands
Commonwealth Scientific and Industrial Research Organisation (CSIRO) 90, 162
comprehensive models 181
computer modelling. *See* modelling
computers for data analysis 9, 207
constraints inventories, during drought 117
controlled climates 6
conversion coefficients for solar radiation 38-41
Cooperative Institute for Meteorological Satellite Studies (Wisconsin) 171
core of the sun 13
corona 14-15
corpuscles (solar) 14
cotton bollworms 134-135
cotton crops
  air temperature and 48
  aphids in 130-131
  chilling injuries 53
  heat stress injuries 52-53
  insect pests 134-135
  soil water management 89
cowpea, chilling injury 53
cranberries, frost control 171-172
CRAS-ALEX 171
Crop Moisture Index 105, 106t
crop pests and diseases. *See* pests and diseases

crop production. *See also* agricultural production
  archives of data on 8
  area used for 247-248
  benefits of screening 240f
  breeding simulation models 182-183
  cardinal temperatures for germination 47t
  fallowing 243-244, 253
  forecasts needed for 217
  global warming effects 282-286
  injury from frost 56
  injury from temperature changes 50-55
  insect pests 130-136
  intercepted radiation and 39, 39t
  rainfed crops 88
  remote sensing of 169
  residues from 245
  rotational 183, 242f, 243f
  simulation models 112, 179-180, 182-186
  solar radiation and 25-30, 38-41
  temperature and 43-68
  water required for 69-70
  water use efficiency 246t
  weather effects on 2, 5, 212
cryosphere 265
CSIRO 90, 162
Culicoides spp. 127, 140-141
cultivars 58, 241
cuticular transpiration 76

day length 26, 65, 129
DECI 110-111
decile method of rainfall ranking 99, 99t
decision support systems (DSS)
  climate information for 237-261
  forecasts needed for 225t-235t, 249-252
  modelling in 183-185, 187-207, 191t-194t
  from remote sensing 170
  software for 259
deficit irrigation 90
defoliator grasshoppers 126-127
DEMs 159
dependable rains 100
desert locusts 128
desertification 119-121
dessication from frozen soil 53
diffuse reflection 147
digital elevation models 159
DigitalGlobe 154
Dirofiliaria immitis 140
diseases. *See* pests and diseases
diversity 119-120, 239, 253
DM 104-105
dog ticks 143
double cropping 251
downscaling forecasts 221-222
drainage lysimeters 74
drought 95-121
  bluetongue infection and 141
  costs of 95
  from frozen soil 53
  grasshoppers affected by 133
  management of 254-255
  monitoring 161-164
  property planning and 239
  thresholds for 98t
dry matter production. *See* crop production
dryland farming 131, 219
DSS. *See* decision support systems
Dubbo (NSW) 200, 201f
dynamic simulation models 181-182

E-RAIN model 75-76
earth
  albedo of 16
  radiative energy budget 13, 18-25
earth satellites. *See* satellites for remote sensing
education and training 10-11
  for agricultural producers 259
  during drought 118
effective heat units 66-68
El Niño events. *See* Southern Oscillation Index
electromagnetic radiation, remote sensing 145-147, 173f
elm bark beetle 124
energy balance 5-6, 19t
Enhanced Thematic Mapper Plus 154
ENSO. *See* Southern Oscillation Index

environment, remote sensing of 7, 145
Environment Protection Australia (EPA) 201-204
environmental degradation 119-121, 237, 240-241, 245
environmental management 9-10
Environmental Resource Information Network (ERIN) 165, 168
environmental temperature. *See* temperature
Epiphyas postvittana 125
ERIN 165, 168
ethephon, for frost control 60
Eudocima salaminia 125
Europe
  chances of 198f, 199f
  evaporation 76, 79-84
  global warming effects 280-281, 285, 288
  lengthening growing season 274
  long-term temperature changes 269-271
  precipitation changes 272
  of rainfall 70
  from reservoirs 91-92
evaporation/precipitation ratio method 72
evapotranspiration 77-84
extended forecasts. *See* long-term forecasts; seasonal forecasts
extension process, limitations of 257-258

faculae 15
fallowing 243-244, 253
FAO-24 methods for measuring evaporation 81-84
FAO CROPWAT program 83, 86-87
farmers. *See* agricultural producers
farming. *See* agricultural production
FARMWEATHER forecasts 214
firebreaks 254
fires 165-168
flexible cropping 248-249
flies 138-139
floating pans 80
flocculi 15
flood management 166, 255-256
flowering 26, 29
flukes 138
fodder, sustainable land use and 253
foliar disease management 172
forecasts. *See also* seasonal forecasts; short-term forecasts
  agricultural 4, 7-8, 211-212, 213f
  decision support from 225t-235t, 249-252
  downscaling 221-222
  frosts 62-64
  long-term 210
  modelling in 184-186
  types of 209-211
  use and benefits 217-220
  users' expectations 223-224
  water use efficiency and 87
freezing. *See* frost damage and control
frost damage and control 48, 54-64, 57t, 63t, 171-172
fruit crops
  agricultural forecasts 215
  chilling and freezing injury 54
  heat stress on 52
fruit flies 135-136
future needs for agrometeorology 8

gamma rays, effect on plants 29-30
GDDs 66-68
general circulation models (GCMs) 185
  global warming projections 275
  seasonal forecasts 210-211
geographic information systems 159-160, 186
Geostationary Operational Environment Satellites (GEOES) 171
germination of seeds
  cardinal temperatures for 47t
  soil temperature and 44-45
  spectral bands and 29
  temperature fluctuations and 65
GISs 159-160, 186
glaciers, shrinkage of 273
global positioning systems 160-161
global warming 8, 269-274, 270f, 277-282, 277f. *See also* climate change; greenhouse gases

Glossina spp. 139
GOES 171
government policy during drought 116-119
GPSs 160-161
grapes, temperature effects 241
GRASP model 111
grass growth 111-112. *See also* rangelands
GrassGro simulation model 112
grasshoppers, temperature effects 132-133
grazing, sustainable 241, 253
green light, effect on plants 29
greenhouse gases 8, 263-264. *See also* global warming
groundcover, reduces water loss 245
groundnuts. *See* peanuts
groundwater 69, 121
growing degree-days 66-68
growth regulators, for frost control 60
growth units 66-68
gypsy moths 129

haemonchosis 136
Haemonchus contortus 137
Hammada elegans 74-75
Hargreaves method of measuring evapotranspiration 81
haze 148
heat stress injury to plants 51-52
heat units 66-68
heaters for frost control 61-62
heaving by frozen soil 53
Heliothis spp. 135
helminth parasites 136-138
Hillston (NSW), MetAccess model 199-200, 200f
history of agrometeorology 5-8
holistic science 1-2
horse flies 139
horse sickness 140-141
houseflies 139
human resources for agrometeorology 223
humidity, effects on pest insects 126, 134-135, 142
Hydrogrow 400 88
hydrological drought 96, 98t
hydrology 273, 277-282
hyperspectral sensing 154-157, 175f

IKONOS satellite 154
illumination geometry 152
image acquisition 152-158
impact assessment committees for drought 117
incident radiation 18-25, 21f, 21t
India
  aphids in 130
  cotton bollworms in 135
  definition of drought 97-98
  global warming effects 283
  integrated watershed development 89
  locusts in 134
  monsoon forecasts 72
  remote sensing satellites 154
infrared thermometry 7
insects 64, 123-126. *See also* pests and diseases
instruments for remote sensing 148-152
integrated pest management (IPM) 185
integrated watershed development 89
interaction processes, remote sensing 147-148
intercepted radiation 31
intercepted rainfall 74
interdisciplinary nature of agrometeorology 1-2
International Society of Biometeorology 1
IPM 185
irrigation
  effect on rivers 121
  for frost control 60-61
  modelling 184
  scheduling of 86-87, 170-171
  sustainable 253
  water required for 70, 84-85, 85f
itch mite 144
Ixodes holocyclus 143

Japanese encephalitis 128
jarrah leaf miners 125
jarrah trees 127

Kenya, armyworms in 127-128
Kirchoff's Law 17
Kondinin Group, forecast review 217
Kytorhinus sharpianus 129

La Niña. *See* Southern Oscillation Index
Lambert's Law 18
land capability 238-240
land degradation 119-121, 245. *See also* sustainable land use
land use 121, 237, 240-241, 265
Landsat Thematic Mapper 154-157, 167
larvae, wind transportation of 129
laser altimeters 150
leaves. *See also* plant canopy
  optimal photosynthesis angle 32f, 33f
  properties affecting transpiration 77
  solar radiation and 26-36, 28f, 28t
  temperature of 50-51
lenticular transpiration 76
lice 141-142
lidar sensors 150
light. *See also* solar radiation; spectral bands
  crop production and 26
  effects on pest insects 129-130
  optimal leaf angle and 32f, 33f
  remote sensing of 145-146
light detection and ranging (lidar) 150
LISS-II sensor 154
Little Ice Age 264
livestock
  benefits from shelter 239-240
  bluetongue in cattle 140-141
  decision lags in managing 4
  forecasting needs for 217
  global warming effects 286-289
  managing during drought 254-255
  overgrazing by 120
  parasites of 136-144
  responses to climate change 10
  weather effects on 212
locusts 125, 133-134
Long Paddock website 215
long-term forecasts 210
longwave radiation 23, 24f, 25t. *See also* radiation, energy budget
low temperature injury to plants. *See* chilling injury; frost damage and control
lucerne 245
Lucilia cuprina 125, 127, 139
lupins 131, 239
Lymantria dispar 129
lymphatic filiarisis 140
lysimeters 80

maize 44-46, 64-65
malaria 140
mangoes, thermoperiodism 65
marine ecosystems, global warming in 274
Maximum Value Composite Differential (MVCD) 162
media use during drought 118
Mei scattering 22
Melanaphis sacchari 130
mesophyll, solar radiation and 26
MetAccess model 197-201
meteorological drought 96
Meteorus trachynotus 124
methane 264, 267-268
microirrigation 84
microwave radiation, remote sensing of 149-150
midges 140-141
migration patterns of pest insects 127-129
milk production, benefits from shelter 239-240
mites 143-144
modelling 9. *See also* general circulation models; simulation models
  agroclimatological management and 179-207
  aphid populations 131
  drought monitoring with 110-113
  effective rainfall 74-76
  summary models 181
moisture effects on pest insects 126-127, 135, 137. *See also* humidity; rainfall
moisture thresholds 99t

Monelliopsis pecanis 124
monitoring committees for drought 117-118
monsoon conditions, determination of 71-73, 99
mosquitoes 140
mulching 59, 88-89, 245
multispectral images 154
Murray-Darling Basin 84, 280
Musca domestica 139
Musca vetusissima 125, 128
MVCD 162

nagana 139
Narromine (NSW), water use efficiency 246-247
Nation Drought Alert Strategic Information System 111
National Climate Centre (AUS) 210, 214
National Drought Alert System (AUS) 161
National Drought Mitigation Center 97
National Drought Policy (AUS) 108
National Farmers' Federation of Australia 217
National Institute of Hydrology, India 98
National Oceanic and Atmospheric Administration (US) 83
National Rainfall Index (Africa) 100
Natural Disaster Relief Section (QLD) 107
NDVI 112, 161, 164-165, 168-169
nematodes 137
Netherlands, decision support in 205-206
netting, protects horticultural crops 253. *See also* screening
New South Wales
  agricultural forecasts 215
  drought monitoring in 106-107
  frost losses in 55
  remote sensing in 165-168
New Zealand, global warming effects 284, 288
Nigeria, rainfall in 71-73
night temperatures, respiration and 48
nitrogen oxides 267
nitrous oxides 264, 267
NOAA 83
noctuid moths 128
Normalized Difference Temperature Index 162-164
Normalized Difference Vegetation Index 112, 161, 164-165, 168-169
North America. *See also* Canada; United States
  global warming effects 281, 286, 289
  precipitation changes 271-272
North Atlantic Oscillation 266
nutrition for frost control 60
Nysius vinitor 128

ocean-atmosphere models. *See* general circulation models
opportunity cropping 245
Orange (NSW) 204, 205t
orbiviruses 140-141
organic matter, soil temperature and 44
Ostertagia spp. 137
ostertagiosis 136
overgrazing 120
oxygen 22. *See also* ozone
ozone
  absorbs solar radiation 22
  accumulation of 264
  tropospheric 268-269

Pacific Ocean 265, 271
palisade parenchyma 26
Palmer Drought Severity Index 101-102, 101t
pan methods for measuring evaporation 79-81, 84
panchromatic images 153
PAR 36-38
paralysis tick 143
parasitic wasps 124
partial root-zone irrigation 90
passive instruments 151
passive protection against frost 56
pasture simulation models 111
path radiance 148

PDSI 101-102, 101t
peaches, temperature fluctuations and 65
peanuts
  air temperature and 49
  chilling injury 54
  solar radiation and 39-40
pears, freezing injuries 54
pecan aphids 124
Pectinophera gossypiella 134-135
Penman method for measuring evaporation 6, 83
Penman-Monteith method 82-83
percent of normal precipitation 97-99
perennial plants 245, 253
personalities of agricultural producers 257
Perthida glyphopa 127
pests and diseases
  climatological management 123-144
  global warming effects 278f, 284
  insects 64, 123-126
  management of 10, 245-246
  modelling in management 185
  weather effects on 212
petrochemicals, reduce water loss 88
Philippines, drought assessment in 99
photoelements, spectral scattering 36f
photoperiodicity 26, 65, 129
photosphere 13-15
photosynthesis
  optimal leaf angle 32f, 33f
  photon flux density 36
  radiation levels for 30, 36-38
  temperature effects 49-50, 50f, 51t
phototropism 29
physiology of plants 183
Piche evaporimeters 80
pine forests 38f
pink bollworm 134
plages 15
Planck's Law 17
plant canopy 30-36, 119-120. *See also* leaves
plant cultivars 58, 241
plant growth
  delay due to chilling 54
  optimal heights of 32
  soil temperature and 45-46
  solar radiation and 13-41
  thermoperiodism 64
plant simulation models. *See* simulation models
plant spectra. *See* vegetation
PMP 238-240, 257
polar stratospheric clouds 268
population dynamics models 67
potatoes
  foliar disease management 172
  soil temperature and 45
  temperature effects on 49f
  thermoperiodism 65
potential evapotranspiration 77-78
PPFD 36
PRD 90
precipitation. *See* rainfall; snow
preliminary models 181
private agroclimatological services 216-217
problem solving cycle 260f
ProFarmer magazine 216
property management planning 238-240, 257
PSCs 268
Psoregates ovis 144
publicity during drought 118

Queensland
  agricultural forecasts 215-216
  bluetongue infection in 140-141
  bushfire monitoring 165
  Department of Primary Industry 211
  drought monitoring 107-108
  remote sensing in 168-169
Queensland fruit fly 135-136
QuickBird satellite 154
quiescence due to heat 52

radar sensing 149-150
radiation
  efficiency of use 38-41
  energy budget 18-25, 20-23, 21f, 21t
  laws of 16-18
radiation frost 55
radiometers 151

rainfall
analysis of 109-110, 113
chances of 195t, 197t
effective use of 70-76
effects on aphids 131
effects on cotton bollworms 134
effects on locusts 134
effects on pest insects 126, 127
effects on Queensland fruit fly 135
effects on sheep lice 142
effects on sheep parasites 137-138
effects on ticks 142-143
efficient use of 242-252
estimating probability of 249
long-term changes in 271-272
measurement of 71
in property planning 238
reduction during drought 97-99
thresholds for 97-98, 98t
rainy seasons 71-73, 99, 138-139
rangelands
forecasts used in 219-220
livestock effects of global warming 288-289
remote monitoring of 164-165
Rayleigh scattering 22
real-time climate information 221
red edge 157
red light, effect on plants 29
reflected radiation 149
reflection from leaves 26-30, 27f, 28t
Regional Review (NSW) 215
regression models 180-181. *See also* statistical methods
remote sensing 7, 145-177
crop water use estimates from 83
drought monitoring with 112-115
research into agrometeorology 6, 9, 221
research into drought 119
reservoirs, water loss reduction 91-93
resilience of agricultural production 237, 239, 252-254
resources inventories 116-117, 120
respiration, temperature effects 48, 49f, 52
reversing layer 14
Rhopalosiphum padi 131
RI 100
rice crops
air temperature and 48
chilling injury 53-54
conversion coefficient of 41
heat stress on 52
irrigation requirements 75
soil temperature and germination 44
windborne insect pests 128
Ricinus communis, conversion coefficient 41
Rift Valley fever 140
Ritchie method of measuring evapotranspiration 81
Riverwatch service (NSW) 215
roots, temperature of 44, 46
Roseworthy (SA) 243f, 244f
roundworms 137
Royal Melbourne Institute of Technology 91
RUE (radiation use efficiency) 38-41
runoff 238, 279-280
Rural Land Protection Boards (NSW) 106

Sahelian region (Africa) 119-120
salinity 121, 245
sap feeders 126-127
Sarcoptes scabei 143-144
sarcoptic mange 143-144
satellites for remote sensing 7, 154, 155t, 156t
orbits for 158-159
solar radiation estimates from 171
scaling of models 7
scattering of EMR 147, 150
Schistocerca gregaria 128
schools, training in 11
science and policy integration during drought 118
Scolytus laevis 124
screening. *See also* windbreaks
for frost control 59
to reflect solar radiation 39-40
sustainable 253
sea levels, long-term rise in 276
sea-surface temperature 185, 265-266
seasonal events
effects on pest insects 126
modelling 186

seasonal events *(continued)*
monsoon conditions 71-73, 99
rainy seasons 70-71
temperature fluctuations 65
seasonal forecasts 210-211, 222
economic policy and 4
estimating probabilities 250f
of frosts 62-64
usefulness of 217-218
seepage from reservoirs 92-93
selective scattering of radiation 22
sensor designs 148-152, 156t
sequence analysis models 183
shading. *See* screening
sheep
blowfly strike 139
lice on 141-142
parasites of 137-138, 144
wool production 220
shelter. *See* screening
short-term forecasts 210
of frosts 64
water use efficiency and 87
shortwave radiation 20-23, 24t
simulation models 6-7
animal models 179
crop water use 82-83
drought monitoring with 111-114
plant models 179-180, 183
simulated annealing 164
SISP 160
site selection, for frost control 58
Sitobion avenae 131
sky radiation 35f
SLATS 169
snow
in determining drought 103-104
effect on plants 53
long-term changes in 271-272
reduction in cover 273
social costs of drought 95-96
sodium carbonate, reduces seepage 93
SOI. *See* Southern Oscillation Index
soil properties
acidity 245
available water content 101
degradation of 120-121
heat storage and 59
moisture levels 243, 248
rainfall and 73
soil properties *(continued)*
to reduce seepage 92
reflectance 152
temperature 43-50, 45f, 164
texture 44
water loss 87-88
when frozen 53
soil-water balance model 73-74
Solanum eleagnifolium 65
solar atmosphere 14
solar constant 15-16
solar flares 14
solar radiation
crop production and 39t
efficiency of use 38-41
plant growth and 13-41
ratio of PAR to 37
sensors for 149
spectral bands 19t
in a spruce forest 34f
sun angle and 35
solar winds 14-15
South America
global warming effects 271, 281, 285-286
hydrological changes 273
livestock effects of global warming 288-289
precipitation changes 272
South Australia, agricultural forecasts 216
Southern Oscillation Index 265-266
African horse sickness and 141
bluetongue infection and 141
decision support and 249-252
energy imbalance and 20
forecasts about 219-220
frost forecasting and 62
modelling 185, 195
mosquito populations and 140
phone hotline 215
seasonal forecasts and 210-211, 222
sowing dates, for frost control 58
soybeans, soil temperature and germination 44
spatial information 145
spatial models 169, 186
spatial resolution, remote sensing 153
spectral bands. *See also* color; light
direct sunlight and sky radiation 35f
effect on plants 29-30

spectral bands *(continued)*
  libraries of 157-158
  in a plant canopy 34-36
  reflectance from vegetation 157
  reflectance of 152, 173f, 175f, 176f
  remote sensing of 153, 173f
  scattering of photoelements 36f
  solar radiation 19t
spectroradiometers 151
spectroscopy 174f
specular reflection 147
SPI 104-105, 104t
Spodoptera exempta 127, 128, 132
spongy parenchyma 26-27
SPOT 154-157
spreading banks 89-90
spring drought 53
sprinklers 84
spruce forests, radiation profile 34f
SST 185, 265-266
stable flies 139
Standardized Precipitation Index 104-105, 104t
Statewide Landcover and Trees Study (QLD) 169
statistical methods 109-111, 180-182
Stefan-Boltzman Law 17-18
stem girdle injury 52
sterility of plants, due to chilling 54
STIN model 185, 216
stomatal transpiration 76-77
Stomoxys calcitrans 139
strategic role of agrometeorology 4, 183-185, 237
strawberries, freezing injuries 54
stream flow, in determining drought 103-104
Stress Index model 185, 216
stubble retention 245
Study of Climate Variability and Predictability 222
Sudan grass, chilling injury 53-54
suffocation by snow 53
sugar beet, soil temperature and growth 46
sugarcane aphids 130
sulfate aerosols 267
summary models 181
summer diapause 123-126
sun scald injury 52
sunflowers, soil temperature and germination 44
Sunken Colorado pan 80
sunspots 15, 264
supercooling of body tissues 124
Surface Water Supply Index 103-104
sustainable land use 237, 240-241. *See also* land degradation
SWSI 103-104
synodic period 14-15
Système Integré de Suivi et Prevision 160
Système Probatoire d'Observation de la Terre 154-157

Tabanidae 139
tactical role of agrometeorology 4, 183-185, 237, 247-249
tailored weather information 4, 223
Tamworth (NSW) 190-196, 195t
tapeworms 268
technology provision to farmers 256-258
temperature. *See also* air temperature; chilling injury to plants; frost damage and control; soil properties, temperature
  climate change and 269-271
  crop production and 43-68
  effect on aphids 130
  effect on cotton bollworms 135
  effect on grasshoppers 132-133
  effect on locusts 134
  effect on Queensland fruit fly 135
  effect on sheep parasites 137
  effect on ticks 142-143
  fluctuations in 65
  heat stress injury to plants 51-52
  plant growth determined by 66-68
  sudden changes in 50-55
  tolerable zones of 123-126
temporal and spatial management of soil water 89
"10 Steps to Drought Preparedness" (US) 115
terrestrial radiation 23
Thailand, drought assessment in 99
thermal infrared 147
  drought monitoring with 162-164
  remote sensing of 149

thermals, insects carried on 129
thermometry 7
thermoperiodism 64-65
Thornthwaite method 6
ticks 142-143
tillage
  to reduce water loss 88
  soil temperature and 43
  sustainable 253
time-lagged influences 123
Timely Satellite Data for Agricultural Management 170, 172
tomatoes 46, 64
Toxoptera aurantii 130
training. *See* education and training
transmission of EMR 147, 174f
  through leaves 26-30, 27f, 28t
transpiration 30, 76-77
tree density 169
trematodes 138
Trichostrongylus spp. 137-138
tropospheric ozone 268-269
trypanosomiasis 139
tsetse flies 139

ultraviolet light 29-30, 35
underground dams 92
United Kingdom 205
United Nations, desertification conventions 121
United States 158, 206. *See also* North America
universities, training at 11
US Department of Agriculture Soil Conservation Service 72
US Weather Bureau Class A pan 79-80
USGS Spectral Library 158
UV light 29-30, 35

validation of models 182
vegetable crops 52, 285
vegetation. *See also* crop production; plant canopy
  global warming and 273-274
  reflectance from 152, 176f
vegetation *(continued)*
  remote sensing of 7, 161-164, 168-169
  spectral reflectance 157
  verification of models 182
volcanic aerosols 265

Walgett (NSW) 196-197, 197t
water balance analysis 164
water bodies, reflectance from 153
water resources 69-93, 280
  for cereal crops 86f
  control during drought 116
  efficiency of use 85-90, 244-246
  long-term changes in supply 277-282
water stress 51, 77
water vapor 22
waterlogging, management of 87
wavelength of radiation 17, 19t, 147
weather conditions 179-180, 209
weather forecasts. *See* forecasts
weather stations 221
weeds, increase frost incidence 59
Wein's Law 18
Western Australia 164-165, 216
western equine encephalitis 140
wet spells. *See* seasonal events
wheat crops
  air temperature and 48
  benefit from shelter 239
  drought monitoring of 110-111
  forecasts used in 219
  frost damage to 55-56
  global warming and 274
  heat stress on 52
  PAR interception by 37
  soil moisture and 248f
  soil temperature and growth 46
  water use efficiency 246-247
WHEATMAN 206
wind-borne pests and diseases 127-129, 134
wind machines for frost control 61-62
windbreaks 239. *See also* screening
Wisconsin, frost control in 171-172

wool production. *See* sheep
World Meteorological Organization 8, 222

Yearly Productivity Index (YPI) 110
yellow fever 140

For Product Safety Concerns and Information please contact our EU representative GPSR@taylorandfrancis.com Taylor & Francis Verlag GmbH, Kaufingerstraße 24, 80331 München, Germany

T - #0035 - 160425 - C6 - 229/152/21 [23] - CB - 9781560229728 - Gloss Lamination